KU-636-759

Contents

Beginning
AutoCAD

Other titles from Bob McFarlane

Beginning AutoCAD   ISBN 0 340 58571 4

Progressing with AutoCAD   ISBN 0 340 60173 6

Introducing 3D AutoCAD   ISBN 0 340 61456 0

Solid Modelling with AutoCAD   ISBN 0 340 63204 6

Assignments in AutoCAD   ISBN 0 340 69181 6

Starting with AutoCAD LT   ISBN 0 340 62543 0

Advancing with AutoCAD LT   ISBN 0 340 64579 2

3D Draughting using AutoCAD   ISBN 0 340 67782 1

Beginning AutoCAD R13 for Windows   ISBN 0 340 64572 5

Advancing with AutoCAD R13 for Windows   ISBN 0 340 69187 5

Modelling with AutoCAD R13 for Windows   ISBN 0 340 69251 0

Using AutoLISP with AutoCAD   ISBN 0 340 72016 6

Beginning AutoCAD R14 for Windows NT and Windows 95   ISBN 0 340 72017 4

Advancing with AutoCAD R14 for Windows NT and Windows 95   ISBN 0 340 74053 1

Modelling with AutoCAD R14 for Windows NT and Windows 95   ISBN 0 340 73161 3

An Introduction to AEC 5.1 with AutoCAD R14   ISBN 0 340 74185 6

Beginning AutoCAD 2002

Bob McFarlane
MSc, BSc, ARCST,
CEng, FIED, RCADDes
MIMechE, MIEE, MIMgt, MBCS, MCSD

*Curriculum Manager CAD and New Media, Motherwell College,
Autodesk Educational Developer*

OXFORD AMSTERDAM BOSTON LONDON NEW YORK PARIS
SAN DIEGO SAN FRANCISCO SINGAPORE SYDNEY TOKYO

Butterworth-Heinemann
An imprint of Elsevier Science
Linacre House, Jordan Hill, Oxford OX2 8DP
225 Wildwood Avenue, Woburn, MA 01801-2041

First published 2002

British Library Cataloguing in Publication Data
A catalogue record for this book is available from the British Library

Library of Congress Cataloguing in Publication Data
A catalogue record for this book is available from the Library of Congress

ISBN 0 7506 5610 7

For information on all Butterworth-Heinemann
publications visit our website at www.bh.com

Produced and typeset by Gray Publishing, Tunbridge Wells, Kent
Printed and bound in Great Britain by Bath Press, Avon

Preface

AutoCAD is probably the most widely used PC-based CAD software package available, and AutoCAD 2002 is the latest release. The program is very similar to AutoCAD 2000, but incorporates several new features including greater Internet accessibility. These new features, combined with the traditional AutoCAD interface will increase the users' draughting skills and improve productivity.

This book is intended for:

a) new users to AutoCAD who have access to AutoCAD 2002
b) experienced AutoCAD users wanting to upgrade their skills from previous releases to AutoCAD 2002.

The objective of the book is to introduce the reader to the essential basic 2D draughting skills required by every AutoCAD user, whether at the introductory, intermediate or advanced level. Once these basic skills have been 'mastered', the user can progress to the more 'demanding' topics such as 3D modelling, customisation and AutoLISP programming.

The book will prove invaluable to any casual AutoCAD user, as well as the student studying any of the City and Guild, BTEC or SQA CADD courses. It will also be useful to undergraduates and postgraduates at higher institutions who require AutoCAD draughting skills. Industrial CAD users will be able to use the book, as both a textbook and a reference source.

As with all my other AutoCAD books, the reader will learn by completing worked examples, and further draughting experience will be obtained by completing the additional activities which complement many of the chapters. All drawing material has been completed using Release 2002 and all work has been checked to ensure there are no errors.

Your comments and suggestions for work to be included in any future publications would be greatly appreciated.

Bob McFarlane

To: Stephen and Amanda.
Many congratulations on your marriage,
from Mum, Dad, Lynda and Ciara.

Chapter 1

What's new in AutoCAD 2002

New features

AutoCAD 2002 has several new and enhanced features including those listed below.

True associative dimensioning

AutoCAD now supports geometry-driven associative dimensioning and trans-spatial dimensioning, providing a new method of dimensioning that eliminates: (*a*) calculating dimension scales and (*b*) creating special annotation layers.

New text features

Several new text utilities include an enhanced spell checker, text scaling, text justifying and matching text between model space and paper space. The spell checker supports all text objects included in block definitions.

CAD standards

Allow a set of common properties to be defined such as layers and text styles. This allows for consistency in interpreting drawings when a large number of users contribute to a drawing.

Design XML

Defines a structure for the efficient delivery of geometric model information over the web. This is an advanced feature of AutoCAD 2002.

Layer translator

Allows layers in the current drawing to be changed to match layers in another drawing or in CAD standard files.

Block attribute manager

Allows attribute data in block definitions to be modified easily.

Enhanced DWF file format

DWF now supports raster image formats, thumbnail and preview images as well as support for additional viewer applications and products.

Enhanced features

AutoCAD Today

Allows the user to manage drawing and template files, load symbol libraries, access the Bulletin Board for collaborative work and access the AutoDESK Point A design portal.

Live Object Enablers

Increases the value of designs and reduces time between drawing and data sharing.

Publish to Web

The user can the Publish to Web wizard with template, themes and I-drop options.

Many (but not all) of these new features will be discussed in this book.

Chapter 2

System requirements and installation

The requirements for using AutoCAD 2002 are:

Operating system

- Windows 2000
- Windows Millennium Edition (ME)
- Windows 98
- Windows NT 4.0 with Service Pack 5.0.

Processor

- Pentium 233 minimum
- Pentium 450 or higher recommended
- Any equivalent processor.

RAM

- 32 MB minimum
- 64 MB recommended.

Video

- 800 × 600 VGA with 256 colours minimum
- 1024 × 768 with 64 thousand colours recommended.
  Note: Windows-supported display adapter required.

Hard disk

- 130 MB for installation
- 64 MB of swap space
- 60 MB minimum system folder (75 MB recommended)
- 20 MB shared files.

Pointing device

- Mouse, tracker ball or other device.

CD-ROM

- For installation purposes – any speed.

Optional hardware

- Open GL-compatible 3D video card
- Printer or plotter
- Digitizer
- Modem or access to an Internet connection
- Network interface card.

The installation procedure should follow the instructions given in the AutoCAD 2002 Users' Manual.

Chapter 3

Using the book

The aim of the book is to assist the reader on how to use AutoCAD 2002 with a series of interactive exercises. These exercises will be backed up with activities, thus allowing the reader to 'practice the new skills' being demonstrated. While no previous CAD knowledge is required, it would be useful if the reader knew how to use:
– the mouse to select items from the screen
– Windows packages, e.g. maximise/minimise screens.

Concepts for using the book

There are several simple concepts with which the reader should become familiar, and these are:

1 Menu selection will be in bold type, e.g. **Draw**

2 A menu sequence will be in bold type and be either:
   a) **Draw**   *or b*) **Draw-Circle-3 Points**
      Circle
      3 Points

3 User keyboard entry will also be highlighted in bold type, e.g.
   a) coordinate entry – **125,36; @100,50; @200<45**
   b) command entry – **LINE; MOVE; ERASE**
   c) response to a prompt – **15**

4 Button/icon selection will be displayed as a small picture of the icon where appropriate – usually the first time the icon is used.

5 The AutoCAD 2002 prompts will be in typewriter face, e.g.
   a) *prompt*   `Specify first point`
   b) *prompt*   `Specify second point of displacement`

6 The symbol **<R>** or **<RETURN>** will be used to signify pressing the RETURN or ENTER key. Pressing the mouse right-button will also give the <RETURN> effect – called right-click.

7 The term **pick** is continually used with AutoCAD, and refers to the selection of a line, circle, text item, dimension, etc. The mouse left button is used to **pick an object** – called left-click.

8 Keyboard entry can be **LINE** or **line**. Both are acceptable.

Saving drawings

All work should be saved for recall at some later time, and drawings can be saved:
– on a formatted floppy disk
– in a named folder in the hard drive.

It is the user's preference as to which method is used, but for convenience purposes only I will assume that a named folder is being used. This folder is named **BEGIN** and when a drawing is being saved or opened, the terminology used will be:
a) save drawing as **BEGIN\WORKDRG**
b) open drawing **BEGIN\EXER_1**.

Chapter 4

The AutoCAD 2002 graphics screen

In this chapter we will investigate the graphics screen and discuss some of the terminology associated with it.

Starting AutoCAD 2002

AutoCAD 2002 is started:
a) from the Windows 'Start screen' with a double left-click on the AutoCAD 2002 icon
b) by selecting the windows taskbar sequence:
Start-Programs-AutoCAD 2002-AutoCAD 2002

Both methods briefly display the AutoCAD 2002 logo and then the AutoCAD 2002 Today screen/dialogue box as Fig. 4.1. The Today screen is divided into three distinct section, these being:
a) My Drawings: Open Drawings, Create Drawings and Symbol Libraries options
b) Bulletin Board: for communication between company 'sites'
c) Autodesk Point A: for Internet access.

At present we will not discuss the Today screen in any detail, so cancel it by picking (left-click) the X button on the right in the title bar. The AutoCAD 2002 graphics screen will then be displayed.

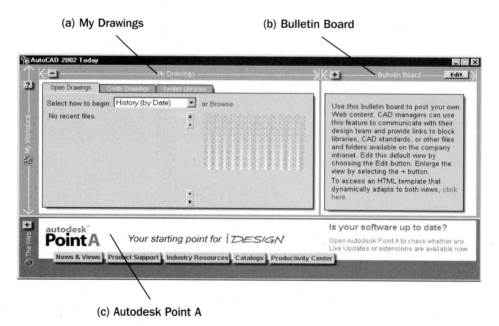

(a) My Drawings   **(b) Bulletin Board**

(c) Autodesk Point A

Figure 4.1   The AutoCAD Today screen.

The graphics screen

The AutoCAD 2002 graphics screen (Fig. 4.2) displays the following:

1 The title bar

2 The 'windows buttons'

3 The menu bar

4 The Standard toolbar

5 The Object Properties toolbar

6 The Windows taskbar

7 The Status bar

8 The Command prompt window area

9 The coordinate system icon

10 The drawing area

11 The on-screen cursor

12 The grips and/or pickfirst box

13 Scroll bars

14 The Layout tabs

15 The Modify toolbar

16 The Draw toolbar.

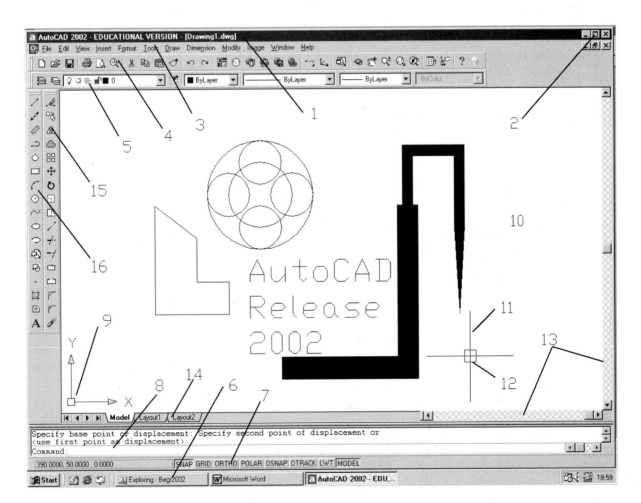

Figure 4.2 The AutoCAD 2002 graphics screen.

Title bar

The title bar is positioned at the top of the screen and displays the AutoCAD 2002 icon, the AutoCAD Release version and the current drawing name.

The Windows buttons

The Windows buttons are positioned to the right of the title bar, and are:
a) left button: minimise screen
b) centre button: maximise screen
c) right button: close current application.

The menu bar

The menu bar displays the default AutoCAD menu headings. By moving the mouse into the menu bar area, the cursor cross-hairs change to a **pick arrow** and with a left-click on any heading, the relevant 'pull-down' menu will be displayed. The full menu bar headings are:

File Edit View Insert Format Tools Draw Dimension Modify Image Windows Help

Figure 4.3 displays the full menu pull-down selections for a sample of menu bar headings, i.e. File, Format, Draw and Modify.

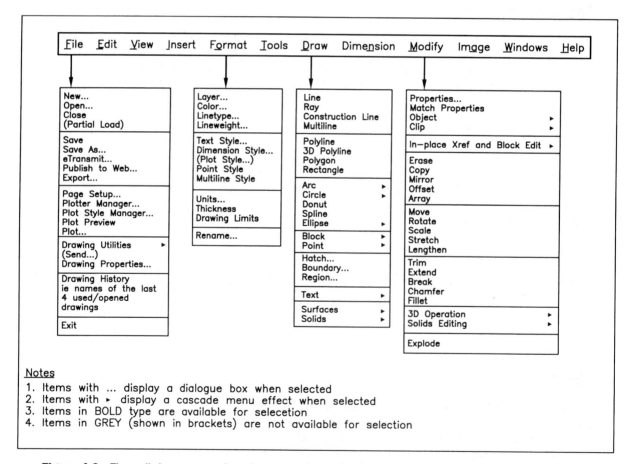

Figure 4.3   The pull-down menus from four menu bar selections.

Menu bar notes

1 Pull-down menu items which display '...' result in a screen dialogue box when the item is selected, i.e. left-clicked

2 Pull-down menu items which display ▶ result in a further menu when selected. This is termed the cascade menu effect.

3 Menu items in BOLD type are available for selection.

4 Menu items in GREY type are not available for selection.

5 Menu bar and pull-down menu items can be selected (picked) with the mouse or by using the **Alt** key with the letter which is underlined, e.g.
a) Alt with M, activates the Modify pull-down menu
b) then Alt with Y, activates the Copy command.

6 Certain items can be activated using the control (**Ctrl**) key with a letter or number. The most common items are:
a) Ctrl with N – New drawing
b) Ctrl with O – Open drawing
c) Ctrl with S – Save drawing
d) Ctrl with P – Plot drawing
e) Ctrl with 1 – Properties dialogue box.

7 In this book, items will be generally selected from the menu bar with the mouse.

The Standard toolbar

The Standard toolbar is normally positioned below the menu bar and allows the user access to 30 button icon selections including New, Open, Save, Print, etc. By moving the cursor pick arrow onto an icon and 'leaving it for about a second', the icon name will be displayed in yellow (default). The standard toolbar can be positioned anywhere on the screen or 'turned off' if required by the user.

The Object Properties toolbar

Normally positioned below the Standard toolbar, this allows a further seven button icon selections. The icons in this toolbar are Make Object's Layer Current, Layers, Layer Control, Layer Previous, Color Control, Linetype control and Lineweight Control.

The Windows taskbar

This is at the bottom of the screen and displays:
a) the Windows 'Start button' and icon
b) the name of any application which has been opened, e.g. AutoCAD
c) the time and the sound control icon
d) perhaps some other icons depending on the user's system.

By left-clicking on 'Start', the user has access to the other Programs which can be run 'on top of AutoCAD', i.e. multi-tasking.

The Status bar

Positioned above the Windows taskbar, the status bar gives useful information to the user:
a) on-screen cursor X, Y and Z coordinates at the left
b) drawing aid buttons, e.g. SNAP, GRID, ORTHO, POLAR, OSNAP, OTRACK, LWT
c) MODEL/PAPER space toggle

Command prompt window area

The command prompt area is where the user 'communicates' with AutoCAD 2002 to enter:
a) a command, e.g. LINE, COPY, ARRAY
b) coordinate data, e.g. 120,150, @15<30
c) a specific value, e.g. a radius of 25.

The command prompt area is also used by AutoCAD to supply the user with information, which could be:
a) a prompt, e.g. from point
b) a message, e.g. object does not intersect an edge.

The command area can be increased in size by 'dragging' the bottom edge of the drawing area upwards. I generally have a two- or three-line command area.

The coordinate system icon

This is the X–Y icon at the lower left corner of the drawing area. This icon gives information about the coordinate system in use. The default setting is the traditional Cartesian system with the origin (0,0) at the lower left corner of the drawing area. The coordinate icon will be discussed in detail later.

The drawing area

This is the user's drawing sheet and can be any size required. In general we will use A3 sized paper, but will also investigate very large and very small drawing paper sizes.

The cursor cross-hairs

Used to indicate the on-screen position, and movement of the pointing device will result in the coordinates in the status bar changing. The 'size' of the on-screen cursor can be increased or decreased to suit user preference and will be discussed later.

The Grips/Pickfirst box

This is the small box which is normally 'attached' to the cursor cross-hairs. It is used to select objects for modifying and will be discussed in detail in a later chapter.

Scroll bars

Positioned at the right and bottom of the drawing area and are used to scroll the drawing area. They are very useful for larger sized drawings and can be 'turned-off' if they are not required.

Layout tabs

Allows the user to 'toggle' between model and paper space for drawing layouts. The layout tabs will be discussed in a later chapter.

Modify and Draw toolbars

By default, Release 2002 displays these two toolbars at the left of the screen. Toolbars will be discussed later in this chapter.

Terminology

AutoCAD 2002 terminology is basically the same as previous releases, and the following gives a brief description of the items commonly encountered by new users to AutoCAD.

Menu

A menu is a list of options from which the user selects (picks) the one required for a particular task. Picking a menu item is achieved by moving the mouse over the required item and left-clicking. There are different types of menus, e.g. pull-down, cascade, screen, toolbar button icon.

Command

A command is an AutoCAD function used to perform some task. This may be to draw a line, rotate a shape or modify an item of text. Commands can be activated by:
a) selection from a menu
b) selecting the appropriate icon from a toolbar button
c) entering the command from the keyboard at the command line
d) entering the command abbreviation
e) using the Alt key as previously described.

Only the first three options will be used in this book.

Objects

Everything drawn in AutoCAD 2002 is termed an **object** or **entity**, e.g. lines, circles, text, dimensions, hatching, etc. are all objects. The user 'picks' the appropriate entity/object with a mouse left-click when prompted.

Default setting

All AutoCAD releases have certain values and settings which have been 'preset' and are essential for certain operations. These default settings are displayed with **< >** brackets, but can be altered by the user as and when required. For example:

1 From the menu bar select **Draw-Polygon** and:
   prompt   `_polygon Enter number of sides<4>`
   respond **press the ESC key** to cancel the command
   Note: *a*) <4> is the default value for the number of sides
            b) _polygon is the active command.

2 At the command line enter **LTSCALE <R>** and:
   prompt   `Enter new linetype scale factor<1.0000>`
   enter   **0.5 <R>**
   Note: *a*) <1.0000> is the LTSCALE default value
            b) we have altered the LTSCALE value to 0.5

The escape (Esc) key

This is used to cancel any command at any time. It is very useful, especially when the user is 'lost in a command'. Pressing the Esc key will cancel any command and return the command prompt line.

Icon

An icon is a menu item in the form of a picture contained on a button within a named toolbar. Icons will be used extensively throughout the book, especially when a command is being used for the first time.

Cascade menu

A cascade menu is obtained when an item in a pull-down menu with ▶ after its name is selected, e.g. by selecting the menu bar sequence **Draw-Circle**, the cascade effect shown in Fig. 4.4 will be displayed. Cascade menus can be cancelled by:

1   moving the pick arrow to any part of the screen and left-clicking

2   pressing the Esc key – cancels the 'last' cascade menu.

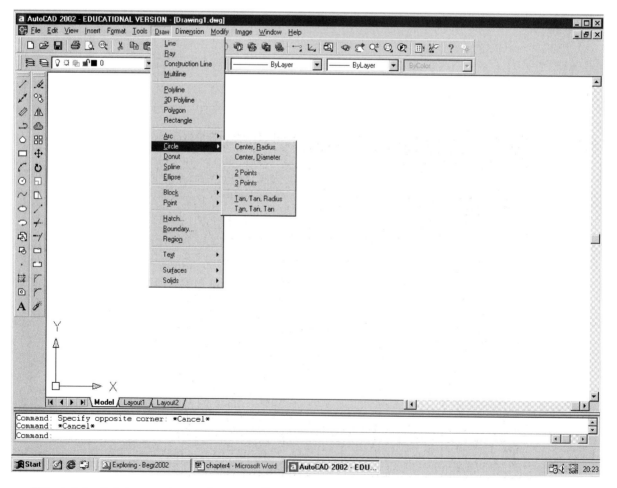

Figure 4.4 Pull-down and cascade menu effect.

Dialogue boxes

A dialogue box is always displayed when an item with '...' after its name is selected, e.g. when the menu bar sequence **Format-Units** is selected, the Drawing Units dialogue box (Fig. 4.5) will be displayed. Dialogue boxes allow the user to alter parameter values or toggle an aid ON/OFF.

Most dialogue boxes display the options On, Cancel and Help which are used as follows:

OK: accept the values in the current dialogue box
Cancel: cancel the dialogue box without any alterations
Help: gives further information in Windows format. The Windows can be cancelled with File-Exit or using the Windows Close button from the title bar (the right-most button).

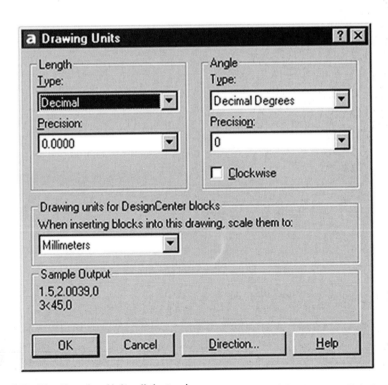

Figure 4.5   The Drawing Units dialogue box.

Toolbars

Toolbars are aids for the user. They allow the Release 2002 commands to be displayed on the screen in button icon form. The required command is activated by picking (left-click) the appropriate button. The icon command is displayed as a **tooltip** in yellow (default colour) by moving the pick arrow onto an icon and leaving it for a second. There are 26 toolbars available for selection, and four are normally displayed by default when AutoCAD 2002 is started, these being the Standard, Object Properties, Modify and Draw toolbars. Toolbars can be:

a) displayed and positioned anywhere in the drawing area

b) customised to the user preference.

To activate a toolbar, select from the menu bar **View-Toolbars** and the Customize dialogue box will be displayed allowing the user access to four tabs: Commands, Toolbars (active), Properties and Keyboard. To display a toolbar, pick the box by the required name. Figure 4.6 displays the Toolbar tab of the Customize dialogue box with the Dimension and Object Snap toolbars toggled on, as well as the default Draw, Modify and Object Properties toolbars active. When toolbars are positioned in the drawing area as the Object Snap and Dimension toolbars in Fig. 4.6, they are called **FLOATING** toolbars. Figure 4.6 also displays the Tooltip from the Snap to Perpendicular object snap icon.

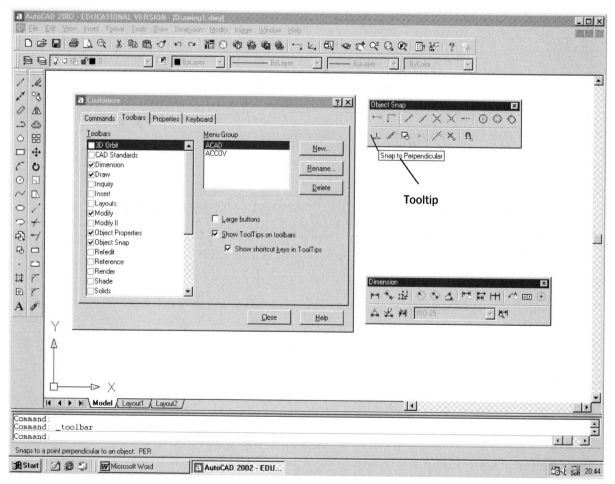

Figure 4.6   The Customize dialogue box with the Toolbars tab active and displaying floating and docked toolbars.

Toolbars can be:

1   Moved to a suitable position on the screen by the user. This is achieved by moving the pick arrow into the blue title area of the toolbar and holding down the mouse left button. Move the toolbar to the required position on the screen and release the left button.

2   Altered in shape by 'dragging' the toolbar edges sideways or downwards.

3   Cancelled at any time by picking the 'Cancel box' at the right of the toolbar title bar.

It is the user's preference as to what toolbars are displayed at any one time. In general I always display the Draw, Modify, Dimension and Object Snap toolbars and activate others as and when required.

Toolbars can be **DOCKED** at the edges of the drawing area by moving them to the required screen edge. The toolbar will be automatically docked when the edge is reached. Figure 4.6 displays two floating and four docked toolbars:
a) Docked: Standard and Object Properties at the top of the screen; Draw and Modify at the left of the screen. These four toolbars 'were set' by default
b) Floating: Object Snap and Dimension. These two toolbars were 'activated' by me.

Toolbars **do not have to be used** – they are an aid to the user. All commands are available from the menu bar, but it is recommended that toolbars are used, as they greatly increase draughting productivity.

When used, it is the user's preference as to whether they are floating or docked.

Fly-out menu

When an button icon is selected an AutoCAD command is activated. If the icon has a ◢ at the lower right corner of the icon box, and the left button of the mouse is held down, a **FLY-OUT** menu is obtained, allowing the user access to other icons. The following fly-out menus are available from the Standard toolbar:

Temporary Tracking Point: object snap icons
UCS: UCS options in icon form as Fig. 4.7
Named view: the viewpoint preset icons
Zoom: the various zoom options in icon form.

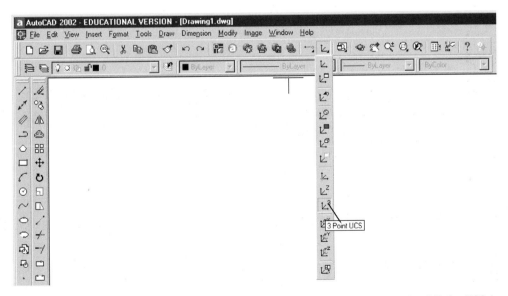

Figure 4.7   The fly-out menu from the UCS button icon with the tooltip from the 3 Point UCS icon.

Wizard

Wizard allows the user access to various parameters necessary to start a drawing session, e.g. units, paper size, etc. There are two Wizard options, these being **Quick Setup** and **Advanced Setup**. We will investigate how to use Wizard in later chapters.

Template

A template allows the user access to different drawing standards with different sized paper, each template having a border and title box. AutoCAD 2002 supports the following standards, the number of templates available for user selection being listed with the standard name:

ANSI: 19; DIN: 10; Gb: 14; ISO: 10; JIS: 12; M: 1

The use of templates will be investigated later in the book.

Toggle

This is the term used when a drawing aid is turned ON/OFF and usually refers to:
a) pressing a key
b) activating a parameter in a dialogue box, i.e. a tick/cross signifying ON, no tick/cross signifying OFF.

Function keys

Several of the keyboard function keys can be used as aids while drawing, these keys being:

F1	accesses the AutoCAD 2002 Help menu
F2	flips between the graphics screen and the AutoCAD Text window
F3	toggles the object snap on/off
F4	toggles the tablet on/off (if attached)
F5	toggles the isoplane top/right/left – for isometric drawings
F6	coordinates on/off toggle
F7	grid on/off toggle
F8	ortho on/off toggle
F9	snap on/off toggle
F10	polar on/off toggle
F11	toggles object snap tracking off
F12	not used.

Help menu

AutoCAD 2002 has an 'on-line' help menu which can be activated at any time by selecting from the menu bar **Help-Help** or pressing the F1 function key. The Help dialogue box will be displayed as two distinct sections:
a) Left: with five tab selections – Contents, Index, Search, Favourites, Ask Me
b) Right: details about the topic.

File types

When a drawing has been completed it should be saved for future recall and all drawings are called *files*. AutoCAD 2002 supports different file formats, including:
.dwg: AutoCAD 2002 drawing
.dws: AutoCAD 2002 Drawing Standard
.dwt: AutoCAD 2002 Template Drawing template file.

AutoCAD 2002 drawings can be saved in other formats as well as in pre-AutoCAD 2002 formats.

Saved drawing names

Drawing names should be as simple as possible. While operating systems support file names which contain spaces and fullstops, I would not recommend this practice. The following are typical drawing file names which I would recommend be used:

EX1; EXER-1; EXERC_1; MYEX-1; DRG1, etc.

When drawings have to be saved during the exercises in the book, I will give the actual named to be used.

Adapting the graphics screen

The graphics screen can be 'customised' to user requirements, i.e. screen colour, scroll bar, screen menu, etc. There are several 'settings' which we will now investigate, but the user should decide for themselves whether they want to customise their graphics screen to my settings. This is **now your personal decision**.

From the menu bar select **Tools-Options** and:

prompt   Options dialogue box with nine tab selections
respond   **by picking the named tab and alter as described.**

A   Display tab
  a) Window elements
    1. Display scroll bars in drawing area active, i.e. tick
    2. Display screen menu not active, i.e. blank
    3. Text lines in command window area: 3
    4. Colors: pick and set Model tab background to white or black then Apply & Close
          (note 1).
  b) Layout elements
    1. Display Layout and Model tabs active
    2. Display margin active
    3. Display paper background and paper shadow both active
    4. Show Page Setup dialog for new layouts active
    5. Create viewport in new layouts active.
  c) Crosshair size
    Default: 5. Set to own size (note 2).
  d) Display resolution: leave as given.
  e) Display performance: leave as given.

B   Open and Save tab
  a) File Safety Precautions
    1. Automatic save active
    2. Minutes between saves: set as required, e.g. 30, 60 or similar
    3. Create backup copy with each save active.
  b) Leave rest as given.

C   System tab
  a) General options
    1. Start up: Show TODAY startup dialog
       Scroll and pick: Show traditional startup dialog (note 3).

D   Other tabs: leave at present.

E   Pick OK to return to the drawing screen.

Notes
1   Allows the user to set a background screen colour.
2   Sets the on-screen cursor size. 100 gives a full screen cursor.
3   This will 'stop' the full TODAY window being shown at start up and will display the New Drawing dialogue box of previous releases. Selecting the TODAY icon from the Standard toolbar at any time will display the 'full' TODAY window.

Other changes

There are a few other alterations which will be discussed before leaving this chapter. These can also be considered as 'customising the system' to user requirements.

The coordinate system icon

Displayed at the lower left of the drawing area, and can be 'set' to display a 2D or 3D icon.

From the menu bar select **View-Display-UCS Icon** and:
a) On and Origin both active (tick)
b) pick Properties and:

    prompt   UCS Icon Properties dialogue box
    respond   1. UCS icon style: 2D
             2. UCS icon color: black
             3. Layout tab icon color: black or pick to suit
             4. dialogue box as Fig. 4.8
             5. pick OK
    and   the traditional AutoCAD 2D icon with the X, Y and W axes will be displayed.

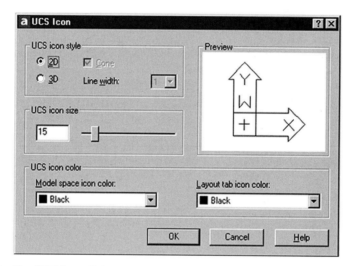

Figure 4.8 The UCS Icon dialogue box.

The Grips/Pickfirst box

The small box attached to the cursor cross-hairs is an aid to the user, but can cause confusion to new AutoCAD users. We will use these aids in later chapters, but at the start 'will turn them off'. This can be achieved with the following keyboard entries:

Enter	*Prompt*	*Enter*
GRIPS<R>	Enter new value for GRIPS	0<R>
PICKFISRT <R>	Enter new value for PICKFIRST	0<R>

We have spent some time discussing the graphics screen and terminology in this rather long chapter and are now ready to start drawing, but before this, select from the menu bar **File-Exit** and pick **No** to Save changes if the AutoCAD message dialogue box is displayed – more on this later.

We have thus customised our drawing screen and quit AutoCAD.

Chapter 5

Drawing, erasing and the selection set

In this chapter we will investigate how lines and circles can be drawn and then erased. When several line and circle objects have been created by different methods, we will investigate the selection set – a powerful aid when modifying a drawing.

Starting a new drawing with Wizard

1 Start AutoCAD with:
    a) a double left-click on the AutoCAD 2002 icon
    b) select from the Windows taskbar the sequence:
       Start-Programs-AutoCAD 2002-AutoCAD2002.

2 When either of the above options is selected, the AutoCAD 2002 logo will be displayed for a short time and if the instructions in Chapter 4 were followed:
    prompt   Startup dialogue box with four selections:
               Open a Drawing; Start from Scratch; Use a Template; Use
               a Wizard
    respond   *a*) pick Use a Wizard icon (right-most icon)
               b) pick Quick Setup – Fig. 5.1
               c) pick OK
    prompt   Quick Setup (Units) dialogue box
    respond   *a*) Select Decimal Units – Fig. 5.2
               b) pick Next>
    prompt   Quick Setup (Area) dialogue box
    respond   *a*) enter Width: 420
               b) enter Length: 297 – Fig. 5.3
               c) pick Finish.

3 The AutoCAD 2002 drawing screen will be displayed and should display the Standard and Object Properties toolbars at the top of the screen, and the Modify and Draw toolbars docked at the left of the screen.

4 *Note*: The toolbars which are displayed will depend on how the last user 'left the system'. If you do not have the Draw and Modify toolbars displayed then:
    a) select from the menu bar **View-Toolbars**
    b) activate the Draw and Modify toolbars with a cross
    c) close the Toolbars dialogue box
    d) position the toolbars to suit, i.e. floating or docked.

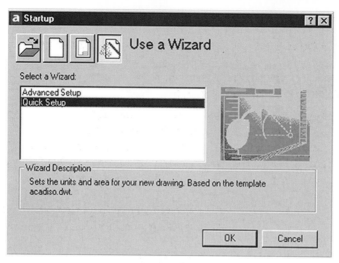

Figure 5.1 The Use a Wizard Startup dialogue box.

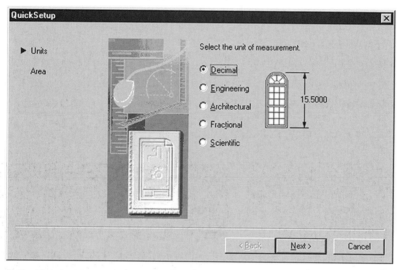

Figure 5.2 The Quick Setup (Units) dialogue box.

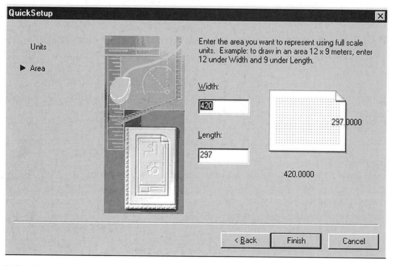

Figure 5.3 The Quick Setup (Area) dialogue box.

Drawing line and circle objects

1 Activate (pick) the LINE icon from the Draw toolbar and the following should have happened:
   a) the command prompt displays: _line Specify first point
   b) the Active Assistance for LINE is displayed – Fig. 5.4(a)

2 You now have to pick a **start** point for the line, so move the pointing device and pick (left-click) any point within the drawing area. Several things should happen:
   i)   a small cross **may** appear at the selected start point – if it does not, don't panic
   ii) as you move the pointing device away from the start point a line will be dragged from this point to the on-screen cursor position. This drag effect is termed **RUBBERBAND**
   iii) as the pointing device is moved, a small coloured box **may** be displayed with text similar to *Polar: 80.00<0*. If it does, don't panic and if it does not don't worry. We will discuss this in the next chapter
   iv) the prompt becomes: Specify next point or [Undo].

3 Move the pointing device to any other point on the screen and left-click. Another cross may appear at the selected point and a line will be drawn between the two 'picked points'.
   This is your first AutoCAD 2002 object

4 The line command is still active with the rubberband effect and the prompt line is still asking you to specify the next point.

5 Continue moving the mouse about the screen and pick points to give a series of 'joined lines'.

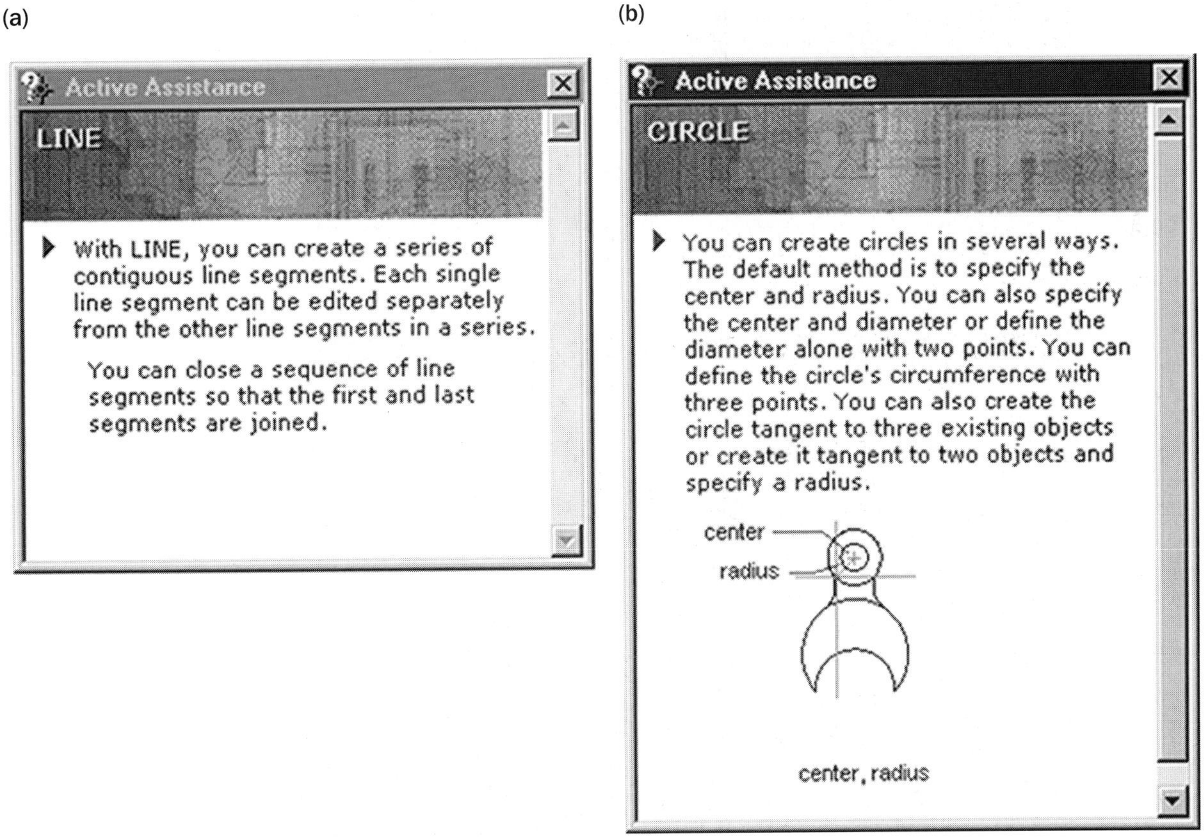

(a)

(b)

Figure 5.4 Active Assistance for (a) LINE and (b) CIRCLE.

6 Finish the LINE command with a right-click on the mouse and:
  a) a pop-up menu will be displayed as Fig. 5.5
  b) pick Enter from this dialogue to end the LINE command and the command line will be returned blank
  c) the Active Assistance for LINE disappears.

Figure 5.5 The command right-click pop-up menu.

7 From the menu bar select **Draw-Line** and the Specify first point prompt will again be displayed in the command area. Draw some more lines and end the command by pressing the RETURN/ENTER key. The LINE command will be 'stopped', but no pop-up menu will have appeared.

8 At the command line enter **LINE <R>** and draw a few more lines. End the command with a right-click and pick Enter from the dialogue box.

9 Right-click on the mouse to display a pop-up menu as Fig. 5.6 and pick **Repeat LINE**. Draw some more lines then end with by pressing the RETURN/ENTER key.

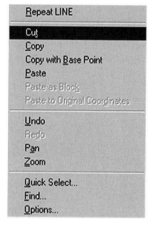

Figure 5.6 The right-button (command) pop-up menu.

10 *Note*
  a) The different ways of activating the LINE command:
    – with the LINE icon from the Draw toolbar
    – from the menu bar with Draw-Line
    – by entering LINE and the command line
    – with a right-click of the mouse (if the LINE command was the last command used).
  b) The two ways to 'exit' a command:
    – with a right click of the mouse which gives a pop-up menu
    – by pressing the RETURN/ENTER key which does not give a pop-up menu.
  c) Cancelling a command with a mouse right-click, **MAY** display the dialogue box similar to Fig. 5.5.
  d) When a command has been finished, a mouse right-click on will display a dialogue box similar to Fig. 5.6 with the **LAST COMMAND** available for selection, e.g. Repeat LINE.
  e) The pop-up menu displayed with the mouse right-click is called a **shortcut menu**. It is a useful aid to the CAD user.

11 From the Draw toolbar activate the CIRCLE icon and:

    prompt   `_circle Specify center point for circle or [3P/2P/Ttr`
                    `(tan tan radius)]`
    and   `Active Assistance for circle displayed` – Fig. 5.4(b)
    respond **pick any point on the screen as the circle centre**
    prompt   `Specify radius of circle or [Diameter]`
    respond **drag out the circle and pick any point for radius.**

12 From the menu bar select **Draw-Circle-Center,Radius** and pick a centre point and drag out a radius.

13 At the command prompt enter **CIRCLE <R>** and create another circle anywhere on the screen.

14 Using the icons, menu bar or keyboard entry, draw some more lines and circles until you are satisfied that you can activate and end the two commands.

15 Figure 5.7(a) displays some line and circle objects.

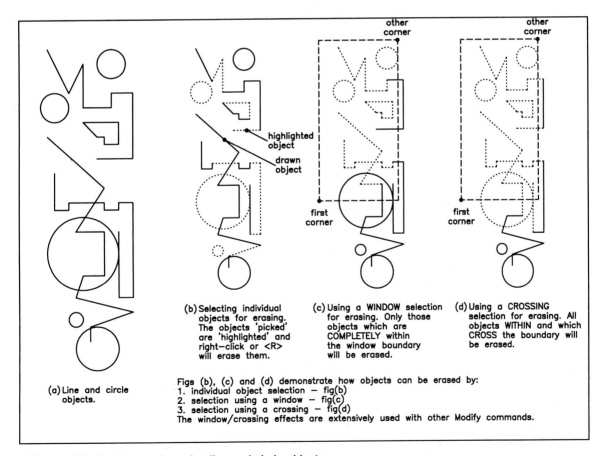

(a) Line and circle objects.

(b) Selecting individual objects for erasing. The objects 'picked' are 'highlighted' and right–click or <R> will erase them.

(c) Using a WINDOW selection for erasing. Only those objects which are COMPLETELY within the window boundary will be erased.

(d) Using a CROSSING selection for erasing. All objects WITHIN and which CROSS the boundary will be erased.

Figs (b), (c) and (d) demonstrate how objects can be erased by:
1. individual object selection – fig(b)
2. selection using a window – fig(c)
3. selection using a crossing – fig(d)
The window/crossing effects are extensively used with other Modify commands.

Figure 5.7 Drawing and erasing line and circle objects.

✓ Blips

Several users may have small crosses at the end of the lines drawn on the screen and at the circle centre points. These crosses are called **BLIPS** and are used to identify the start and end point of lines, circle centres, etc. The are *NOT OBJECTS/ENTITIES* and will not be plotted out on a final drawing. Personally I find them a nuisance and always turn them off. This can be achieved by entering **BLIPMODE <R>** at the command line and:

prompt   `Enter mode [ON/OFF]`
enter   **OFF <R>**

If you do not want to turn the blips off, then by selecting from the menu bar **View-Redraw or View-Regen** the drawing screen is regenerated (refreshed) and the blips are removed.

Active Assistance

The Active Assistance is a dialogue box with information about the command which has been selected. When the LINE and CIRCLE commands were activated, the corresponding Active Assistance dialogue box was displayed. This Active Assistance is a new concept with AutoCAD 2002. It is a very useful aid to the new AutoCAD user, and for commands being used for the first time. I will not refer to it again, and will let the user decide for themselves whether they want to have it displayed with every command or not.

Should you decide to have the dialogue box displayed with every command, then **do not** complete the following, but proceed to the erasing objects section. If you decide to not display the dialogue box, then:

1   Activate the LINE command and Active Assistance (LINE) displayed.

2   Move the cursor into the Active Assistance dialogue box.

3   Right-click in the dialogue box to display a shortcut menu.

4   Pick Settings from this shortcut menu and the Active Assistance Settings dialogue box will be displayed.

5   Deactivate Show on start and Hover Help (both blank).

6   Activate On demand – black dot as Fig. 5.8.

7   Pick OK.

8   Complete your line.

9   Activate the LINE command again and draw another line – no Active Assistance dialogue box should be displayed.

The Active Assistance dialogue box can be displayed at any time by:
a) Entering ASSIST at the command line before a command is activated. This will display the dialogue box at all times until it is closed.
b) Picking the Active Assistance icon from the Standard toolbar. This has the same effect as (*a*).
c) Right-click the Active Assistance icon from the Windows taskbar and selecting Show Active Assistance. This also has the same effect as (*a*).
d) By right-clicking the Active Assistance icon from the Windows taskbar and picking Settings, the Active Assistance Settings dialogue box will be displayed, allowing the user to alter the display of the dialogue box.

Remember: using the Active Assistance is your decision.

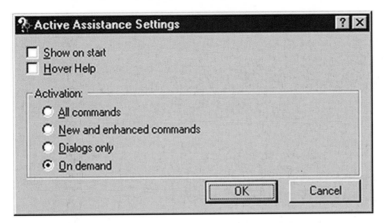

Figure 5.8 The Active Assistance dialogue box.

Erasing objects

Now that we have drawn some lines and circles, we will investigate how they can be erased – seems daft? The erase command will be used to demonstrate different options available to us when it is required to modify a drawing. The actual erase command can be activated by one of three methods:

a) picking the ERASE icon from the Modify toolbar
b) with the menu bar sequence **Modify-Erase**
c) entering **ERASE <R>** at the command line.

Before continuing with the exercise, select from the menu bar the sequence **Tools-Options** and:

prompt	Options dialogue box
respond	pick the Selection tab and ensure:

      a) Noun/verb selection not active, i.e. no tick in box
      b) Use shift to add to selection not active
      c) Press and drag not active
      d) Implied windowing active, i.e. tick in box
      e) Object Grouping active
      f) Pickbox size: set to suit (about 1/4 distance from left)
      g) pick OK when complete

respond	pick OK

Now continue with the erase exercise.

1   Ensure you still have several lines and circles on the screen. Figure 5.7(a) is meant as a guide only.

2   From the menu bar select **Modify-Erase** and:

prompt	Select objects
and	cursor cross-hairs replaced by a 'pickbox' which moves as you move the mouse
respond	**position the pickbox over any line and left-click**
and	the following will happen:

      a) the selected line will 'change appearance', i.e. be 'highlighted'
      b) the prompt displays Select objects: 1 found and then: Select objects.

3   Continue picking lines and circles to be erased (about six) and each object will be highlighted.

4   When enough objects have been selected, right-click the mouse.

5   The selected objects will be erased, and the Command prompt will be returned blank.

6   Figure 5.7(b) demonstrates the individual object selection erase effect.

Oops

Suppose that you had erased the wrong objects. Before you **do anything else**, enter **OOPS <R>** at the command line.

The erased objects will be returned to the screen. Consider this in comparison to a traditional draughtsman who has rubbed out several lines/circles – they would have to redraw each one.

OOPS must be used **immediately** after the last erase command and must be entered from the keyboard.

Erasing with a Window/Crossing effect

Individual selection of objects is satisfactory if only a few objects (e.g. lines/circles) have to be modified (we have only used the erase command so far). When a large number of objects require to be modified, the individual selection method is very tedious, and AutoCAD overcomes this by allowing the user to position a 'window' over an area of the screen which will select several objects 'at the one pick'.

To demonstrate the window effect, ensure you have several objects (about 20) on the screen and refer to Fig. 5.7(c).

1   Select the ERASE icon from the Modify toolbar and:

    prompt   Select objects
    enter   **W <R>** (at the command line) – the window option
    prompt   Specify first corner
    respond   **position the cursor at a suitable point and left-click**
    prompt   Specify opposite corner
    respond   **move the cursor to drag out a window (rectangle) and left-click**
    prompt   ??? found and certain objects highlighted
    then   Select objects i.e. any more objects to be erased?
    respond   **right-click or <R>**

2   The highlighted objects will be erased.

3   At the command line enter **OOPS <R>** to restore the erased objects

4   From the menu bar select **Modify-Erase** and:

    prompt   Select objects
    enter   **C <R>** (at the command line) – the crossing option
    prompt   Specify first corner
    respond   **pick any point on the screen**
    prompt   Specify opposite corner
    respond   **drag out a window and pick the other corner**
    prompt   ??? found and highlighted objects
    respond   **right-click.**

5   The objects highlighted will be erased – Fig. 5.7(d).

Note on window/crossing

1   The window/crossing concept of selecting a large number of objects will be used extensively with the modify commands, e.g. erase, copy, move, scale, rotate, etc. The objects which are selected when **W** or **C** is entered at the command line are as follows:
window: all objects *completely within* the window boundary are selected
crossing: all objects *completely within and also which cross* the window boundary are selected.

2   The window/crossing option **is entered from the keyboard**, i.e. W or C.

3   Figure 5.7 demonstrates the single object selection method as well as the window and crossing methods for erasing objects.

4   *Automatic window/crossing*

In the example used to demonstrate the window and crossing effect, we entered a W or a C at the command line. AutoCAD allows the user to activate this window/crossing effect automatically by picking the two points of the 'window' in a specific direction. Figure 5.9 demonstrates this with:
a) the window effect by picking the first point anywhere and the second point either upwards or downwards to the right
b) the crossing effect by picking the first point anywhere and the second point either upwards or downwards to the left.

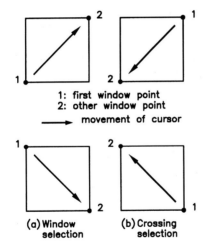

Figure 5.9   Automatic window/crossing selection.

The selection set

Window and crossing are only two options contained within the selection set, the most common selection options being:

```
Crossing, Crossing Polygon, Fence, Last, Previous, Window and
Window Polygon.
```

During the various exercises in the book, we will use all of these options but will only consider three at present.

1 Erase all objects from the screen – individual selection or window option

2 Refer to Fig. 5.10(a) and draw some new lines and circles – the actual layout is not important, but try and draw some objects 'inside' others

3 Refer to Fig. 5.10(b), select the ERASE icon from the Modify toolbar and:
   prompt Select objects
   enter **F <R>** – the fence option
   prompt First fence point
   respond **pick a point (pt 1)**
   prompt Specify endpoint of line or [Undo]
   respond **pick a suitable point (pt 2)**
   prompt Specify endpoint of line or [Undo]
   respond **pick point 3, then points 4 and 5 then right-click**
   prompt Shortcut menu
   respond **pick Enter**
   prompt ??? found and certain objects highlighted
   respond **right-click or <R>**

4 The highlighted objects will be erased.

5 Enter OOPS <R> to restore these erased objects.

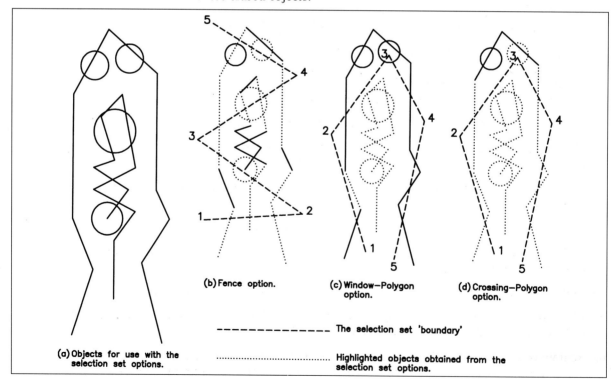

(b) Fence option.

(c) Window–Polygon option.

(d) Crossing–Polygon option.

(a) Objects for use with the selection set options.

– – – – – – – – – – – – The selection set 'boundary'

.... Highlighted objects obtained from the selection set options.

Figure 5.10 Further selection set options.

6   Menu bar with **Modify-Erase** and referring to Fig. 5.7(c):
    prompt   Select objects
    enter   **WP <R>** – the window-polygon option
    prompt   First polygon point
    respond   **pick a point (pt 1)**
    prompt   Specify endpoint of line or [Undo]
    respond   **pick points 2,3,4,5 then right-click and pick Enter**
    prompt   ??? found and objects highlighted
    respond   **right-click** to erase the highlighted objects.

7   OOPS to restore the erased objects.

8   *a*) activate the ERASE command
    b) enter **CP <R>** at command line – crossing polygon option
    c) pick points in order as Fig. 5.10(d) then right-click and pick Enter
    d) right-click to erase the highlighted objects.

9   The fence/window polygon/crossing polygon options of the selection set are very useful when the 'shape' to be modified does not permit the use of the normal rectangular window. The user can 'make their own shape' for selecting objects to be modified.

Activity

Spend some time using the LINE, CIRCLE and ERASE commands and become proficient with the various selection set options for erasing – this will greatly assist you in later chapters. Read the summary and proceed to the next chapter. Do not exit AutoCAD if possible.

Summary

1   The LINE and CIRCLE draw commands can be activated:
    a) by selecting the icon from the Draw toolbar
    b) with a menu bar sequence, e.g. Draw-Line
    c) by entering the command at the prompt line, e.g. LINE <R>

2   The ERASE command can be activated:
    a) with the ERASE icon from the Modify toolbar
    b) from the menu bar with Modify-Erase
    c) by entering ERASE <R> at the command line.

3   All modify commands (e.g. ERASE) allow access to the Selection Set.

4   The selection set has several options including Window, Crossing, Fence, Window-Polygon and Crossing-Polygon.

5   The appropriate selection set option can be activated from the command line by entering the letters W,C,F,WP,CP.

6   The term WINDOW refers to all objects completely contained in the window boundary.

7   A CROSSING includes all objects which cross the window boundary and are also completely within the window.

8   OOPS is a useful command that 'restores' objects erased with the last erase command.

9   Blips are small crosses used to display the start and endpoints of lines. They are **not objects** and I would advise keeping them turned off.

10   Redraw is a command which will 'refresh' the drawing screen and remove both blips and any 'ghost image' from the screen. The command is best used from the icon in the Standard toolbar.

Chapter 6

The 2D drawing aids

Now that we know how to draw and erase lines and circles, we will investigate the aids which are available to the user. AutoCAD 2002 has several drawing aids which include:

Grid   allows the user to place a series of imaginary dots over the drawing area. The grid spacing can be altered by the user at any time while the drawing is being constructed. As the grid is imaginary, it does **not** appear on the final plot.

Snap   allows the user to set the on-screen cursor to a pre-determined point on the screen, this usually being one of the grid points. The snap spacing can also be altered at any time by the user. When the snap and grid are set to the same value, the term **grid lock** is often used.

Ortho   an aid which allows only horizontal and vertical movement

Polar tracking allows objects to be drawn at specific angles along an alignment path. The user can alter the polar angle at any time.

Object Snap   the user can set a snap relative to a pre-determined geometry. This drawing aid will be covered in detail in a later chapter.

Getting ready

1   Still have some line and circle objects from Chapter 5 on the screen?

2   Menu bar with **File-Close** and:
     prompt   AutoCAD Message dialogue box with Save changes options
     respond **pick No** – more on in next chapter.

3   Begin a new drawing with the menu bar sequence **File-New** and:
     prompt   Create New Drawing dialogue box
     respond *a)* pick Use a Wizard
                b) pick Quick Setup
                c) pick OK
     prompt   Quick Setup (Units) dialogue box
     respond **pick Decimal then Next>**
     prompt   Quick Setup (Area) dialogue box
     respond   *a)* set Width: 420 and Length: 297
                b) pick Finish.

4   A blank drawing screen will be displayed.

5   Menu bar with **Draw-Rectangle** and:
     prompt   Specify first corner point and enter: **0,0 <R>**
     prompt   Specify other corner point and enter: **420,297 <R>**

6   Menu bar with **View-Zoom-All** and the rectangle shape will 'fill the screen'. This rectangle will be 'our drawing paper'.

Grid and Snap setting

The grid and snap spacing can be set by different methods and we will investigate setting these aids from the command line and from a dialogue box.

1   At the command line enter **GRID <R>** and:
     prompt   Specify grid spacing (X) or ...
     enter   **20 <R>**

2   At the command line enter **SNAP <R>** and:
    prompt   Specify snap spacing or ...
    enter   **20 <R>**

3   Refer to Fig. 6.1 and use the LINE command to draw the letter H using the grid and snap settings of 20.

4   Using keyboard entry, change the grid and snap spacing to 15.

5   Use the LINE command and draw the letter E.

6   From the menu bar select **Tools-Drafting Settings** and:
    prompt   Drafting Settings dialogue box
    respond   activate the Snap and Grid tab
    and   *a*) Snap on with X and Y spacing 15
          b) Grid on with X and Y spacing 15
          c) These values are from our previous step 4 entries.
    respond   1. alter the Snap X spacing to 10 by:
              a) click to right of last digit
              b) back-space until all digits removed
              c) enter 10
              d) left click at Snap Y spacing – alters to 10.
          2. alter the Grid X spacing by:
              a) position pick arrow to left of first digit
              b) hold down left button and drag over all digits – they wil be highlighted
              c) enter 10
              d) left click at Grid Y spacing – alters to 10 as Fig. 6.2.
          3. pick OK.

7   Use the LINE command to draw the letter L.

8   Use the Drafting Settings dialogue box to set both the grid and snap spacing to 5 and draw the letter P.

9   *Note:* the Drafting Settings dialogue box allows the user access to the following drawing aids:
    a) the grid and snap settings
    b) polar tracking
    c) object snap settings.

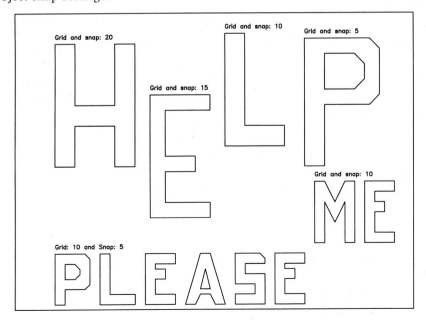

Figure 6.1 Using the GRID and SNAP drafting aids.

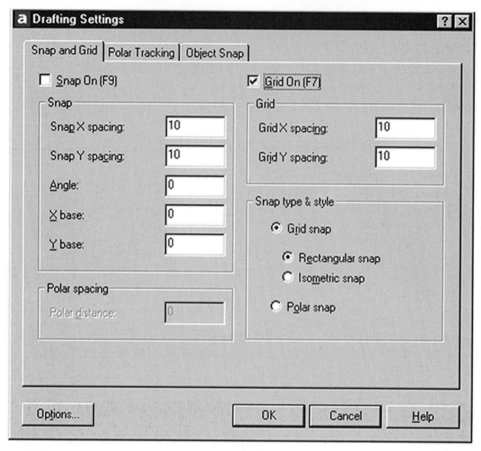

Figure 6.2 Drafting Settings (Snap and Grid) dialogue box.

Toggling the grid/snap/ortho

1   The drawing aids can be toggled ON/OFF with:
    a) the function keys, i.e. F7 – grid; F8 – ortho; F9 – snap
    b) the Drafting Settings dialogue where a tick in the box signifies that the aid is on, and
        a blank box means the aid is off
    c) the status bar with a left-click on Snap, Grid, Ortho.

2   My preference is to set the grid and snap spacing values from the dialogue box or
    command line then use the function keys to toggle the aids on/off as required.

3   Take care if the ortho drawing aid is on. Ortho only allows horizontal and vertical
    movement and lines may not appear as expected. I tend to ensure that ortho is off.

4   The Drafting Settings dialogue box can be activated:
    a) from the menu bar with Tools-Drafting Settings
    b) with a right-click on Snap or Grid from the Status bar and then picking Settings.

Task

Refer to Fig. 6.1 and:
a) with the grid and snap set to 10, draw ME
b) with the grid set to 10 and the snap set to 5, complete 'PLEASE' to your own design
    specification.
c) When complete, do not erase any of the objects.

Drawing with the Polar Tracking aid

1 The screen should still display 'HELP ME PLEASE'?

2 Menu bar with **File-New** and:
   prompt   `Create New Drawing dialogue box`
   respond   *a*) pick Start from Scratch icon (second left)
            b) pick Metric
            c) pick OK
            d) blank drawing screen returned.

3 Set the grid and snap on with settings of 20.

4 Right click on POLAR in the Status bar, pick Settings and:
   prompt   `Drafting Settings dialogue box` with Polar Tracking tab active
   respond   *a*) ensure Polar Tracking On (F10)
            b) scroll at Incremental angle and pick 30
            c) ensure Track using all polar angle settings is active
            d) ensure Absolute active
            e) dialogue box as Fig. 6.3
            f) pick OK.

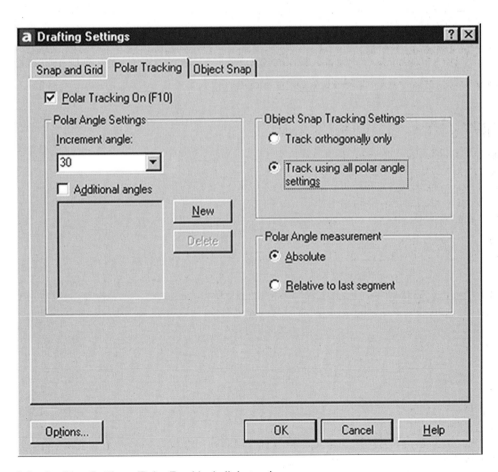

Figure 6.3 Drafting Settings (Polar Tracking) dialogue box.

5 Activate the LINE command and pick a suitable grid/snap start point towards the top of the screen.

6 Move the cursor horizontally to the right and observe the polar tracking information displayed. Move until the tracking data is Polar: 100.0000<0 as Fig. 6.4(a) then left click. This is a line **segment** drawn using the polar tracking drawing aid.

7 Now move the cursor vertically downwards until 40.0000<270 is displayed as Fig. 6.4(b) then left click.

8 Move the cursor downwards and to the right until a 300 degree angle is displayed as Fig. 6.4(c) and enter 50 from the keyboard. The entered value of 50 is the length of the line segment.

9 Move upwards to right until a 30 degree angle is displayed in the polar tracking tip box as Fig. 6.4(d) and enter 80 from the keyboard.

10 Complete the polar tracking line segments with:
   a) an angle of 270 and a keyboard entry of 50 – Fig. 6.4(e)
   b) a line segment length of 150 at an angle of 180 – Fig. 6.4(f)
   c) end the line command with right-click/enter.

11 *Note:* the polar tracking aid displays information of the format 100.0000<90, i.e. **a length and an angle**.

12 When this exercise is complete, proceed to the next chapter but try not to exit AutoCAD.

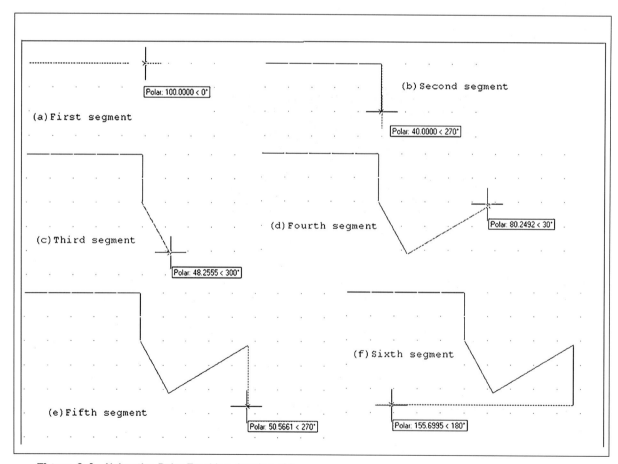

Figure 6.4 Using the Polar Tracking drawing aid.

Chapter 7

Saving and opening drawings

It is essential that *all* users know how to save and open a drawing, and how to exit AutoCAD correctly. These operations can cause new users to CAD a great deal of concern, and as AutoCAD 2002 allows multiple drawings to be opened during a drawing session, it is important to know the correct procedure for saving/opening drawings and how to exit AutoCAD. In this and all the following chapters, all drawing work will be saved to the named folder **BEGIN**.

Saving a drawing and exiting AutoCAD

1   If work has been followed correctly, the user has two drawings opened:
    a) the line segments drawn with polar tracking – active
    b) the HELP ME PLEASE drawing created using grid/snap.

2   Menu bar with **File-Exit** and:
    prompt   `AutoCAD message dialogue box` – similar to Fig. 7.1.

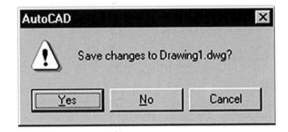

Figure 7.1   The AutcoCAD message dialogue box.

3   This dialogue box is informing the user that since starting the current drawing session, changes have been made and that these drawing changes have not yet been saved. The user has to respond to one of the three options which are:
    Yes   picking this option will save a drawing with the name displayed, i.e. Drawing1.dwg or similar
    No   selecting this option means that the alterations made will not be saved
    Cancel   returns the user to the drawing screen.

4   At this time pick Cancel, as we want to investigate how to save a drawing.

5 Select from the menu bar **File-Save As** and:
   prompt   `Save Drawing As dialogue box`
   respond   *a*) Scroll at Save in by picking the arrow at right of name box
             b) pick (left-click) the C: drive to display folder names
             c) double left-click on the **Begin** folder
             d) alter the File name to DRG2
             e) note the file type name and extension
             f) dialogue box as Fig. 7.2
             g) pick Save.

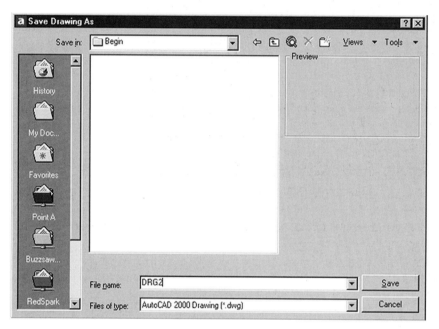

Figure 7.2   The Save Drawing As dialogue box.

6 The screen drawing will be saved to the floppy disc, but will still be displayed on the screen.

7 Menu bar with File-Close and the line segments drawing will disappear from the screen and the HELP ME PLEASE drawing will be displayed

8 Menu bar with **File-Save As** and using the Save Drawing As dialogue box:
   a) ensure the Begin folder is current
   b) alter File name to MYFIRST
   c) pick Save.

9 Now menu bar with **File-Exit** to exit AutoCAD.

10 *Note*
    a) when multiple drawings have been opened in AutoCAD 2002, the user is prompted to save changes to each drawing before AutoCAD can be exited
    b) the Save Drawing As dialogue box displays other options, e.g. History, My Doc, Favorites, etc. These are typical of Windows terminology.

Opening, modifying and saving existing drawings

While AutoCAD is used to create drawings, it also allows existing drawings to be displayed and modified/altered. To demonstrate this:

1 Start AutoCAD and:
   prompt   Startup dialogue box
   respond **pick the Open a Drawing tab**
   prompt   dialogue box similar to Fig. 7.3
   with   a list of the last few opened drawings with their paths.

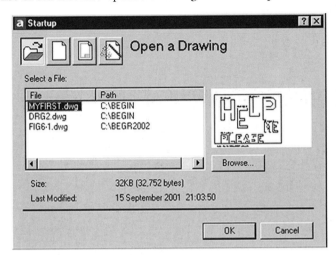

Figure 7.3 The Startup (Open a Drawing) dialogue box.

2 It may be that the drawing you want to open is displayed in the dialogue box at this
   stage, but we will suppose that it is not, so:
   respond **pick Browse**
   prompt   Select File dialogue box
   respond *a*) scroll at Look in and pick the C: drive
              b) double left-click on Begin folder to display all saved drawings
              c) pick MYFIRST
              d) preview displayed
              e) pick Open.

3 The 'HELP ME PLEASE' drawing will be displayed.

4 Set the grid to 10 and snap to 5 then:
   a) erase the 'ME' letters
   b) draw lines around each of the H, E, L and P letters
   c) draw lines around the 'PLEASE' word – Fig. 7.4.

Figure 7.4 Modified MYFIRST drawing.

5 Menu bar with **File-Save As** and:
prompt   Save Drawing As dialogue box
with   MYFIRST.dwg as the File name
respond   pick Save
prompt   Save Drawing As message dialogue box – Fig. 7.5
with   **C:\BEGIN\MYFIRST.dwg already exists**
                Do you want to replace it?
respond   Do nothing at present.

Figure 7.5 The Save Drawing As message dialogue box.

6 This dialogue box is very common with AutoCAD and it is important that you understand what the three options will give if they are selected:
Cancel   does nothing and returns the dialogue box
No   returns the dialogue box allowing the user to alter the file name which should be highlighted
Yes   will overwrite the existing file name and replace the original drawing with any modifications made.

7 At this stage, respond to the message with:
a) pick No
b) alter the file name to MYFIRST1
c) pick Save.

8 What have we achieved?
a) we opened drawing MYFIRST from the C:\BEGIN folder
b) we altered the drawing layout
c) we saved the alterations as MYFIRST1
d) the original MYFIRST drawing is still available and has not been modified.

9 Menu bar with **File-Open** and:
prompt   Select File dialogue box
respond   *a*) pick DRG2 and note the preview
           b) pick Open.

10 The screen will display the line segments drawn with Polar tracking.

11 We now have two opened drawings – DRG2 and MYFIRST1 (or is it MYFIRST?).

Closing files

We will use the two opened drawings to demonstrate how AutoCAD should be exited when several drawings have been opened in the one drawing session.

1 Erase the line segments with a window selection – easy?

2 Menu bar with **File-Close** and:
   prompt AutoCAD Message dialogue box
   with   Save changes to A:\DRG2.dwg massage
   respond **pick No** – can you reason out why we picked No?

3 The screen will display the MYFIRST1 (HELP PLEASE) modified drawing

4 Menu bar with **File-Close** and a blank screen will be displayed with a short menu bar
   – File, View, Window, Help.

5 Select **File** from the menu bar to display a pull down menu similar to Fig. 7.6 with the
   last four opened drawings listed.

6 Respond to the pull down menu by picking **Myfirst1** to display the modified HELP
   PLEASE drawing.

7 *a*) Menu bar with **File** and pick Myfirst
   b) menu bar with **File** and pick Drg2.

8 We have now opened three drawings with Drg2 displayed.

9 Menu bar with:
   a) File-Close to close DRG2 and display MYFIRST
   b) File-Close to close MYFIRST and display MYFIRST1
   c) File-Close to close MYFIRST1 and display a blank screen
   d) File-Exit to exit AutoCAD.

10 All drawings having been saved correctly and AutoCAD has been exited properly.

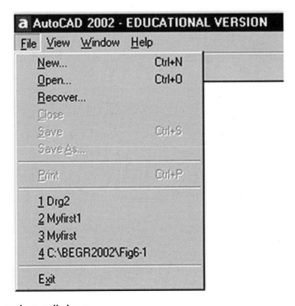

Figure 7.6 Menu bar pull-down.

Save and Save As

The menu bar selection of **File** allows the user to pick either **Save** or **Save As**. The new AutoCAD user should be aware of the difference between these two options.

Save: will save the current drawing with the same name with which the drawing was opened. No dialogue box will be displayed. The original drawing will be **automatically overwritten** if alterations have been made to it.

Save As: allows the user to enter a drawing name via a dialogue box. If a drawing already exists with the entered name, a message is displayed in a dialogue box.

It is strongly recommended that the SAVE AS selection be used at all times.

Assignment

You are now in the position to try a drawing for yourself, so:

1   Start AutoCAD and select Start from Scratch-Metric-OK.

2   Refer to Activity drawing 1 (all activity drawings are grouped together at the end of the book).

3   Set a grid and snap spacing to suit, e.g. 10 and/or 5.

4   Use only the LINE and CIRCLE commands (and perhaps ERASE if you make a mistake) to draw some simple shapes. The size and position are not really important at this stage, the objective being to give you a chance to practice drawing using the drawing aids.

5   When you have completed the drawing, save it as C:\BEGIN\ACT1.

Summary

1   The **recommended** procedure for saving a drawing is:
    a) menu bar with **File-Save As**
    b) select your named folder
    c) enter drawing name in File name box
    d) pick Save.

2   The procedure to open a drawing from **within AutoCAD** is:
    a) menu bar with **File-Open**
    b) select named folder
    c) pick drawing name from list
    d) preview obtained
    e) pick Open.

3   The procedure to open a drawing **from start** is:
    a) start AutoCAD
    b) pick **Open a Drawing** tab from the Start Up dialogue box
    c) either 1. pick drawing name if displayed then OK
        or   2. pick Browse then:
                a) scroll at Look in
                b) select folder name
                c) pick the drawing name
                d) note preview then pick Open.

4   The **recommended** procedure to end an AutoCAD session is:
    a) complete the drawing
    b) save the drawing to your named folder
    c) menu bar with **File-Close** to close all opened drawings
    d) menu bar with **File-Exit** to quit AutoCAD.

5   The save and open command can be activated:
    a) by menu bar selection – recommended method
    b) by keyboard entry
    c) from icon selection in the Standard toolbar.

6   The menu bar method for saving a drawing is recommended as it allows the user the facility to enter the drawing name. This may be different from the opened drawing name.

7   The Save icon from the Standard toolbar and entering SAVE at the command line is a **quick save** option, and does not allow an different file name to be entered. These methods save the drawing with the opened name, and therefore overwrite the original drawing. This may not be what you want?

Chapter 8

Standard sheet 1

Traditionally one of the first things that a draughtsperson does when starting a new drawing is to get the correct size sheet of drawing paper. This sheet will probably have borders, a company logo and other details already printed on it. The drawing is then completed to 'fit into' the pre-printed layout material. A CAD drawing is no different from this, with the exception that the user does not 'get a sheet of paper'. Companies who use AutoCAD will want their drawings to conform to their standards in terms of the title box, text size, linetypes being used, the style of the dimensions, etc. Parameters which govern these factors can be set every time a drawing is started, but this is tedious and against CAD philosophy. It is desirable to have all standard requirements set automatically, and this is achieved by making a drawing called a **standard sheet** or **prototype drawing** – you may have other names for it. Standard sheets can be 'customised' to suit all sizes of paper, e.g. A0, A1, etc. as well as any other size required by the customer. These standard sheets will contain the companies settings, and the individual draughtsman can add their own personal settings as required. It is this standard sheet which is the CAD operators 'sheet of paper'.

We will create an A3 standard sheet, save it, and use it for all future drawing work. At this stage, the standard sheet will not have many 'settings', but we will continue to refine it and add to it as we progress through the book.

The A3 standard sheet will be created using the Advanced Wizard so:

1 Start AutoCAD and:
   prompt   Startup dialogue box
   respond   *a*) pick Use a Wizard
               b) pick Advanced Setup
               c) pick OK
   prompt   Advanced Setup dialogue box
   respond   to each dialogue box with the following selections:
               a) Units: Decimal with 0.00 precision then Next>
               b) Angle: Decimal degrees with 0.0 precision then Next>
               c) Angle Measure: East for 0 degrees then Next>
               d) Angle Direction: Counter-Clockwise then Next>
               e) Area: Width of 420, Length of 297 then Finish.

2 A blank drawing screen will be returned.

3 Menu bar with **Tools-Drafting Settings** and select:
   a) Snap and Grid tab with:
       1. Snap on with X and Y spacing set to 5
       2. Grid on with X and Y spacing set to 10
       3. Rectangular snap style active
   b) Polar Tracking tab with:
       1. Polar Tracking off
   c) Object Snap tab with:
       1. Object Snap off
       2. Object Snap Tracking off
       3. All snap modes off
   d) pick OK.

4 *Note*: The object snap drawing aid will be considered in a later chapter.

5 Menu bar with **Format-Drawing Limits** and:
   prompt   Specify lower left corner and enter: **0,0 <R>**
   prompt   Specify upper right corner and enter: **420,297 <R>**

6 Menu bar with **View-Zoom-All** and the grid will 'fill the screen'.

7 At the command line enter:
   a) **GRIPS <R>** and set to 0
   b) **PICKFIRST <R>** and set to 0.

8 Display toolbars to suit. I would suggest:
   a) Standard and Object Properties docked at the top as default
   b) Draw and Modify docked to left as default
   c) Other toolbars will be displayed as required.

9 Menu bar with **Draw-Rectangle** and:
   prompt   Specify first corner point
   enter   **0,0 <R>**
   prompt   Specify other corner point
   enter   **420,297 <R>**

10 This rectangle will represent our 'drawing area'.

11 Menu bar with **File-Save As** and:
    a) scroll and pick your named folder, e.g. **C:\BEGIN**
    b) enter the file name as **A3PAPER**
    c) pick Save.

Note

1 This completes our standard sheet (at this stage). We have created an A3 sized sheet of paper which has the units and screen layout set to our requirements. The Status bar displays the coordinates to two decimal places with both the Snap and Grid ON. The A3PAPER standard sheet has been saved to the C:\BEGIN folder as a drawing (.dwg) file.

2 Although we have activated several toolbars in our standard sheet, the user should be aware that these may not always be displayed when your standard sheet drawing is opened. AutoCAD displays the screen toolbars which were active when the system was 'shut down'. If other CAD operators have used 'your machine', then the toolbar display may not be as you left it.

   If you are the only user on the machine, then there should not be a problem.

   Anyway you should know how to display toolbars?

3 We will discuss AutoCAD's template files in more detail in a later chapter.

4 Do not confuse my Standard sheet idea with the CAD Standards in AutoCAD 2002. The standard sheet idea is a reference name only.

Chapter 9

Line creation and coordinate input

The line and circle objects so far created were drawn at random on the screen without any attempt being made to specify position or size. To draw objects accurately, coordinate input is required and AutoCAD 2002 allows different 'types' of coordinate entry including:

1 Absolute, i.e. from an origin point.

2 Relative (or incremental), i.e. from the last point entered.

In this chapter we will use our A3PAPER standard sheet to create several squares by different entry methods. The completed drawing will then be saved for future work.

Getting started

1 Still in AutoCAD?
   A. **Answer YES**: then menu bar with **File-Open** and:
      prompt   Select File dialogue box
      respond *a*) scroll at Look in
              b) pick C: drive
              c) double left-click on BEGIN
              d) pick A3PAPER and note Preview
              e) pick Open
   B. **Answer NO:** then Start AutoCAD and:
      prompt   Startup dialogue box
      respond *a*) select Open a Drawing icon
              b) pick Browse – Select File dialogue box displayed
              c) scroll at Look in and pick C: drive
              d) double left-click on BEGIN
              e) pick A3PAPER
              f) pick Open.

   Note: you may have the C:BEGIN\A3PAPER listed when the Open a Drawing icon is selected. If so then open it at this stage.

2 The A3PAPER standard sheet will be displayed, i.e. a black border with the grid and snap spacing set to 10.

3 Display the Draw and Modify toolbars and position to suit and decide if you want to use polar tracking. Ensure that the Object Snap modes are off – they should be.

4 Refer to Fig. 9.1.

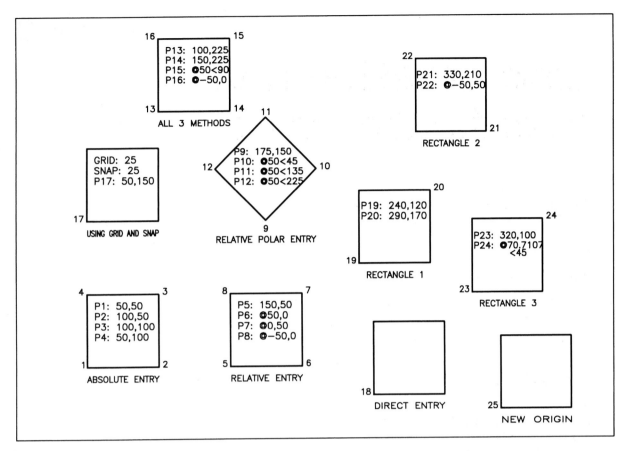

Figure 9.1 Line creation using different entry methods.

Absolute coordinate entry

This is the traditional X-Y Cartesian system, where the origin point is (0,0) at the lower left corner of the drawing area. This origin point can be 'moved' by the user as you will discover later in this chapter.

Select the LINE icon from the Draw toolbar and:
		Fig. 9.1 ref
prompt	Specify first point and enter: **50,50 <R>**	start pt 1
prompt	Specify next point and enter: **100,50 <R>**	pt 2
prompt	Specify next point and enter: **100,100 <R>**	pt 3
prompt	Specify next point and enter: **50,100 <R>**	pt 4
prompt	Specify next point and enter: **50,50 <R>**	end pt 1
prompt	Specify next point and right-click	
then	Pick Enter to end the line command.	

Relative (absolute) coordinate entry

Relative coordinates are from the **last point** entered and the @ symbol is used for the incremental entry.

From the menu bar select **Draw-Line** and enter the following X-Y coordinate pairs, remembering <R> after each entry.

prompt	Specify first point and enter: **150,50 <R>**	start pt 5
prompt	Specify next point and enter: **@50,0 <R>**	pt 6
prompt	Specify next point and enter: **@0,50 <R>**	pt 7
prompt	Specify next point and enter: **@-50,0 <R>**	pt 8
prompt	Specify next point and enter: **@0,-50 <R>**	end pt 5
prompt	Specify next point and right-click then pick Enter.	

The @ symbol has the following effect:

a) @50,0 is 50 units in the positive X direction and 0 units in the Y direction from the last point, which is 150,50

b) @0,–50 is 0 units in the X direction and 50 units in the negative Y direction from the last point on the screen

c) thus an entry of @30,40 would be 30 in the positive X direction and 40 in the positive Y direction from the current cursor position

d) similarly an entry of @–80,–50 would be 80 units in the negative X direction and 50 units in the negative Y direction.

Relative (polar) coordinate entry

This also allows coordinates to be specified relative to the last point entered and uses the @ symbol as before, but also introduces angular entry using the < symbol.

Activate the LINE command (icon or menu bar) and enter the following coordinates:

Specify first point	**175,150 <R>**	start pt 9
Specify next point	**@50<45 <R>**	pt 10
Specify next point	**@50<135 <R>**	pt 11
Specify next point	**@50<225 <R>**	pt 12
Specify next point	**C <R>** to close the square and end line command.	

Note

1 The relative polar entries can be read as:
   a) @50<45 is 50 units at an angle of 45 degrees from the last point which is 175,150
   b) @50<225 is 50 units at an angle of 225 degrees from the current cursor position.

2 The entry **C <R>** is the **CLOSE** option and:
   a) closes the square, i.e. a line is drawn from the current screen position (point 12) to the start point (point 9)
   b) ends the sequence, i.e. <R> not needed
   c) the close option works for any straight line shape.

3 There is **NO** comma (,) with polar entries. This is a common mistake with new AutoCAD users, i.e.
   a) @50<45 is correct
   b) @50,<45 is wrong and gives the command line error:
   Point or option keyword required.

Using all three entry methods

The different coordinate entry methods can be 'mixed and matched' when drawing a series of line segments. Activate the LINE command then enter the following:

Specify first point	**100,225 <R>**	start pt 13
Specify next point	**150,225 <R>** : absolute entry	pt 14
Specify next point	**@50<90 <R>** : relative polar entry	pt 15
Specify next point	**@-50,0 <R>** : relative absolute entry	pt 16
Specify next point	**C <R>** to close square and end command.	

Grid and snap method

The grid and snap drawing aids can be set to any value suitable for current drawing requirements, so:

a) set the grid and snap spacing to 25
b) with the LINE command, draw a 50 unit square the start point being at **50,150** which is pt 17 in Fig. 9.1
c) when the 50 unit square has been drawn, reset the grid and snap to original values, i.e. 10.

Direct distance entry

This method uses the position of the on-screen cursor and is very suitable when polar tracking is active – remember that our polar tracking is off. Activate the LINE command and:

prompt	Specify first point
enter	**250,30 <R>** – this is pt 18 in Fig. 9.1
prompt	Specify next point
respond	move cursor horizontally to right and enter: **50 <R>**
prompt	Specify next point
respond	move cursor vertically upwards and enter: **50 <R>**
prompt	Specify next point
respond	move cursor horizontally to left and enter: **50 <R>**
prompt	Specify next point
enter	**C <R>**

Rectangles

Rectangular shapes (in our case, squares) can be created from coordinate input by specifying two points on a diagonal of the rectangle, and the command can be used with absolute or relative input.

1   From the menu bar select **Draw-Rectangle** and:

prompt	Specify first corner point	
enter	**240,120 <R>**	pt 19
prompt	Specify other corner point	
enter	**290,170 <R>**	pt 20

2   Select the rectangle icon from the Draw toolbar and:

prompt	Specify first corner point	
enter	**330,210 <R>**	pt 21
prompt	Specify other corner point	
enter	**@-50,50 <R>**	pt 22

3   At the command line enter **RECTANG <R>** and:

prompt	Specify first corner point and enter: **320,100 <R>**	pt 23
prompt	Specify other corner point and enter: **@70,7107<45 <R>**	pt 24

Question: why the 70.7107 length entry?

Moving the origin

All the squares created so far have been with the origin point (0,0) at the lower left corner of 'our drawing paper'. If you move the cursor onto this position (snap on) and observe the status bar, it will display 0.00, 0.00, 0.00. These are the x, y and z coordinates of this point. At present we are drawing in 2D and thus the third coordinate will always be 0.00. The origin can be moved to any point on the screen. We will reset the origin and draw a 50 unit square from this new origin position, so:

1   Menu bar with **View-Display-UCS Icon** and ensure that both On and Origin are active (tick at name).

2   Menu bar with **Tools-New UCS-Origin** and:
prompt   Specify new origin point
enter   **340,20 <R>**

3   The UCS icon should move to this position. If it does not, repeat step 1

4   Move the cursor (snap on) to the + in the icon and observe the status bar. The coordinates read 0.00, 0.00, 0.00, i.e. we have reset the origin point – point 25 in Fig. 9.1

5   Now draw a square of side 50 from this new origin using any of the methods previously described.

6   At the command line enter **UCS <R>** and:
prompt   Enter an option [New/Move. ...
enter   **P <R>** – the previous option.

7   The UCS icon will be 'returned' to its original origin position at the lower left corner of out drawing page rectangle.

Saving the squares

The drawing screen should now display ten squares positioned as Fig. 9.1 but without the text. This drawing must be saved as it will be used in other chapters, so from the menu bar select **File-Save As** and:
prompt   Save Drawing As dialogue box
respond 1. scroll at Save in and pick C: drive
          2. double left-click on BEGIN folder
          3. enter file name as **DEMODRG**
          4. pick Save.

Conventions

When using coordinate input, the user must know the positive and negative directions for both linear and angular input. The two conventions are as follows:

1   *Coordinate axes*
The X-Y axes convention used by AutoCAD is shown in Fig. 9.2(a) and displays four points with their coordinate values. When using the normal X-Y coordinate system:
a) a positive X direction is to the right, and a positive Y direction is upwards
b) a negative X direction is to the left, and a negative Y direction is downwards.

2   *Angles*
When angles are being used:
a) positive angles are anti-clockwise
b) negative angles are clockwise.

Figure 9.2(b) displays the angle convention with four points and their polar coordinates.

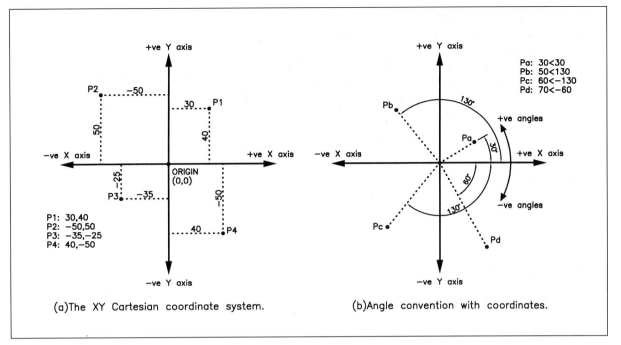

Figure 9.2 Coordinate and angle convention.

Task

Before leaving this exercise, try the following:

1 Make sure you have saved the squares as C:\BEGIN\DEMODRG.

2 Erase a square created with a Draw-Rectangle sequence. It is erased with a 'single pick' as it is a polyline – more on this in a later chapter.

3 Now erase all the squares from the screen.

4 Draw a line sequence using the following entries:
```
Specify first point    30,40
Specify next point    -100,100
Specify next point    -150,-200
Specify next point     80,-100
Specify next point     C <R>
```

5 Draw four lines, entering the following:
a) Specify first point: 0,0; Specify next point: 100<30
b) Specify first point: 0,0; Specify next point: 200<150
c) Specify first point: 0,0; Specify next point: 250<-130
d) Specify first point: 0,0; Specify next point: 90<-60.

6 When these eight line segments have been drawn, not all are completely visible on the screen.

7 Menu bar with **View-Zoom-All** and:
a) all eight lines are visible, i.e. we can draw 'off the screen'?
b) move the cursor to the intersection of the four polar lines, and with the snap on, the status bar displays 0,0,0 as the coordinates.
c) now **View-Zoom-Previous** from the menu bar.

Notes

1   When using coordinate input with the LINE command, it is very easy to make a mistake with the entries. If the line 'does not appear to go in the direction it should', then either:
    a) enter **U <R>** from the keyboard to 'undo' the last line segment drawn, or
    b) right-click and pick **undo** from the right button dialogue box
    c) this 'undo effect' can be used until all segments are erased.

2   The @ symbol very is useful if you want to 'get to the last point referenced on the screen'. Try the following:
    a) draw a line and cancel the command with a <RETURN>
    b) re-activate the line command and enter @ **<R>**
    c) the cursor 'snaps to' the endpoint of the drawn line.

3   A <RETURN> keyboard press will always activate the last command.

4   A right-click on the mouse will activate a 'pop-up' dialogue box, allowing the last command to be activated.

Assignment

This activity only uses the LINE command (and ERASE?), but requires coordinate entry (and some 'sums') for you to complete the drawing.

1   Close all existing drawings then open your A:A3PAPER standard sheet.

2   Refer to Activity 2 and draw the three template shapes using coordinate input. Any entry method can be used, but I would recommend that:
    a) position the start points with absolute entry
    b) use relative entry as much as possible.

3   When the drawing is complete, save it as **C:\BEGIN\ACT2.**

4   Read the summary then proceed to the next chapter.

Summary

1 Coordinate entry can be **ABSOLUTE** or **RELATIVE.**

2 ABSOLUTE entry is from an origin – the point (0,0). Positive directions are UP and to the RIGHT, negative directions are DOWN and to the LEFT. The entry format is **X,Y**, e.g. 30,40.

3 RELATIVE entry refers the coordinates to the last point entered and uses the @ symbol. The entry format is:
   a) relative absolute: **@X,Y**, e.g. @50,60
   b) relative polar: **@X<A**, e.g. @100<50 and note – no comma.

4 An angle of –45° is the same as an angle of +315°.

5 The following polar entries are the same:

   (*a*) @–50<30; (*b*) @50<210; (*c*) @50<–150.

   Try them if you are not convinced.

6 All entry methods can be used in a line sequence.

7 A line sequence is terminated with:
   a) the <RETURN> key
   b) a mouse right-click and pick Enter
   c) 'closing' the shape with a C <R>

8 The rectangle command is useful, but it is a 'single object' and not four 'distinct lines'.

9 The LINE command can be activated by:
   a) icon selection from the Draw toolbar
   b) menu bar selection with Draw-Line
   c) entering LINE <R> at the command line.

   It is the user's preference what method is to be used.

Chapter 10

Circle creation

In this chapter we will investigate how circles can be created by adding several to the squares created in the previous chapter, so:

1   Open you C:\BEGIN\DEMODRG to display the ten squares created in Chapter 9.

2   Refer to Fig. 10.1 and ensure the Draw and Modify toolbars are displayed.

AutoCAD 2002 allows circles to be created by six different methods and the command can be activated by icon selection, menu bar selection or keyboard entry. When drawing circles, absolute coordinates are usually used to specify the circle centre, although the next chapter will introduce the user to the Object Snap modes. These Object Snap modes allow greater flexibility in selecting existing entities for reference.

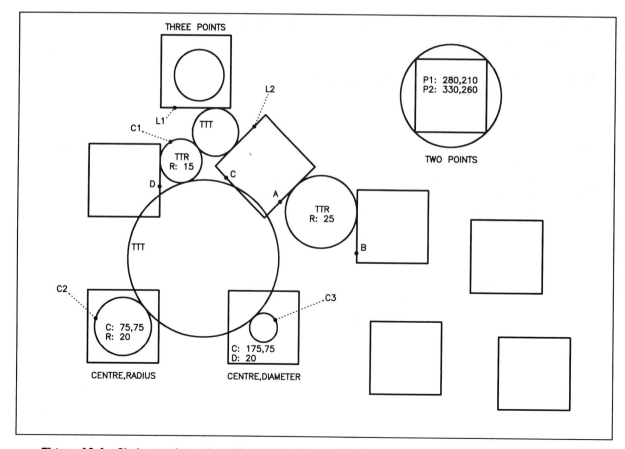

Figure 10.1   Circle creation using different selection methods.

Centre-radius

Select the CIRCLE icon from the Draw toolbar and:

prompt Specify center point for circle or [3P/2P/Ttr (tan tan radius)]:
enter **75,75 <R>** – the circle centre point
prompt Specify radius of circle or [Diameter]
enter **20 <R>** – the circle radius.

Centre-Diameter

From the menu bar select **Draw-Circle-Center,Diameter** and:

prompt Specify center point for circle or ..
enter **175,75 <R>**
prompt Specify diameter of circle
enter **20 <R>**

Two points on circle diameter

At the command line enter **CIRCLE <R>** and:

prompt Specify center point for circle or ..
enter **2P <R>** – the two point option
prompt Specify first end point on circle's diameter
enter **280,210 <R>**
prompt Specify second end point on circle's diameter
enter **330,260 <R>**

Three points on circle circumference

Menu bar selection with **Draw-Circle-3 Points** and:

prompt Specify first point on circle
respond **pick any point within the top left square**
prompt Specify second point on circle
respond **pick another point within the top left square**
prompt Specify third point on circle
respond **drag out the circle and pick a point.**

TTR: tangent-tangent-radius

a) Menu bar with **Draw-Circle-Tan,Tan,Radius** and:

prompt	Specify point on object for first tangent of circle
respond	**move cursor on to line A and leave for a second**
and	1. a small marker is displayed
	2. Deferred Tangent tooltip displayed
respond	**pick line A**, i.e. left click on it
prompt	Specify point on object for second tangent of circle
respond	**pick line B**
prompt	Specify radius of circle
enter	**25 <R>**
and	a circle is drawn as tangent to the two selected lines.

b) At the command line enter **CIRCLE <R>** and:

prompt	Specify center point for circle or..
enter	**TTR <R>** – the tan,tan,radius option
prompt	first tangent point prompt and: **pick line C**
prompt	second tangent point prompt and: **pick line D**
prompt	radius prompt and enter: **15 <R>**
and	a circle is drawn tangential to the two selected lines, line C being assumed extended.

TTT: tangent-tangent-tangent

a) Menu bar with **Draw-Circle-Tan,Tan,Tan** and:

prompt	Specify first point on circle and: **pick line L1**
prompt	Specify second point on circle and: **pick line L2**
prompt	Specify third point on circle and: **pick circle C1.**

b) Activate the **Draw-Circle-Tan,Tan,Tan** sequence and:

prompt	Specify first point on circle and: **pick circle C1**
prompt	Specify second point on circle and: **pick circle C2**
prompt	Specify third point on circle and: **pick circle C3.**

Circles have been drawn tangentially to selected objects, these being:
a) two lines and a circle
b) three circles.

Questions

1 How long would it take to draw a circle as a tangent to three other circles by conventional draughting methods, i.e. drawing board, T square, set squares, etc.?

2 Can a circle be drawn as a tangent to two circles and a line, or to three inclined lines?

Saving the drawing

Assuming that the CIRCLE commands have been entered correctly, your drawing should resemble Fig. 10.1 (without the text) and is ready to be saved for future work.

From the menu bar select **File-Save As** and:

prompt	Save Drawing As dialogue box
with	1. Begin folder name active
	2. File name: DEMODRG
respond	**pick Save**
prompt	Drawing already exists message
respond	**pick Yes** – obvious?

Task

The two Tan,Tan,Tan circles have been created without anything being known about their radii.

1   From the menu bar select **Tools-Inquiry-List** and:
    prompt   Select objects
    respond   **pick the smaller TTT circle**
    prompt   1. found and Select objects
    respond   **right-click**
    prompt   AutoCAD Text window with information about the circle.

2   Note the information then cancel the text window by picking the right (X) button from the title bar

3   Repeat the **Tools-Inquiry-List** sequence for the larger TTT circle

4   The information for my two TTT circles is as follows:

	smaller	*larger*
Centre point	139.52, 208.53	131.88, 122.06
Radius	16.47	53.83
Circumference	103.51	338.21
Area	852.67	9102.42

5   Could you calculate these figures manually as easily as has been demonstrated?

Assignment

1   Open your A:STDA3 standard sheet.

2   Refer to the Activity 3 drawing which can be completed with only the LINE and CIRCLE commands.

3   The method of completing the drawings is at your discretion.

4   Remember that absolute coordinates are recommended for circle centres and that the TTR method is very useful.

5   You may require some 'sums' for certain circle centres, but the figures are relatively simple.

6   When the drawing is complete, save it as C:\BEGIN\ACT3.

Summary

1   Circles can be created by six methods, the user specifying:
    a) a centre point and radius
    b) a centre point and diameter
    c) two points on the circle diameter
    d) any three points on the circle circumference
    e) two tangent specification points and the circle radius
    f) three tangent specification points.

2   The TTR and TTT options can be used with lines, circles, arcs and other objects.

3   The centre point and radius can be specified by:
    a) coordinate entry
    b) picking a point on the screen
    c) referencing existing entities – next chapter.

Chapter 11

Object snap

The lines and circles drawn so far have been created by coordinate input. While this is the basic method of creating objects, it is often desirable to 'reference' existing objects already displayed on the screen, e.g. we may want to:
a) draw a circle with its centre at the midpoint of an existing line
b) draw a line, from a circle centre perpendicular to another line.

These types of operations are achieved using **the object snap modes** – generally referred to as **OSNAP** – and are one of the most useful (and powerful) draughting aids.

Object snap modes are used **transparently**, i.e. whilst in a command, and can be activated:
a) from the Object Snap toolbar
b) by direct keyboard entry.

While the toolbar method is the quicker and easier to use, we will investigate both methods.

Getting ready

1   Open your C:\BEGIN\DEMODRG of the squares and circles.
2   Erase the two TTT circles and the lower right square.
3   Display the Draw, Modify and Object Snap toolbars and position them to suit.
4   Refer to Fig. 11.1.

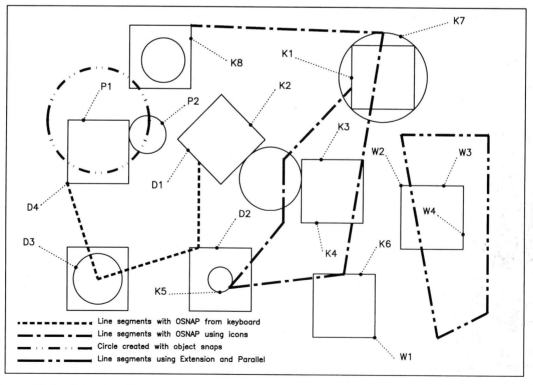

Figure 11.1   Using the object snap modes with C:\BEGIN\DEMODRG.

Using object snap from the keyboard

Activate the LINE command and:

prompt	Specify first point
enter	**MID <R>**
prompt	of
respond	1. move cursor to line D1 and leave for few seconds
	2. coloured triangular marker at line midpoint
	3. Midpoint tooltip displayed in colour
now	**pick line D1, i.e. left-click**
and	line 'snaps to' the midpoint of D1
prompt	Specify next point
enter	**PERP <R>**
prompt	to
respond	**pick line D2** – note coloured Perpendicular marker
prompt	Specify next point
enter	**CEN <R>**
prompt	of
respond	**pick circle D3** – note coloured Center marker
prompt	Specify next point
enter	**INT <R>**
prompt	of
respond	**pick point D4** – note blue Intersection marker
prompt	Specify next point
respond	**right-click and pick Enter** to end the line sequence.

Using object snap from the toolbar

Activate the LINE command and:

prompt	Specify first point	
respond	**pick the Snap to Nearest icon**	
prompt	nea to	
respond	**pick any point on line K1**	
prompt	Specify next point	
respond	**pick the Snap to Apparent Intersection icon**	
prompt	appint of	
respond	pick line K2	
prompt	and	
respond	**pick line K3**	
prompt	Specify next point	
respond	**pick the Snap to Perpendicular icon**	
prompt	per to	
respond	**pick line K4**	
prompt	Specify next point	
respond	**pick the Snap to Tangent icon**	
prompt	tan to	
respond	**pick circle K5**	
prompt	Specify next point	
respond	**pick the Snap to Midpoint icon**	
prompt	mid of	
respond	**pick line K6**	
prompt	Specify next point	
respond	**pick the Snap to Quadrant icon**	
prompt	qua of	
respond	**pick circle K7**	
prompt	Specify next point	
respond	**pick the Snap to Endpoint icon**	
prompt	endp of	
respond	**pick line K8**	
prompt	Specify next point	
respond	**right-click and pick Enter** to end the line sequence.	

Object snap with circles

Select the CIRCLE icon from the Draw toolbar and:

prompt   Specify center point for circle or..

respond   **pick the Snap to Midpoint icon**

prompt   mid of

respond   **pick line P1**

prompt   Specify radius of circle or..

respond   **pick the Snap to Center icon**

prompt   cen of

respond   **pick circle P2.**

Note

1   Save your drawing at this stage as C:\BEGIN\DEMODRG.

2   The endpoint 'snapped to' depends on which part of the line is 'picked'. The coloured marker indicates which line endpoint.

3   A circle has four quadrants, these being at the 3, 12, 9, 6 o'clock positions. The coloured marker indicates which quadrant will be snapped to.

The extension and parallel object snap modes

The object snap modes selected so far should have been self-explanatory to the user, i.e. endpoint will snap to the end of a line, center will snap to the centre of a circle, etc. The extension and parallel modes are used as follows:

a) Extension: used with lines and arcs and gives a temporary extension line as the cursor is passed over the endpoint of an object

b) Parallel: used with straight line objects only and allows a vector to be drawn parallel to another object.

To demonstrate these two object snap modes:

1 Continue with the DEMODRG and turn on the grid and snap. Set the spacing to 10 for both.

2 Activate the LINE command and:

prompt   Specify first point

respond **pick the Snap to Extension icon**

prompt   ext of

respond **move cursor over point W1 then drag to right**

and   1. highlighted extension line dragged out

         2. information displayed about distance from endpoint as a tooltip

respond 1. move cursor until Extension: 50.00<0.0 displayed

         2. left-click

prompt   Specify next point

respond **pick the Snap to Extension icon**

prompt   ext of

respond 1. move cursor over point W2

         2. move cursor vertically up until Extension:40.00<90.0 is displayed

         3. left-click

prompt   Specify next point

respond **pick the Snap to Parallel icon**

prompt   par to

respond 1. move cursor over line W3, leave for a few seconds and note the display

         2. move cursor to right of last pick point and:

           a) highlighted line

           b) information about distance and angle displayed

         3. move cursor to right until 70.00<0.0 displayed

         4. left-click

prompt   Specify next point

respond **pick the Snap to Parallel icon**

prompt   par to

respond 1. move cursor over line W4

         2. move cursor vertically below last pick point

         3. move cursor until 140.00<270.0 displayed

         4. left-click

prompt   Specify next point

enter   **C <R>** to close the shape and end the line command

3 The line segments should be as Fig. 11.1

4 Save your layout as **C:\BEGIN\DEMODRG**, updating the existing DEMODRG.

Running object snap

Using the object snap icons from the toolbar will increase the speed of the draughting process, but it can still be 'tedious' to have to pick the icon every time an ENDpoint (for example) is required. It is possible to 'preset' the object snap mode to ENDpoint, MIDpoint, CENter, etc., and this is called a **running object snap**. Pre-setting the object snap does not preclude the user from selecting another mode, i.e. if you have set an ENDpoint running object snap, you can still pick the INTersection icon.

The running object snap can be set:

1 From the menu bar with **Tools-Drafting Settings** and pick the Object Snap tab

2 Entering **OSNAP <R>** at the command line

3 Picking the Object Snap Settings icon from the Object Snap toolbar

4 With a right-click on OSNAP in the Status bar and picking Settings

Each method displays the Drafting Settings dialogue box with the Object Snap tab active.

Task

1 Select **Tools-Drafting Settings** from the menu bar and:
   prompt   Drafting Settings dialogue box
   respond   1. ensure the Object Snap tab is active
             2. ensure Object Snap On (F3) is active
             3. activate Endpoint, Midpoint and Nearest by picking the appropriate box –
                tick means active
             4. dialogue box as Fig. 11.2
             5. pick OK.

2 Now activate the LINE command and move the cursor cross-hairs onto any line and
   leave it.

3 A coloured marker will be displayed at the Nearest point of the line – it may be that you
   have the midpoint marker displayed.

4 Press the **TAB** key to cycle through the set running object snaps, i.e. the line should
   display the cross, square and triangular coloured markers for the Nearest, Endpoint and
   Midpoint object snap settings.

5 Cancel the line command with **ESC.**

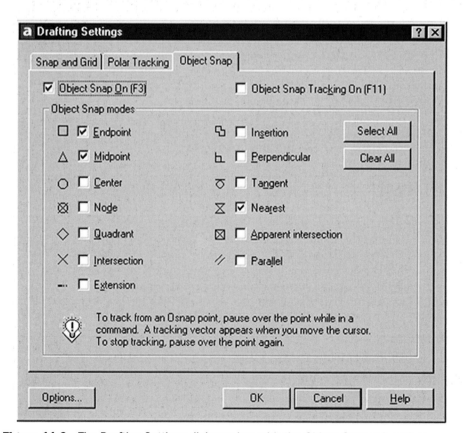

Figure 11.2   The Drafting Settings dialogue box with the Object Snap tab active.

AutoSnap and AutoTrack

The Object Snap dialogue box allows the user to select **Options** which displays the Drafting tab dialogue box – Fig. 11.3. With this dialogue box, the user can control both the AutoSnap and the AutoTrack settings which can be simply defined as:

a) AutoSnap: a visual aid for the user to see and use object snaps efficiently, i.e. a marker and tooltip displayed.

b) AutoTrack: an aid to the user to assist with drawing at specific angles, i.e. polar tracking.

The settings which can be altered with AutoSnap and AutoTrack are:

AutoSnap	*AutoTrack*
Marker	Display polar tracking vector
Magnet	Display full screen tracking vector
Display tool tip	Display autotrack tooltip
Display aperture box	Alter alignment point acquisition
Alter marker colour	Alter aperture size
Alter marker size.	

The various terms should be self-explanatory at this stage(?):

a) Marker: is the geometric shape displayed at a snap point.

b) Magnet: locks the aperture box onto the snap point.

c) Tool tip: is a flag describing the name of the snap location.

The rest of the options should be apparent. It is normal to have the marker, magnet and tool tip active (ticked). The colour of the marker and the aperture box sizes are at the user's discretion, as is having polar tracking 'on'.

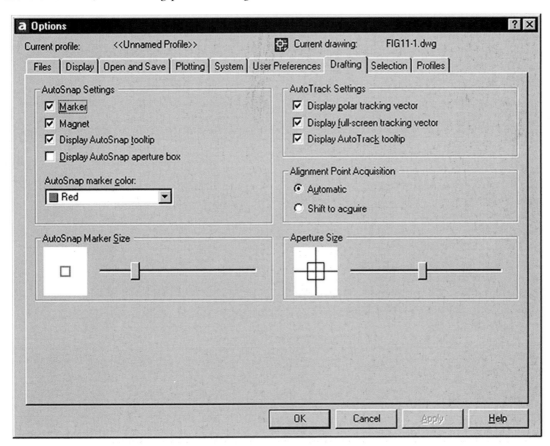

Figure 11.3   The Drafting tab of the Options dialogue box.

Cancelling a running object snap

A running object snap can be left 'active' once it has been set, but this can cause problems if the user 'forgets' about it. The running snap can be cancelled:

1 From the Object Settings dialogue box with **Clear All.**

2 By entering **–OSNAP <R>** at the command line and:
   prompt   Enter list of object snap modes
   enter   **NONE <R>**

3 *Note*:
   a) Selecting the Snap to None icon from the Object Snap toolbar will turn off the object snap running modes for the next point selected.
   b) Using the Object Snap dialogue box is the recommended way of activating and de-activating object snaps.

The Snap From object snap

This is a very useful object snap, allowing the user the reference points relative to existing objects.

1 Using the squares and circles which should still be displayed, erase the lines and circle created with the object snaps. This is to give 'some space'. Ensure the Object Snap toolbar is displayed.

2 Activate the circle command and:
   prompt   Specify center point for circle
   respond **pick Snap From icon from the Object Snap toolbar**
   prompt   from Base Point
   respond **pick Snap to Midpoint icon**
   prompt   mid of
   respond **pick line K8**
   prompt   <Offset>
   enter   **@50,0 <R>**
   prompt   Specify radius of circle
   enter   **20 <R>**

3 A circle is drawn with its centre 50mm horizontally from the midpoint of the selected line

4 Select the LINE icon and:
   prompt   Specify first point
   respond **pick Snap From icon**
   prompt   from Base point
   respond **pick Intersection icon**
   prompt   int of
   respond **pick point W1**
   prompt   <Offset>
   enter   **@25,25 <R>**
   prompt   Specify next point
   respond **pick Snap From icon**
   prompt   from Base point
   respond **pick Snap to Centre icon**
   prompt   cen of
   respond **pick circle K7**
   prompt   <Offset>
   enter   **@80<–20 <R>**
   prompt   Specify next point
   respond **right-click and Enter.**

5 A line is drawn between the specified points. The endpoints of this line have been 'offset' from the selected objects by the entered coordinate values.

6 Do not save these additions to your drawing layout.

Object Snap Tracking

Object snap tracking allows the user to 'acquire coordinate data' from the object snap modes which have been set. To demonstrate this drawing aid:

1 Erase all squares, circles, etc. from the screen to leave the basic rectangular outline.

2 Refer to Fig. 11.4 and draw a line from 50,50 to 100,150 and a circle, centre at 200,200 with radius 50 – fig. (a).

3 Right-click OSNAP from the Status bar, pick Settings and:
prompt   Drafting Settings dialogue box with Object Snap tab active
respond   a) Object Snap active
            b) Endpoint and Center object snap modes active
            c) Object Snap Tracking active
            d) pick OK.

4 Activate the LINE command and:
   a) move the cursor over the top end of the drawn line and the endpoint marker (square) will be displayed
   b) move the cursor vertically upwards to display object snap tracking information, similar to fig. (b)
   c) enter **80 <R>** and the start point of the line will be obtained
   d) move the cursor to the centre of the circle and the centre marker (circle) will be displayed
   e) move the cursor horizontally to the right and object snap tracking information displayed similar to fig. (c)
   f) enter **78 <R>** and a line segment is drawn – fig. (d)
   g) move cursor onto bottom end of first line to acquire the endpoint marker then move horizontally to the right to display object snap tracking data similar to fig. (e)
   h) enter **100 <R>** then right-click/enter to end the line command
   i) the second line segment is complete – fig. (f).

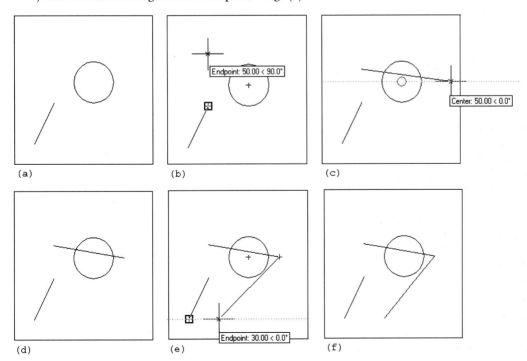

Figure 11.4 Using Object Snap Tracking to draw line segments.

5   *Task 1*

    a) from the Object Snap Settings, deactivate the Endpoint and Center snap modes and activate the Midpoint snap mode

    b) from the Polar Tracking Settings, set the incremental angle to 30

    c) activate the circle command and acquire the midpoint of the second line drawn by object snap tracking

    d) move downwards to the right until object snap tracking data displays an angle of 330 degrees

    e) enter **50 <R>** then **20 <R>** for circle radius

    f) circle drawn at selected point.

6   *Task 2*

    a) erase the last circle and the two line segments drawn with object snap tracking to leave the original line and circle

    b) set the polar tracking angle to 90

    c) turn off the midpoint object snap mode, and activate the endpoint and center modes

    d) with the circle command:

        1. acquire the endpoint of lower end of line

        2. acquire the centre point of the circle

        3. move cursor vertically downwards until it is horizontally in line with the lower end of line

        4. the two acquired point object snap tracking data should be displayed similar to Fig. 11.5

        5. pick this point as the circle centre

        6. enter a radius value, e.g. 30

        7. think of the benefits of this type of operation, i.e. acquiring centre point data without any coordinate input.

7   This completes the object snap tracking exercise. Do not save.

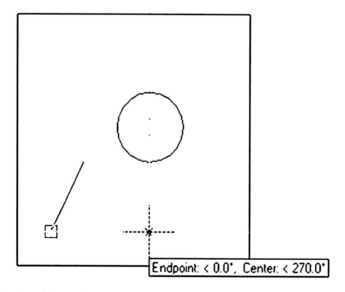

Figure 11.5   Acquiring a circle centre point using object snap tracking.

Assignment

1 Open your C:\BEGIN\A3PAPER standard sheet.

2 Refer to Activity 4 and draw the three components using lines and circles.

3 The Object snap modes will require to be used and hints are given.

4 When complete, save as C:\BEGIN\ACT4.

5 Read the summary then progress to the next chapter.

Summary

1 Object snap (OSNAP) is used to reference existing objects.

2 The object snap modes are invaluable aids to draughting and should be used whenever possible.

3 The user can 'pre-set' a running objects snap.

4 Geometric markers will indicate the snap points on objects.

5 The Drafting Settings dialogue box allows the user to 'control' the geometric markers.

6 Object snap is an example of a **transparent** command, as it is activated when another command is being used.

7 The object snap modes can be set and cancelled using the dialogue box or toolbar.

8 Object snap tracking allows the user to 'acquire' data for selected points 'set' by the object snap modes.

9 *Drawing aids*
At this stage the user now has knowledge about the basic 2D draughting aids, these being:
a) Grid
b) Snap
c) Ortho
d) Object snap modes
e) Polar tracking
f) Object snap tracking.

Chapter 12

Arc, donut and ellipse creation

These three drawings commands will be discussed in turn using our square and circle drawing. Each command can be activated from the toolbar, menu bar or by keyboard entry and both coordinate entry and referencing existing objects (OSNAP) will be demonstrated.

Getting started

1 Open your C:\BEGIN\DEMODRG to display the squares, circles and object snap lines, etc.

2 Erase the objects created during the object snap exercise.

3 Refer to Fig. 12.1 and activated the Draw, Modify and Object Snap toolbars.

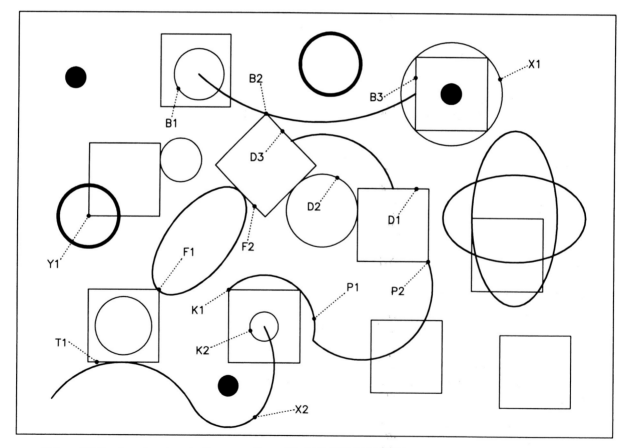

Figure 12.1 Arc, donut and ellipse creation with C:\BEGIN\DEMODRG.

Arcs

There are ten different arc creation methods. Arcs are normally drawn in an anti-clockwise direction with combinations of the arc start point, end point, centre point, radius, included angle, length of arc, etc. We will investigate four different arc creation methods as well as continuous arcs. You can try the others for yourself.

Start,Center,End

From the menu bar select **Draw-Arc-Start,Center,End** and:

prompt   Specify start point of arc
respond **Snap to Midpoint icon and pick line D1**
prompt   Specify center point of arc
respond **Snap to Center icon and pick circle D2**
prompt   Specify end point of arc
respond **Snap to Midpoint icon and pick line D3.**

Start,Center,Angle

Menu bar with **Draw-Arc-Start,Center,Angle** and:

prompt   Specify start point of arc
respond **Snap to Intersection icon and pick point K1**
prompt   Specify center point of arc
respond **Snap to Center icon and pick circle K2**
prompt   Specify included angle
enter   **–150 <R>**

Note that negative angle entries draw arcs in a clockwise direction.

Start,End,Radius

Menu bar again with **Draw-Arc-Start,End,Radius** and:

prompt   Start point and **snap to Endpoint of arc P1**
prompt   End point and **snap to Intersection of point P2**
prompt   Radius and enter **50 <R>**

Three points (on arc circumference)

Activate the 3 Points arc command and:

prompt   Start point and **snap to Center of circle B1**
prompt   Second point and **snap to Intersection of point B2**
prompt   End point and **snap to Midpoint of line B3**

Continuous arcs

1   Activate the 3 Points arc command again and:
    prompt   Start point and enter **25,25 <R>**
    prompt   Second point and **snap to Midpoint of line T1**
    prompt   End point and enter **@50,–30 <R>**

2   Select from the menu bar **Draw-Arc-Continue** and:
    prompt   End point, and cursor snaps to end point of the last arc drawn
    enter   **@50,0 <R>**

3   Repeat the **Arc-Continue** selection and:
    prompt   End point
    respond **snap to Center of circle K2.**

Donut

A donut (doughnut) is a 'solid filled' circle or annulus (a washer shape), the user specifying the inside and outside diameters and then selecting the donut centre point.

1 Menu bar with **Draw-Donut** and:
  prompt   Specify inside diameter of donut and enter: **0 <R>**
  prompt   Specify outside diameter of donut and enter: **15 <R>**
  prompt   Specify center of donut and enter: **40,245 <R>**
  prompt   Specify center of donut
  respond   **Snap to Center of circle X1**
  prompt   Specify center of donut
  respond   **Snap to Center of arc X2**
  prompt   Specify center of donut and **right-click.**

2 Repeat the donut command and:
  prompt   Specify inside diameter of donut and enter: **40 <R>**
  prompt   Specify outside diameter of donut and enter: **45 <R>**
  prompt   Specify center of donut and enter: **220,255 <R>**
  prompt   Specify center of donut
  respond   **Snap to Intersection of point Y1**
  prompt   Specify center of donut and **right-click.**

3 *Note*: the donut command allows repetitive entries to be made by the user, while the circle command only allows one circle to be created per command – I don't know why this is!

Ellipse

Ellipses are created by the user specifying:
a) either the ellipse centre and two axes endpoints
b) or three points on the axes endpoints.

1 Select from the menu bar **Draw-Ellipse-Center** and:
  prompt   Specify center of ellipse and enter: **350,150 <R>**
  prompt   Specify endpoint of axis and enter: **400,150 <R>**
  prompt   Specify distance to other axis and enter: **350,120 <R>**

2 Select the ELLIPSE icon from the Draw toolbar and:
  prompt   Specify axis endpoint of ellipse or (Arc/Center)
  enter   **C <R>** – the center option
  prompt   Specify center of ellipse
  respond   **Snap to Center icon and pick the existing ellipse**
  prompt   Specify endpoint of axis and enter: **@30,0 <R>**
  prompt   Specify distance to other axis and enter: **@0,60 <R>**

3 Menu bar with Draw-Ellipse-Axis,End and:
  prompt   Specify axis endpoint of ellipse
  respond   **Snap to Intersection of point F1**
  prompt   Specify other endpoint of axis
  respond   **Snap to Midpoint of line F2**
  prompt   Specify distance to other axis or [Rotation]
  enter   **R <R>** – the rotation option
  prompt   Specify rotation around major axis and enter: **60 <R>**

Note

1 At this stage your drawing should resemble Fig. 12.1, but without the text.

2 Save the layout as C:\BEGIN\DEMODRG for future recall if required. Remember that this will 'over-write' the existing C:\BEGIN\DEMODRG file.

3 Arcs, donuts and ellipses have centre points and quadrants which can be 'snapped to' with the object snap modes.

4 It is also possible to use the tangent snap icon and draw tangent lines, etc. between these objects. Try this for yourself.

Solid fill

Donuts are generally displayed on the screen 'solid', i.e. 'filled in'. This solid fill effect is controlled by the FILL system variable and can be activated from the menu bar or the command line.

1 Still with the DEMODRG layout on the screen?

2 Menu bar with **Tools-Options** and:
prompt   Options dialogue box
respond **pick the Display tab**
then   1. Apply solid fill OFF, i.e. no tick in box
        2. pick OK.

3 Menu bar with **View-Regen** and the donuts will be displayed without the fill effect.

4 At the command line enter **FILL <R>**
prompt   Enter mode [ON/OFF] <OFF>
enter   **ON <R>**

5 At the command line enter **REGEN <R>** to 'refresh' the screen and display the donuts with the fill effect.

6 *Note*: this fill effect also applies to polylines, which will be discussed in a later chapter.

Summary

Arcs, donuts and ellipses are Draw commands and can be created by coordinate entry or by referencing existing objects.

General

1   The three objects have a centre point and quadrants.

2   They can be 'snapped to' with the object snap modes.

3   Tangent lines can be drawn to and from them.

Arcs

1   Several different creation options.

2   Normally drawn in an anti-clockwise direction.

3   Very easy to draw in the wrong 'sense' due to the start and end points being selected wrongly.

4   Continuous arcs are possible.

5   A negative angle entry will draw the arc clockwise.

Donuts

1   Require the user to specify the inside and outside diameters.

2   Can be displayed filled of unfilled.

3   An inside radius of 0 will give a 'filled circle'.

4   Repetitive donuts can be created.

Ellipses

1   Two creation methods.

2   Partial ellipses (arcs) are possible.

3   The created ellipses are 'true', i.e. have a centre point.

Chapter 13

Layers and standard sheet 2

All the objects that have been drawn so far have had a continuous linetype and no attempt has been made to introduce centre or hidden lines, or even colour. AutoCAD has a facility called **LAYERS** which allows the user to assign different linetypes and colours to named layers. For example, a layer may for red continuous lines, another may be for green hidden lines, and yet another for blue centre lines. Layers can also be used for specific drawing purposes, e.g. there may be a layer for dimensions, one for hatching, one for text, etc. Individual layers can be 'switched' on/off by the user to mask out drawing objects which are not required.

The concept of layers can be imagined as a series of transparent overlays, each having its own linetype, colour and use. The overlay used for dimensioning could be switched off without affecting the remaining layers. Figure 13.1 demonstrates the layer concept with:
a) Five layers used to create a simple component. Each 'part' of the component has been created on 'its own' layer.
b) The layers 'laid on top of each other'. The effect is that the user 'sees' one component.

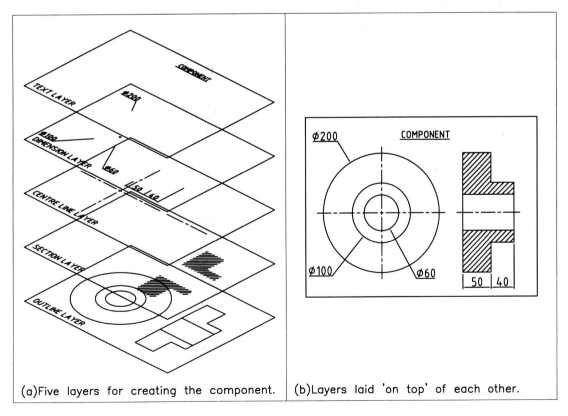

(a)Five layers for creating the component. (b)Layers laid 'on top' of each other.

Figure 13.1   Layer concept.

The following points are worth noting when considering layers:

1 All objects are drawn on layers.

2 Layers should be used for each 'part' of a drawing, i.e. dimensions should not be on the same layer as centre lines (for example).

3 New layers must be 'created' by the user, using the Layer Properties Manager dialogue box.

4 Layers are one of the most important concept in AutoCAD.

5 Layers are essential for good and efficient draughting.

As layers are very important, and as the user must have a sound knowledge of how they are used, this chapter is therefore rather long (and perhaps boring). I make no apology for this, as all CAD operators must be able to use layers correctly.

As a point of interest, try and complete this chapter at 'the one sitting' as it is important for all future drawing work.

Getting started

Several different aspects of layers will be demonstrated in this chapter. Once these concepts have been discussed, we will modify our existing standard sheet, so:

1 Open your C:\BEGIN\A3PAPER standard sheet.

2 Draw a horizontal and vertical line each of length 200, and a circle of radius 75 anywhere on the screen.

The Layer Properties Manager dialogue box

1 From the menu bar select **Format-Layer** ... and:
   prompt Layer Properties Manager dialogue box
   respond Study the layout of the dialogue box.

2 The Layer Properties Manager dialogue box has a number of distinct 'sections' as follows:
   a) Named layer filter with options:
      1. Show all layers – usually active
      2. Show all used layers – very useful to the user
      3. Show all Xref dependent layers
      4. Invert filter and Apply to Object Property toolbar toggle.
   b) Six selections: New, Delete, Current, Show details, Save state, Restore state.
   c) Current Layer: 0 – at present.
   d) Layer information display with:
      Name: only 0 at present
      Three layer states: On, Freeze, (L)ocked
      Color: White (with a black box)
      Linetype: Continuous
      Lineweight: Default
      Plot Style   Color_7 (probably?)
      Plot   Printer icon.

3 Layer 0 is the layer on which all objects have so far been drawn, and is 'supplied' with AutoCAD. It is the **current** layer and is displayed in the Objects Properties toolbar with the layer state icons.

4 Certain areas of the dialogue box are 'greyed out', i.e. inactive.

5 Move the pointing device arrow onto the 0 named line and:
   a) pick with a left-click
   b) the 'complete line' is highlighted in colour
   c) pick Show Details to reveal other information about the selected layer as Fig. 13.2.

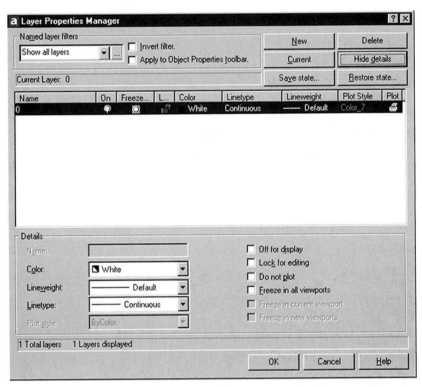

Figure 13.2   The Layer Properties Manager dialogue box.

6 Move the cursor along the highlighted area and pick the On/Off icon which is a yellow light bulb (indicating ON) and:
   prompt   AutoCAD layer message box – Fig. 13.3
   and   On/Off icon now a blue light bulb, indicating OFF
   respond   1. pick OK from the message box
                2. pick OK from the Layer Properties Manager dialogue box
   and   the drawing screen will be returned and no objects are displayed – they were all drawn on layer 0 which has been turned off.

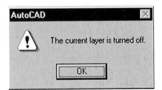

Figure 13.3   Layer Message dialogue box.

7 Re-activate the Layer Properties Manager dialogue box and:
   a) pick the 0 layer name – highlighted in colour
   b) pick the 'blue light' and it changes to yellow (ON)
   c) pick OK.

8 The line and circle objects displayed – layer 0 is on.

Linetypes

AutoCAD allows the user to display objects with different linetypes, e.g. continuous, centre, hidden, dotted, etc. Until now, all objects have been displayed with continuous linetype.

1 Activate the menu selection **Format-Layer** and:

prompt	Layer Properties Manager dialogue box
respond	**pick Continuous from layer 0 'line'**
prompt	Select Linetype dialogue box
with	Loaded linetypes:

Linetype	Appearance	Description
Continuous	_____	Solid line

respond	**pick Load ...**
prompt	Load or Reload Linetype dialogue box
with	*a*) Filename: **acadiso.lin**
	b) a list of all linetypes in the acadiso.lin file
respond	1. scroll (at right) and **pick CENTER** – Fig. 13.4
	2. hold down the **Ctrl** key
	3. scroll and **pick HIDDEN**
	4. pick OK from Load or Reload Linetypes dialogue box
prompt	Select Linetype dialogue box
with	Loaded Linetypes:

Linetype	Appearance	Description
CENTER	__ – __ __ – __	Center__ – __ __ – __
Continuous	_____	Solid line
HIDDEN	____ ____ __	Hidden__ __ __ __ __

result	Figure 13.5, i.e. we have loaded the CENTER and HIDDEN linetypes from the acadiso.lin file into our drawing and they are now ready to be used
respond	1. pick CENTER – it becomes highlighted
	2. pick OK from Select Linetype dialogue box
prompt	Layer Properties Manager dialogue box
with	layer 0 having CENTER linetype
respond	pick OK from Layer Properties Manager dialogue box.

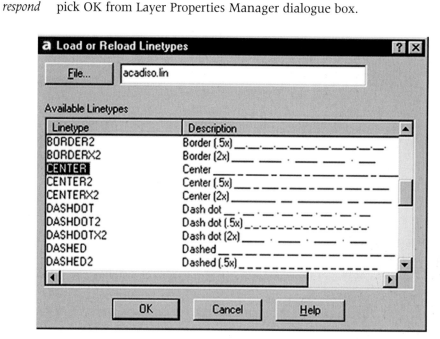

Figure 13.4 Load or Reload Linetypes dialogue box.

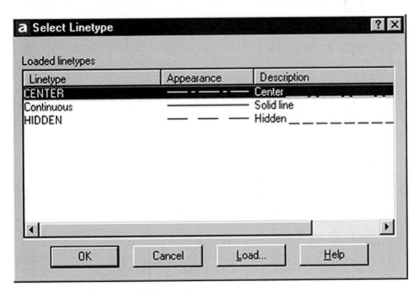

Figure 13.5 Select Linetype dialogue box.

2   The drawing screen will display the three objects and the border with center linetype appearance. Remember that AutoCAD is an American package – hence center, and not centre.

3   *Note*

Although we have used the Layer Properties Manager dialogue box to 'load' the CENTER and HIDDEN linetypes, linetypes can also to loaded by selecting from the menu bar **Format-Linetype** and the Linetype Manager dialogue box will be displayed. The required linetypes are loaded 'into the current drawing' in the same manner as described.

Colour

Individual objects can be displayed on the screen in different colours, but I prefer to use layers for the colour effect. This is achieved by assigning a specific colour to a named layer and will be discussed later in the chapter.

For this exercise:

1   Activate the Layer Properties Manager dialogue box with Show Details active.

2   *a*) Pick line 0 to activate the layer information
     b) pick the scroll arrow at Details Color: White to display the seven standard colours
     c) pick Red and:
        i)   White alters to Red at Details Color
        ii) layer 0 line displays Color: red square Red
     d) pick OK.

3   The screen displays the objects with red centre lines.

4   Note the Object Properties toolbar which displays:
     a) layer 0 with a red square
     b) the ByLayer colour is a red square
     c) the ByLayer linetype has a center linetype appearance.

Task

By selecting the Layers icon from the Object Properties toolbar, use the Layer Properties Manager dialogue box to:
a) set layer 0 linetype to Continuous
b) set layer 0 colour to White/Black
c) screen should display the original continuous white/black objects
d) note the Object Properties toolbar.

Note on colour

1 The default AutoCAD colour for objects is dependent on your screen configuration, but is generally:
either   *a*) white background with black lines
or   *b*) black background with white lines.

2 The white/black linetype can be confusing?

3 All colours in AutoCAD are numbered. There is a total of 255 colours, but the seven standard colours are:
1 red   2 yellow   3 green   4 cyan   5 blue   6 magenta   7 black/white.

4 The number or colour name can be used to select a colour. The numbers are associated with colour pen plotters.

5 The complete 255 'colour palette' can be activated from:
a) Layer Program Manager dialogue box by picking the coloured square under Color
b) from the Details: Color by scrolling and picking Others …

6 Generally only the seven standard colours will be used in the book, but there may be the odd occasion when a colour from the palette will be selected.

Creating new layers

Layers should be made to suit individual or company requirements, but for our purposes the layers which will be made for all future drawing work are:

Usage	Layer name	Layer colour	Layer linetype
General	0	white/black	continuous
Outlines	OUT	red	continuous
Centre lines	CL	green	center
Hidden detail	HID	number 12	hidden
Dimensions	DIMS	magenta	continuous
Text	TEXT	blue	continuous
Hatching	SECT	number 74	continuous
Construction	CONS	to suit	continuous

These seven layers must be 'made' by us (remember that layer 0 is given to us) so:

1 Close the existing drawing (no to save changes) then open your A3PAPER standard sheet

2 Menu bar with **Format-Layer** and:
prompt   Layer Properties Manager dialogue box
respond **pick New**
and   Layer1 added to layer list with the same properties as layer 0, i.e. white colour and continuous linetype
respond pick New until 7 new layers are added to the layer list, i.e. 8 layers in total:
     a) the original layer 0 and
     b) the added new layers, named layer1-layer7
a) *naming the new layers*
   1. move pick arrow onto layer1 and pick it – highlighted
   2. Show details active
   3. at Details, highlight Layer1 name and enter OUT
   4. Layer1 is renamed OUT in the layer list
   5. repeat the above steps and rename layer2-layer7 with the following names:
       layer2: CL   layer3: HID   layer4: DIMS
       layer5: TEXT   layer6: SECT   layer7: CONS

b) assigning the *linetypes*
1. pick layer line CL
2. move pick arrow to Linetype-Continuous and pick it
3. from the Select Linetype dialogue box, pick Load and load the CENTER and HIDDEN linetypes from the acadiso.lin file as discussed earlier
4. when the CENTER and HIDDEN linetype are displayed in the Select Linetype dialogue box, pick CENTER then OK
5. pick layer line HID
6. move pick arrow to Linetype-Continuous and pick it
7. the Select Linetype dialogue box should display HIDDEN
8. pick HIDDEN then OK
9. Layer Properties Manager dialogue box and layer CL should have a CENTER linetype, and layer HID a HIDDEN linetype

c) assigning the *colours*
1. pick layer line OUT
2. scroll at Details-Color and pick Red
3. pick the other layer lines and set the following colours:

    CL: green   HID: number 12   DIMS: magenta
    TEXT: blue   SECT: number 74   CONS: colour to suit.

3   At this stage the Layer Properties Manager dialogue box should resemble Fig. 13.6.

4   Pick OK from the Layer Properties Manager dialogue to save the layers (with linetypes and colours) which have been created.

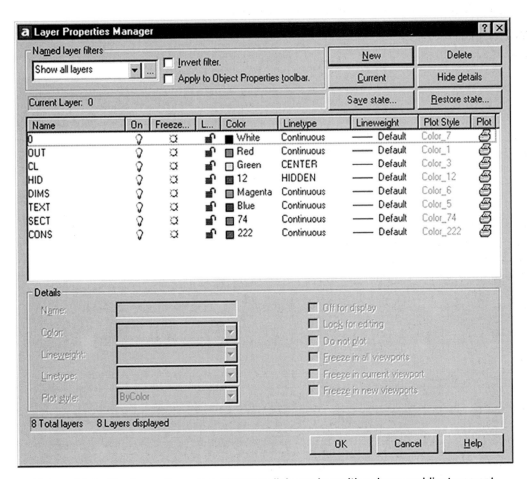

Figure 13.6 The Layer Properties Manager dialogue box with colours and linetypes set.

The current layer

The current layer is the one on which all objects are drawn. The current layer name appears in the Layer Properties Manager section of the Object Properties toolbar, which is generally docked below the Standard toolbar at the top of the screen. The current layer is also named in the Layer Properties Manager dialogue box when it is activated. The current layer is 'set' by the user.

1 Activate the Layer Properties Manager dialogue box and:
   prompt Layer Properties Manager dialogue box
   with   the layers created earlier displayed in numeric then alphabetical order, i.e. 0, CL, CONS ... TEXT
   respond 1. pick layer line OUT – becomes highlighted
         2. pick Current
         3. pick OK.

2 The drawing screen will be returned, and the Object Properties dialogue box will display OUT with a red box, and ByLayer will also display a red box.

   The Layer Properties Manager dialogue box is used to create new layers with their colours and linetypes. Once created, the current layer can be set:
   a) from the Layer Properties Manager dialogue box
   b) from the Object Properties toolbar by scrolling at the right of the layer name and selecting the name of the layer to be current.

   I generally start a drawing with layer OUT current, but this is a personal preference. Other users may want to start with layer CL or 0 as the current layer, but it does not matter, as long as the objects are eventually 'placed' on their correct layers.

   Having created layers it is now possible to draw objects with different colours and linetypes, simply by altering the current layer. All future work should be completed with layers used correctly, i.e. if text is to be added to a drawing, then the TEXT layer should be current.

Saving the layers to the standard sheet

1 Use the Layer Properties Manager dialogue box and make layer OUT current.

2 Select one of the following:
   a) the Save icon from the Standard toolbar
   b) menu bar with **File-Save.**

3 This selection will automatically update the **C:\BEGIN\A3PAPER** standard sheet drawing opened earlier.

4 The standard sheet has now been saved as a drawing file with:
   a) units set to metric
   b) sheet size A3
   c) grid, snap, etc. set as required
   d) several new layers
   e) a border effect on layer 0.

5 With the layers having been saved to the A3PAPER standard sheet, the layer creation process does not need to be undertaken every time a drawing is started. Additional layers can be added to the standard sheet at any time – the process is fairly easy?

Layer states

Layers can have different 'states', for example, they could be:
a) ON or OFF
b) THAWED or FROZEN
c) LOCKED or UNLOCKED.

The layer states are displayed both in the Layer Control box of the Objects Properties toolbar as well as in the Layer Properties Manager dialogue box itself. In both, the layer states are displayed in icon form. These icon states are:
a) yellow – ON, THAWED
b) bluey grey – OFF, FROZEN
c) lock and unlock should be obvious?

The following exercise will investigate layers:

1 Your A3PAPER standard sheet should be displayed with the black border and layer OUT current.

2 Using the seven layers (our created six and layer 0):
   a) make each layer current in turn
   b) draw a 50 radius circle on each layer anywhere within the border
   c) make layer 0 current
   d) toggle the grid off to display eight coloured circles.

3 The green circle will be displayed with center linetype and the hidden linetype circle displayed with colour number 12.

4 From the Layer Properties Manager dialogue box ensure that Show details is active and:
   a) pick On/Off icon on CL layer line – layer line highlighted
   b) note the Details – tick at *Off for display*
   c) pick OK – no green circle.

5 Layer Properties Manager dialogue box and:
   a) pick Freeze/Thaw icon on DIMS layer line – highlighted line
   b) note the Details – tick at *Freeze in all viewports*
   c) pick OK – no magenta circle.

6 Pick the scroll arrow at Layer in Object Properties toolbar and:
   a) note CL and DIMS display information!
   b) pick Lock/Unlock icon on HID layer – note icon appearance
   c) left-click to side of pull-down menu
   d) hidden linetype circle still displayed.

7 From the Object Properties layer pull-down:
   a) pick Freeze and Lock icons for SECT
   b) left-click to side
   c) no dark green circle.

8 Using the pull down layer menu effect:
   a) turn off, freeze and lock the TEXT layer
   b) no blue circle.

9 Activate the Layer Properties Manager dialogue box and:
   a) note the icon display
   b) make layer OUT current
   c) pick Freeze icon
   d) Warning message – **Cannot freeze the current layer**
   e) pick OK
   f) turn layer OUT OFF
   g) Warning message – **The current layer is turned off**
   h) pick OK from this message dialogue box
   i) pick OK from Layer Properties Manager dialogue box
   j) no red circle displayed.

10 The screen should now display:
   a) a black circle – drawn on layer 0
   b) a hidden linetype circle (colour 12) – on locked layer HID
   c) a circle on layer CONS coloured to your own selection
   d) the black border – drawn on layer 0.

11 Thus objects will **not be displayed** if their layer is **OFF or FROZEN**.

12 Make layer 0 current – easy for you?

13 Erase the hidden linetype circle – you cannot
   The prompt line displays: *1 was on a locked layer.*

14 Using the CEN object snap, draw a line from the centre of the hidden linetype circle to the centre of the black circle. You can reference the hidden linetype circle, although it is on a locked layer.

15 *a*) Make layer OUT current
   b) Warning message? – pick OK
   c) Draw a line from 50,50 to 200,200.
   d) No line is displayed – the reason should be obvious?

16 Erase the coloured circle on layer CONS then using the Layer Properties Manager dialogue box, investigate the Named layer filters:
   a) Show all layers: eight displayed
   b) Show all used layers: seven displayed – no CONS layer
   c) Show all Xref dependent layers – no layer names displayed
   d) display Show all layers then pick OK.

17 Using the Layer Properties Manager dialogue box:
   a) ensure all layers are: ON, THAWED and UNLOCKED
   b) pick OK
   c) seven circles and two lines displayed
   d) the CONS layer circle was erased.

18 The layer states can be activated using the Layer Properties Manager dialogue box, or the Object Properties pull-down menu.

Saving layer states

Layer states can be saved so that the user can restore the original layer 'settings' at any time. To demonstrate this:

1 Erase any lines to leave the circles.

2 At the command line enter **LAYER <R>** and:
   prompt   `Layer Properties Manager dialogue box`
   respond **pick Save State**
   prompt   `Save Layer State dialogue box`
   respond 1. enter A3PAPER as new layer state name
             2. make the following layer states active (tick):
                On/Off; Frozen/Thawed; Locked/Unlocked
             3. make the following later properties active:
                Color; Linetype; Lineweight
             4. dialogue box as Fig. 13.7
             5. pick OK
   prompt   `Layer Properties Manager dialogue box`
   respond pick OK.

3 Scroll at layer information from the Object Properties toolbar and:
   a) turn off and lock all layers
   b) message about current layer displayed – pick OK
   c) screen display is blank – obviously?

4 Pick the Layers icon from the Object Properties dialogue box and:
   prompt   `Layer Properties Manager dialogue box`
   with   all layers off and locked icons
   respond **pick Restore state**
   prompt   `Layer States Manager dialogue box`
   with   A3PAPER highlighted. If it is not, pick it
   then   *pick Restore*
   prompt   `Layer Properties Manager dialogue box`
   with   all layers on and unlocked icons
   respond **pick OK.**

5 The original objects will be displayed.

6 It may be useful to have the A3PAPER saved layer state added to our A3PAPER standard sheet. I will let you decide for yourself if this would be a worthwhile exercise.

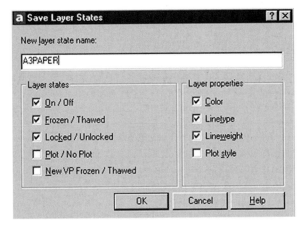

Figure 13.7 The Save Layer States dialogue box.

Make Object's Layer Current icon

Layers icon

Layers Previous icon

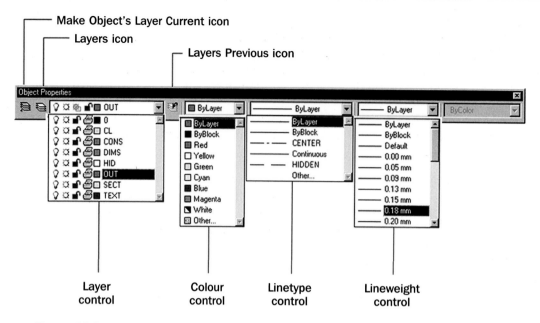

| Layer control | Colour control | Linetype control | Lineweight control |

Figure 13.8 The Object Properties toolbar information for C:\BEGIN\A3PAPER.

The Object Properties toolbar

The Object Properties toolbar is generally displayed at all times and is usually docked below the Standard toolbar at the top of the drawing screen under the title bar. This toolbar gives information about layers and their states, and Fig. 13.8 displays the complete toolbar details for our current drawing. The various 'areas' of the toolbar are:

a) *Make Object's Layer Current icon*
Select this icon and:
prompt   Select object whose layer will become current
respond **pick the green circle**
prompt   CL is now the current layer
and   CL (green) displayed in the Object Properties toolbar.

b) *Layers icon*
Selecting this icon will display the Layer Properties Manager dialogue box, i.e. an alternative to the menu bar selection Format-Layer.

c) *Pull down layer information*
Allows the user to quickly set a new current layer, or activate one of the layer states, e.g. off, freeze, lock, etc.

d) *Layer Previous icon*
Select this icon and:
prompt   Restored previous layer status
and   Layer OUT will again be current.

e) *Color control*
By scrolling at the colour control arrow, the user can select one of the standard colours. By selecting *Other* from the colour pull down menu, the Select Color dialogue box is displayed allowing access to all colours available in AutoCAD. Selecting a colour by this method will allow objects to be drawn with that particular colour, irrespective of the current layer colour. This is **not recommended** but may be useful on certain occasions, i.e. coloured objects should (ideally) be created on their own layer. It is **recommended** that the Color control always displays ByLayer.

f) *Linetype control*

Similar to colour control, the scroll arrow display all linetypes loaded in the current drawing. By selecting *Other* the Linetype Manager dialogue box will be displayed, allowing the user to load any other linetype from a named file – usually **acadiso.lin.** By selecting a linetype from the pull down menu, the user can create objects with this linetype, independent of the current layer linetype. Again this is **not recommended**. Layers should be used to display a certain linetype, but it may be useful to have different linetypes displayed on the one layer occasionally. It **is recommended** that the Layer control always displays **ByLayer.**

g) *Lineweight control*

Lineweight allows objects to be displayed (and plotted) with different thicknesses from 0 to 2.11 mm and the lineweight control scroll arrow allows the user to select the required thickness. We will be investigating this topic in greater detail in a later chapter, and will not discuss at this stage. The Lineweight control should display ByLayer.

Renaming and deleting layers

Unwanted or wrongly named layers can easily be deleted or renamed in AutoCAD 2002.

1   Still have the A3PAPER drawing on the screen with 7 coloured circles (the circle on layer CONS was erased. Erase any lines still displayed.

2   Make layer OUT current.

3   Activate the Layer Properties Manager dialogue box by picking the Layer icon from the Object Properties toolbar and:
   a) pick New twice to add two new layers to the list – Layer1 and Layer2
   b) rename Layer1 as NEW1 and Layer2 as NEW2
   c) pick **Show all used layers** and CONS, NEW1 and NEW2 will not be listed
   d) pick **Show all layers** then pick OK.

4   Making each new layer current in turn, draw a circle anywhere on the screen on each new layer.

5   Make layer OUT current.

6   *a)* Activate the Layer Properties Manager dialogue box and:
   respond   **pick NEW1 layer line then Delete**
   prompt   AutoCAD message – The selected layer was not deleted.
   b) Pick OK from this message box (Fig. 13.9) then pick OK from the Layer Properties Manager dialogue box.

Figure 13.9   The Layer warning message box.

7   *a)* Erase the two added circles
   b) activate the Layer Properties Manager dialogue box
   c) pick NEW1-Delete
   d) pick NEW2-Delete
   e) pick OK.

8   This completes this chapter, so exit AutoCAD (do **not** save changes). Hopefully your C:\BEGIN\A3PAPER standard sheet was saved with the layers and the border earlier in the chapter?

Assignment

There is no activity specific to layers, but all future drawings will be started by opening the C:\BEGIN\A3PAPER drawing file.

Summary

1 Layers are one of the most important concept in AutoCAD. Perhaps even the most important?

2 Layers allow objects to be created with different colours and linetypes.

3 Layers are created using the Layer Properties Manager dialogue box – often called layer control.

4 There are 255 colours available, but the seven Standard colours should be sufficient for most of the users needs.

5 Linetypes are loaded by the user as required.

6 Layers saved to a standard sheet need only be created once.

7 New layers can easily be added as and when required.

8 The layer states are:
   ON: all objects are displayed and can be modified
   OFF: objects are not displayed
   FREEZE: similar to OFF but the screen regenerates faster.
   THAW: undoes a frozen layer
   LOCK: objects are displayed but **cannot** be modified
   UNLOCK: undoes a locked layer.

9 Layer states are displayed and activated in icon form from the Layer Properties Manager dialogue box or using Layer Control from the Object Properties toolbar.

10 Care must be taken when modifying a drawing with layers which are turned off or frozen. More on this later.

11 Layers can be renamed at any time.

12 Unused layers can be deleted at any time.

13 Layer states can be saved and restored at any time.

Chapter 14

User exercise 1

By now you should have the confidence and ability to create line and circle objects by various methods, e.g. coordinate entry, referencing existing objects, etc.

Before proceeding to other draw and modify commands, we will create a working drawing which will be used to introduce several new concepts, as well as reinforcing your existing draughting skills. The exercise will also demonstrate how to:

a) open an existing drawing file
b) complete a drawing exercise
c) save a completed drawing with a new file name.

1   Start AutoCAD and:
    prompt   Startup dialogue box
    respond   **pick Open a Drawing icon**
    then   *a*) if A3PAPER displayed in file list:
                    1. pick A3PAPER.dwg
                    2. pick OK
    or   *b*) if A3PAPER not displayed in file list:
                    1. pick Browse – Select File dialogue box displayed
                    2. scroll at Look in and pick C: drive
                    3. double left-click the Begin folder
                    4. pick A3PAPER
                    5. pick Open.

2   Either selection method (a) or (b) will display the A3 standard sheet with layer OUT current

3   Refer to Fig. 14.1 and:
    a) draw full size the component given. Only layer OUT is used.
    b) a start point is given – **use it as it is important for future work**
    c) *do not attempt to add the dimensions*
    d) use absolute coordinates for the (50,50) start point then relative coordinates for the outline
    e) use absolute coordinates for the circle centres – some 'sums' are required, but these should give you no trouble?

4   When the drawing is complete, menu bar with **File-Save As** and:
    prompt   Save Drawing As dialogue box
    with   File name: A3PAPER
    respond   1. ensure C:\BEGIN is current folder to Save in
                2. alter file name to: **WORKDRG**
                3. pick Save

5   We have now opened our A3PAPER standard sheet, completed a drawing exercise and saved this drawing with a different name to that which was opened.

6   This is (at present) the method which will be used to complete all new drawing exercises.

Now continue to the next chapter.

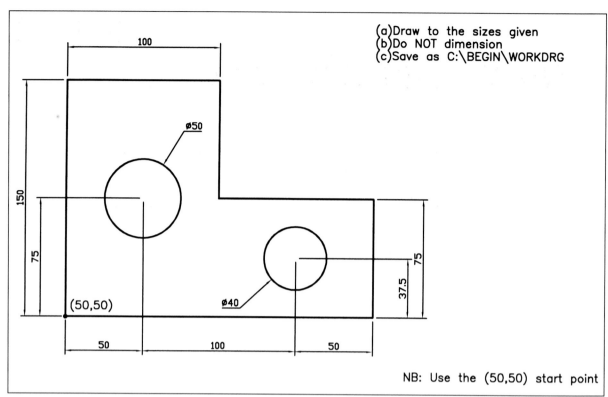

Figure 14.1 User exercise 1.

Chapter 15

Fillet and chamfer

In this chapter we will investigate how the fillet and chamfer commands can be used to modify an existing drawing. Both commands can be activated by icon, menu bar selection or keyboard entry.

1   Open the **C:\BEGIN\WORKDRG** created in the previous chapter or simply continue from the previous chapter.

2   Ensure layer OUT is current and display the Draw and Modify toolbars.

3   Refer to Fig. 15.1.

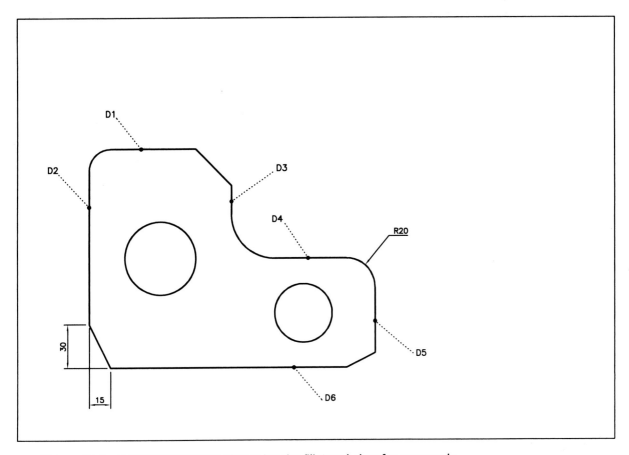

Figure 15.1   C:BEGIN\WORKDRG after using the fillet and chamfer commands.

Fillet

A fillet is a radius added to existing line/arc/circle objects. The fillet radius must be specified before the objects to be filleted can be selected.

1   Select the FILLET icon from the Modify toolbar and:
    prompt   Current settings: MODE=TRIM, Radius=??
               Select first object or [Polyline/Radius/Trim]
    enter   **R <R>** – the radius option
    prompt   Specify fillet radius<??>
    enter   **15 <R>**
    prompt   Select first object or [Polyline/Radius/Trim]
    respond **pick line D1**
    prompt   Specify second object
    respond **pick line D2.**

2   The corner selected will be filleted with a radius of 15, and the two 'unwanted line portions' will be erased and the command line will be returned.

3   From the menu bar select **Modify-Fillet** and:
    prompt   Select first object or [Polyline/Radius/Trim]
    enter   **R <R>** – the radius option
    prompt   Specify fillet radius<15.00>
    enter   **30 <R>**
    prompt   Select first object and: **pick line D3**
    prompt   Select second object and: **pick line D4.**

4   At the command line enter **FILLET <R>** and:
    a) set the fillet radius to 20
    b) fillet the corner indicated.

Chamfer

A chamfer is a straight 'cut corner' added to existing line objects. The chamfer distances must be 'set' prior to selecting the object to be chamfered.

1 Select the CHAMFER icon from the Modify toolbar and: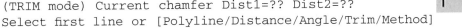
   prompt (TRIM mode) Current chamfer Dist1=?? Dist2=??
         Select first line or [Polyline/Distance/Angle/Trim/Method]
   enter **D <R>** – the Distance option
   prompt Specify first chamfer distance<??>
   enter **25 <R>**
   prompt Specify second chamfer distance<25.00>
   enter **25 <R>**
   prompt Select first line and: **pick line D1**
   prompt Select second line and: **pick line D3.**

2 The selected corner will be chamfered, the unwanted line portions removed and the command line returned.

3 Menu bar selection with **Modify-Chamfer** and:
   prompt Select first line or [Polyline/Distance/Angle/Trim/Method]
   enter **D <R>**
   prompt Specify first chamfer distance and enter: **10 <R>**
   prompt Specify second chamfer distance and enter: **20 <R>**
   prompt Select first line and: **pick line D5**
   prompt Select second line and: **pick line D6.**

4 Note that the pick order is important when the chamfer distances are different. The first line picked will have the first chamfer distance set.

5 At the command line enter **CHAMFER <R>** and:
   a) set first chamfer distance: 15
   b) set second chamfer distance: 30
   c) chamfer the corner indicated.

Saving

When the three fillets and three chamfers have been added to the component, select from the menu bar **File-Save**. This will automatically update the existing **C:\BEGIN\WORKDRG** drawing file.

Error messages

The fillet and chamfer commands are generally used without any problems, but the following error messages may be displayed at the command prompt:

1 Radius is too large.

2 Distance is too large.

3 Chamfer requires 2 lines (not arc segments).

4 No valid fillet with radius ??.

5 Lines are parallel – this is a chamfer error.

6 Cannot fillet an entity with itself.

These error messages should be self-evident to the user.

Fillet and chamfer options

Although simple to use, the fillet and chamfer commands have several options. It is in your own interest to attempt the following exercises, so:

1 Erase the filleted/chamfered component from the screen but ensure that C:\BEGIN\WORKDRG has been saved.

2 Refer to Fig. 15.2.

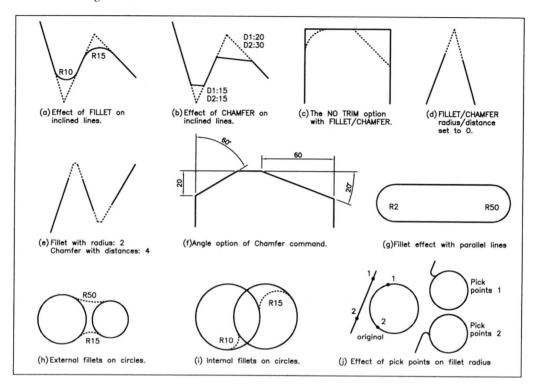

Figure 15.2 The fillet and chamfer options.

3 *Fillet/Chamfer with inclined lines*
   Our working drawing demonstrated the fillet and chamfer commands with lines which were at right angles to each other. This was not deliberate – just the way the component was drawn. Now draw three-four inclined lines then use the fillet and chamfer commands with the radius (R) and distance (D) values given. The effect of using the fillet/chamfer commands with inclined lines is displayed:
   a) fillet effect – fig. (a)
   b) chamfer effect – fig. (b).

4 Both the fillet and chamfer commands have a Polyline option and this option will be discussed when we have investigated the polyline command in a later chapter.

5 The two commands have a TRIM option and this effect is displayed in fig. (c). The option is obtained by entering **T <R>** at the prompt line after activated the fillet/chamfer command and setting the radius/distance value. The response is:
   prompt   Enter trim mode option [Trim/No Trim]<Trim>
   enter   *a*) T <R>: corner removed. This is the default
               b) N <R>: corners not removed.

   Note that if the no trim (N) option is used, this effect will always be obtained until the user 'resets' the trim (T) option.

6 The two commands can be used to extend two inclined lines to a point as demonstrated in fig. (d) with:

either   *a*) the fillet radius set to 0

or   *b*) both chamfer distances set to 0.

7 Chamfer with one distance set to 20 and the other set to 0. Think about this effect!

8 Interesting effect with inclined lines if the fillet radius or the chamfer distances are 'small'. The lines are extended if required, and the fillet/chamfer effect 'added at the ends'. This effect is displayed in fig. (e).

9 Chamfer has an angle option and when the command is selected:

prompt   Select first line or [Polyline/Distance/Angle/Trim/Method]

enter   **A <R>** – the angle option

prompt   Specify chamfer length on the first line and enter: **20<R>**

prompt   Specify chamfer angle from the first line and enter: **60<R>**

The required lines can now be chamfered as fig. (f) which displays:

a) length of 20 and angle of 60

b) length of 60 and angle of 20.

The user has to be careful when selecting the first line as this is used for the distance value entered.

10 The fillet command can be used with parallel lines, the effect being to add 'an arc' to the ends selected. This arc is added independent of the set fillet radius as fig. (g) demonstrates. The fillet radii was set to 2 and 50, but the added 'arc' is the same at both ends. In my fig. (g) example, the actual arc radius is 15. Any idea why?

11 Two circles can be filleted but are not 'trimmed' as lines. The fillet effect on circles is displayed:

a) externally as fig. (h)

b) internally as fig. (i).

12˙ Two circles cannot be chamfered. If any circle is selected:

prompt   Chamfer requires 2 lines (not arc segments).

13 Lines-circles-arcs can be filleted, but the position of the pick points is important as displayed in fig. (j).

14 Chamfer has a Method option and when **M** is entered:

prompt   Enter trim method [Distance/Angle]

and   *a*) entering **D** sets the two distance method (default)

           b) entering **A** sets the length and angle method.

Summary

1 FILLET and CHAMFER are Modify commands, activated from the menu bar, by icon selection or by keyboard entry.

2 Both commands require the radius/distances to be set before they can be used.

3 When values are entered, they become the defaults until altered by the user.

4 Lines, arcs and circles can be filleted.

5 Only lines can be chamfered.

6 A fillet/chamfer value of 0 is useful for extending two inclined lines to meet at a point.

7 Both commands have several useful options.

Chapter 16

The offset, extend, trim and change commands

In this chapter we will investigate OFFSET, EXTEND and TRIM – three of the most commonly used draughting commands. We will also investigate how an object's properties can be altered with the CHANGE command, and finally we will discuss LTSCALE, a system variable.

To demonstrate these new commands:

1 Open C:\BEGIN\WORKDRG – easy by now?

2 Ensure layer OUT is current and display the Draw and Modify toolbars.

Offset

The offset command allows to user to draw objects parallel to other selected objects and lines, circles and arcs can all be offset. The user specifies:
- an offset distance
- the side to offset the selected object.

1 Refer to Fig. 16.1.

2 Pick the OFFSET icon from the Modify toolbar and:
   prompt   Specify offset distance or [Through]
   enter   **50 <R>** – the offset distance
   prompt   Select object to offset or <exit>
   respond **pick line D1**
   prompt   Specify point on side to offset
   respond **pick any point to right of line D1 as indicated**
   and   line D1 will be offset by 50 units to right
   prompt   Select object to offset, i.e. any more 50 offsets
   respond **pick line D2**
   prompt   Specify a point on side to offset
   respond **pick any point to left of line D2 as indicated**
   and   line D2 will be offset 50 units to left
   prompt   Select object to offset, i.e. any more 50 offsets
   respond **right-click** to end command.

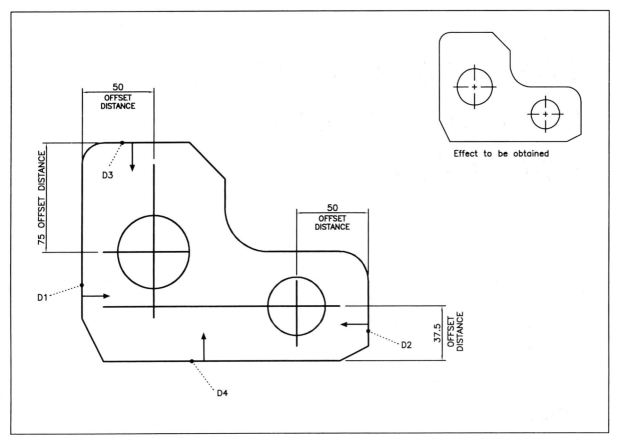

Figure 16.1 C:\BEGIN\WORKDRG after the OFFSET command.

3   Menu bar with **Modify-Offset** and:
    prompt   Specify offset distance or [Through]<50.00>
    enter   **75 <R>**
    prompt   Select object to offset and **pick line D3**
    prompt   Specify a point on side to offset and **pick as indicated.**

4   At the command line enter **OFFSET <R>** and:
    a) set an offset distance of 37.5
    b) offset line D4 as indicated.

5   We have now created lines through the two circle centres and later in the chapter we
    will investigate how these lines can be modified to be 'real centre lines'.

6   Continue to the next part of the exercise.

Extend

This command will extend an object 'to a boundary edge', the user specifying:
a) the actual boundary – an object
b) the object which has to be extended.

1   Refer to Fig. 16.2(a) and with SNAP OFF, select the EXTEND icon from the Modify toolbar and:

prompt   Current settings: Projection=UCS, Edge=None
        Select boundary edges ...
        Select objects
respond **pick line D1**
prompt   1 found
        Select objects, i.e. any more boundary edges
respond **pick line D2**
prompt   1 found, 2 total
        Select objects
respond **right-click** to end boundary edge selection
prompt   Select object to extend or shift-select to trim or
        [Project/Edge/Undo]
respond **pick lines D3, D4 and D5 then right-click-Enter.**

2   The three lines will be extended to the selected boundary edges

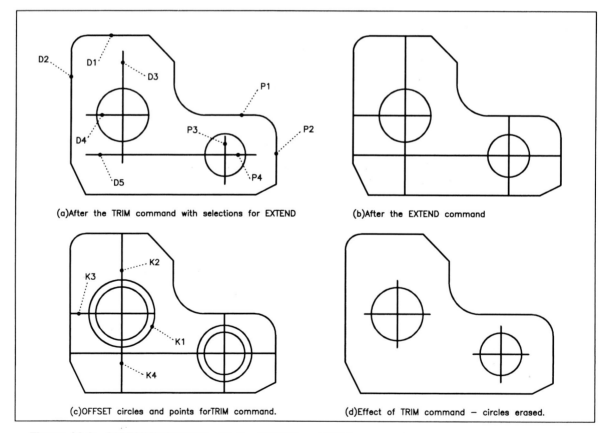

(a)After the TRIM command with selections for EXTEND   (b)After the EXTEND command

(c)OFFSET circles and points forTRIM command.   (d)Effect of TRIM command – circles erased.

Figure 16.2   C:\BEGIN\WORKDRG with the EXTEND and TRIM commands.

3   From the menu bar select **Modify-Extend** and:
   prompt   Select objects, i.e. the boundary edges
   respond   **pick lines P1 and P2 then right-click**
   prompt   Select object to extend
   respond   **pick lines P3 and P4 then right-click an pick Enter.**

4   At the command line enter **EXTEND <R>** and extend the two vertical 'centre lines' to lower horizontal outline.

5   When complete, the drawing should resemble Fig. 16.2(b).

Trim

Allows the user to trim an object 'at a cutting edge', the user specifying:
a) the cutting edge – an object
b) the object to be trimmed.

1   Refer to Fig. 16.2(c)and OFFSET the two circles for a distance of 5 'outwards' – easy?

2   Extend the top horizontal 'circle centre line' to the offset circle – should be obvious why?

3   Select the TRIM icon from the Modify toolbar and:
   prompt   Current settings: Projection=UCS, Edge=None
            Select cutting edges ...
            Select objects
   respond   **pick circle K1 then right-click**
   prompt   Select object to trim or shift-select to extend or
            [Project/Edge/Undo]...
   respond   **pick lines K2, K3 and K4 then right-click-Enter.**

4   From menu bar select **Modify-Trim** and:
   prompt   Select objects, i.e. the cutting edge
   respond   **pick the other offset circle the right-click**
   prompt   Select object to trim
   respond   **pick the four circle centre lines then right-click.**

5   Erase the two offset circles.

6   Now have 'neat lines' through the circle centres as Fig. 16.2(d).

7   At this stage select the Save icon from the Standard toolbar to automatically update C:\BEGIN\WORKDRG. We will recall it shortly.

Additional exercises

Offset, extend and trim are powerful commands and can be used very easily. To demonstrate additional use for the commands, try the examples which follow:

1 Erase all objects from the screen – have you saved WORKDRG?

2 Refer to Fig. 16.3 and attempt the exercise which follow.

3 *Offset for circle centre point*
Using OFFSET to obtain a circle centre point is one of the most common uses for the command. Figure 16.3(a) demonstrates offsets of 18.5 horizontally and 27.8 vertically to position the circle centre point.
Question: what about inclined line offsets?

4 *Offset through*
This is a very useful option of the command, as it allows an object to be offset through a specified point. When the command is activated:

prompt	Specify offset distance or [Through]
enter	**T <R>** – the through option
prompt	Select object to offset or <exit>
respond	pick the required object, e.g. a line
prompt	Specify through point
respond	**Snap to Center icon and pick the circle**
prompt	Specify object to offset, i.e. any more through offsets
respond	**<RETURN>** to end command.

The line will be offset through the circle centre – fig. (b).

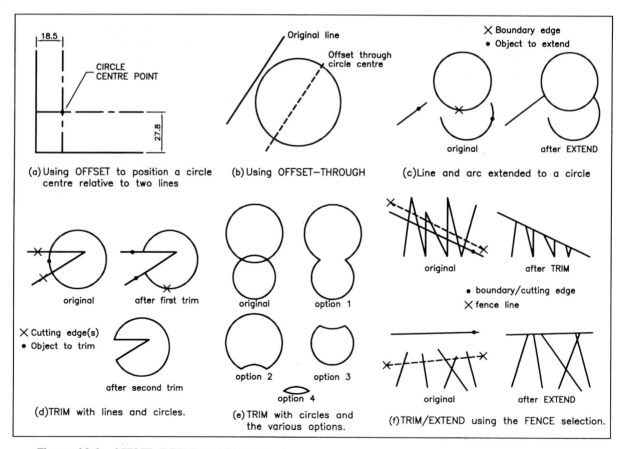

(a) Using OFFSET to position a circle centre relative to two lines

(b) Using OFFSET–THROUGH

(c) Line and arc extended to a circle

(d) TRIM with lines and circles.

(e) TRIM with circles and the various options.

(f) TRIM/EXTEND using the FENCE selection.

Figure 16.3 OFFSET, EXTEND and TRIM exercises.

5 *Extending lines and arcs*
   Lines and arcs can be extended to other objects, including circles – fig. (c).

6 *Trim lines and circles*
   Lines, circles and arcs can be trimmed to 'each other' and fig. (d) demonstrates trimming lines with a circle.

7 *Trimming circles*
   Circles can be trimmed to each other, but the selected objects to be trimmed can give different effects. Can you obtain the various options in fig. (e)?

8 *Trim/Extend with a fence selection*
   When several objects have to be trimmed or extended, the fence selection option can be used – fig. (f). The effect is achieved by:
   a) activating the command
   b) selecting the boundary (extend) or cutting edge (trim)
   c) entering **F <R>** – the fence option
   d) draw the fence line then right-click.

9 *Trim/Extend toggle effect*
   When the trim/extend command is activated and an object selected for the cutting edge/boundary, the prompt line will display:

   ```
   Select object to trim (extend) or shift-select to extend (trim)
   ```

   By using the shift key the user can 'toggle' between the two commands. This is very useful. Try it with a few lines.

10 *The RETURN key*
    If the trim/extend command is activated and the <RETURN> key is pressed at the `Select objects` first prompt, then the user can trim/extend any object without having to select a cutting edge or boundary edge, i.e. every object on the screen is a cutting or boundary edge. Try this for yourself.

    This completes the additional exercises, which do not need to be saved.

Changing the offset centre lines

The WORKDRG drawing was saved with 'centre lines' obtained using the offset, extend and trim commands. These lines pass through the two circle centres, but they are continuous lines and not centre lines. We will modify these lines to be centre lines using the CHANGE command. This command will be fully investigated in a later chapter, but for now:

1   Re-open C:\BEGIN\WORKDRG saved earlier in this chapter.

2   At the command line enter **CHANGE <R>** and:

prompt	Select objects
respond	**pick the four offset centre lines then right-click**
prompt	Specify change point or [Properties]
enter	**P <R>** – the properties option
prompt	Enter property to change [Color/Elev/LAyer/LType/ ltScale/LWeight/Thickness]
enter	**LA <R>** – the layer option
prompt	Enter new layer name<OUT>
enter	**CL <R>**
prompt	Enter property to change, i.e. any more changes?
respond	**right-click-Enter** to end command.

3   The four selected lines will be displayed as green centre lines, as they were changed to the CL layer. This layer was made with centre linetype and colour green. Confirm with Format-Layer if you are not convinced!

4   Although the changed lines are centre lines, their 'appearance' may not be ideal and an additional command is required to 'optimise' the centre line effect. This command is LTSCALE.

5   As stated, the CHANGE command will be investigated in more detail in a later chapter.

LTSCALE

LTSCALE is a system variable used to 'alter the appearance' of non-continuous lines on the screen. It has a default value of 1.0 and this value is altered by the user to 'optimise' centre lines, hidden lines, etc. To demonstrate its use:

1 At the command line enter **LTSCALE <R>** and:
   prompt   Enter new linetype scale factor<1.0000>
   enter   **0.6 <R>**

2 The four centre lines should now be 'better defined'.

3 The value entered for LTSCALE depends on the type of lines being used in a drawing, and can be further refined – more on this in a later chapter.

4 Try other LTSCALE values until you are satisfied with the appearance of the four centre lines, then save the drawing as C:**BEGIN\\WORKDRG.**

5 *Note*
   a) LTSCALE must be entered from the command line. There is no icon or menu bar sequence to activate the command.
   b) The LTSCALE system variable is **GLOBAL**. This means that when its value is altered, all linetypes (centre, hidden, etc.) will automatically be altered to the new value. In a later chapter we will discover how this can be 'overcome'.

Question

In our offset exercise we obtained four circle centre lines using the offset command. These lines were then changed to the CL layer to display then as 'real centre lines'.

The question I am repeatedly asked by new AutoCAD users is: '**Why not use offset with the CL layer current?**' This question is reasonable so to investigate it:

1 C:\\BEGIN\\WORKDRG should still be displayed with the four 'changed' centre lines.

2 Make layer CL current.

3 Set an offset distance of 30 and offset:
   a) any red perimeter line
   b) any green centre line.

4 The effect of the offset command is:
   a) the offset red outline is a red outline
   b) the offset green centre line is a green centre line.

5 The offset command will therefore offset an object 'as it was drawn' and is independent of the current layer.

Match properties

This is a very useful 'tool' to the user, as it does exactly what it says – it matches properties.

1   Erase the two offset lines, still with layer CL current.

2   Select the Match Properties icon from the Standard toolbar and:
    prompt   Select source object
    respond   **pick any green centre line**
    prompt   Select destination object(s) or [Settings]
    respond   **pick the two circles then right-click-Enter.**

3   The circles will now be displayed as green centre lines.

4   Menu bar with **Modify-Match Properties** and:
    a) pick any red line as source
    b) pick the two circles as the destination objects
    c) right-click-enter.

5   The circles will now be displayed as red outlines.

6   *Settings*
    The Match Properties command as a settings option, which allows the user to determine which properties can be matched. To use this option:
    a) activate the Match Properties command
    b) pick any object as the source
    c) at the destination prompt, enter **S <R>** to display the Property Settings dialogue box – Fig. 16.4
    d) generally all the six Basic Properties and the three Special Properties are active at all times, but the user can alter these if they do not want a specific property matched
    e) cancel the dialogue box then ESC to end the command.

This completes the exercises in this chapter. Ensure you have saved the WORKDRG with the four green centre lines.

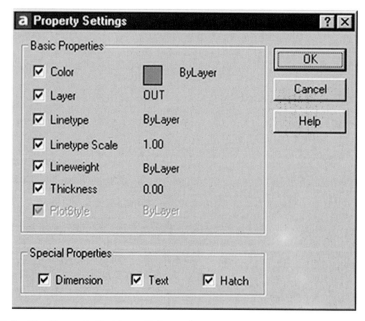

Figure 16.4   The Property Settings dialogue box.

Assignment

It is now some time since you have attempted any activities on your own. Refer to the Activity 5 drawing and create the two template shapes as given. Use the commands already investigated and make particular use of FILLET, CHAMFER, OFFSET and TRIM as much as possible. **Do not attempt to add dimensions.**

Start with your C:\BEGIN\A3PAPER standard drawing file and save the completed drawing as C:\BEGIN\ACT5.

Summary

1 OFFSET, TRIM and EXTEND are modify commands and can be activated by icon selection, from the menu bar or by keyboard entry.

2 Lines, arcs and circles can be offset by:
   a) entering an offset distance and selecting the object
   b) selecting an object to offset through a specific point.

3 Extend requires:
   a) a boundary edge
   b) objects to be extended.

4 Trim requires:
   a) a cutting edge
   b) objects to be trimmed

5 Lines and arcs can be extended to other lines, arcs and circles.

6 Lines, circles and arcs can be trimmed.

7 The extend and trim commands allow the user to 'toggle' between each command by using the shift key.

Chapter 17

User exercise 2

In this chapter we will create another working drawing using the commands discussed in previous chapters. The drawing will be saved and then we will list all drawings so far been completed.

1   Open your C:\BEGIN\A3PAPER standard sheet with layer OUT current and display the Draw and Modify toolbars.

2   Refer to Fig.17.1 and complete the drawing using:
    a) the basic shape from six lines : **USE THE 140,100 START POINT**
    b) three offset lines
    c) four extended lines
    d) trim to give the final shape.

3   When complete save as **C:\BEGIN\USEREX.**

4   Do not exit AutoCAD.

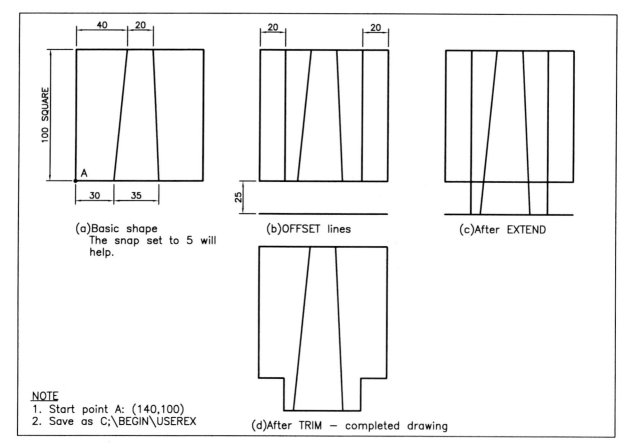

(a)Basic shape
The snap set to 5 will help.

(b)OFFSET lines

(c)After EXTEND

(d)After TRIM — completed drawing

NOTE
1. Start point A: (140,100)
2. Save as C;\BEGIN\USEREX

Figure 17.1   C:\BEGIN\USEREX construction.

So far, so good

At this stage of our learning process, we have completed the following drawings:

1   A demonstration drawing DEMODRG, used in the creation of lines, circles, arcs and for investigating object snap.

2   A standard sheet with layers and other settings, saved as A3PAPER.

3   A working drawing WORKDRG which was used to demonstrate several new topics. This drawing will still be used to investigate other new topics as they are discussed.

4   A new working drawing USEREX, which will also be used to demonstrate new topics.

5   Five activity exercises, ACT1 to ACT5, which you have been completing and saving as the book progresses. Or have you?

And now

Proceed to the next chapter in which we will discuss how text can be added to a drawing.

Chapter 18

Text

Text should be added to a drawing whenever possible. This text could simply be a title and date, but could also be a parts list, a company title block, notes on costing, etc. AutoCAD 2002 has text suitable for:
a) short entries, i.e. a few lines
b) larger entries, i.e. several lines.

In this chapter we will consider the short entry type text, and leave the other type (multi-line entry) to a later chapter, so:

1   USEREX still on the screen – it should be. If it is not, then open the appropriate drawing file.

2   Make layer Text (blue) current and refer to Fig. 18.1.

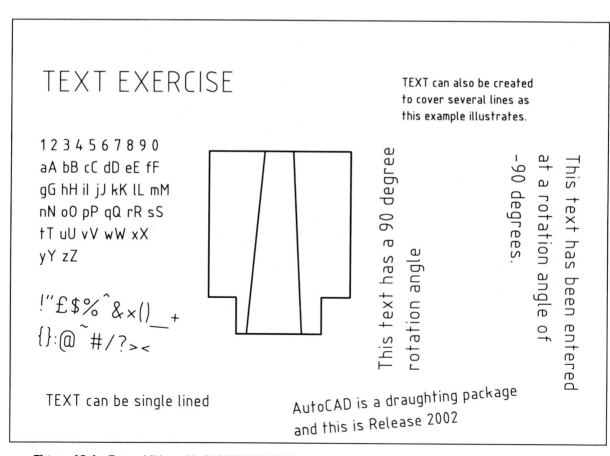

Figure 18.1   Text addition with C:\BEGIN\USEREX.

One line of text

1 Menu bar with **Draw-Text-Single line text** and:

   prompt   `Specify start point of text or [Justify/Style]`
   enter   **25,25 <R>**
   prompt   `Specify height<??>` and enter: **8 <R>**
   prompt   `Specify rotation angle of text<0.0>` and enter: **0 <R>**
   prompt   `Enter text`
   enter   **TEXT can be single lined <R>**
   prompt   `Enter text`, i.e. any more lines of text
   respond   **press the <RETURN> key** to end the command.

2 The entered text item will be displayed at the entered point. The command must be ended with a return key press.

3 Repeat the Draw-Text-Single line text sequence with:
   a) start point: 20,240 <R>
   b) height: 15 <R>
   c) rotation: 0 <R>
   d) text: TEXT EXRECISE <R><R> – two returns
      NB: this spelling is deliberate!

4 *Note*: the text is displayed on the screen as you enter it from the keyboard. This is referred to as *dynamic* text.

Several lines of text

1 Select from the menu bar **Draw-Text-Single Line Text** and:
   prompt   `Specify start point of text` and enter: **275,245 <R>**
   prompt   `Specify height` and enter: **6 <R>**
   prompt   `Specify rotation angle of text` and enter: **0 >R>**
   prompt   `Enter text` and enter: **TEXT can also be created <R>**
   prompt   `Enter text` and enter: **to cover several lines as <R>**
   prompt   `Enter text` and enter: **this example illustrates <R>**
   prompt   `Enter text` and respond: **<RETURN>**

2 At the command line enter **TEXT <R>** and:
   prompt   `Specify start point of text` and enter: **200,20 <R>**
   prompt   `Specify height` and enter: **8 <R>**
   prompt   `Specify rotation angle of text` and enter: **5 <R>**
   prompt   `Enter text` and enter: **AutoCAD is a draughting package<R>**
   prompt   `Enter text` and enter: **and this is Release 2002 <R>**
   prompt   `Enter text` and enter: **<R>** – to end command.

3 Using the menu bar sequence Draw-Text-Single Line Text or the command line entry TEXT, add the other items of text shown in Fig. 18.1. The start point, height and rotation angle are at your discretion.

Editing existing screen text

Hopefully text which has been entered on a drawing is correct, but there may be spelling mistakes and/or alterations to existing text may be required. Text can be edited as it is being entered from the keyboard if the user notices the mistake.

Screen text which needs to be edited requires a command.

1   From the menu bar select **Modify-Object-Text-Edit** and:

    prompt   Select an annotation object or [Undo]
    respond **pick the TEXT EXRECISE item**
    prompt   Edit Text dialogue box – Fig. 18.2 with the text phrase highlighted
    either   1. retype the highlighted phrase correctly then pick OK
    or   2. *a*) left-click at right of the text item
                b) backspace to remove error
                c) retype correctly
                d) pick OK
    or   3. *a*) move cursor to TEXT EXR|ECISE and left-click
                b) backspace to give TEXT EX|ECISE
                c) move cursor to TEXT EXE|CISE and left-click
                d) enter R to give TEXT EXER|CISE
                e) pick OK
    prompt   Select an annotation object, i.e. any more selections
    respond **right-click and Enter.**

2   The text item will now be displayed correctly.

3   *Note* You could always erase the text item and enter it correctly?

4   AutoCAD has a built-in spell checker which 'uses' a dictionary to check the spelling. This dictionary can be changed to suit different languages and must be 'loaded' before it can be used. We will use a British English dictionary so at the command line:

    enter   **DCTMAIN <R>**
    prompt   Enter new value for DCTMAIN (main dictionary)
    enter   **ens <R>** – British English.

5   From the menu bar select **Tools-Spelling** and:

    prompt   Select object
    respond **pick AutoCAD is a draughting package then right-click**
    prompt   Check Spelling dialogue box
    with   Current word: AutoCAD and suggestions
    respond **pick Ignore**
    then   Current word: Draughting
    with   Suggestion: draughtiness (and draught) – Fig. 18.3(a)
    respond **pick Ignore**
    prompt   AutoCAD Message box – Fig. 18.3(b)
    respond pick OK – spell check is complete.

6   Save the screen layout if required, **But not as C:\BEGIN\USEREX.**

Figure 18.2 The Edit Text dialogue box.

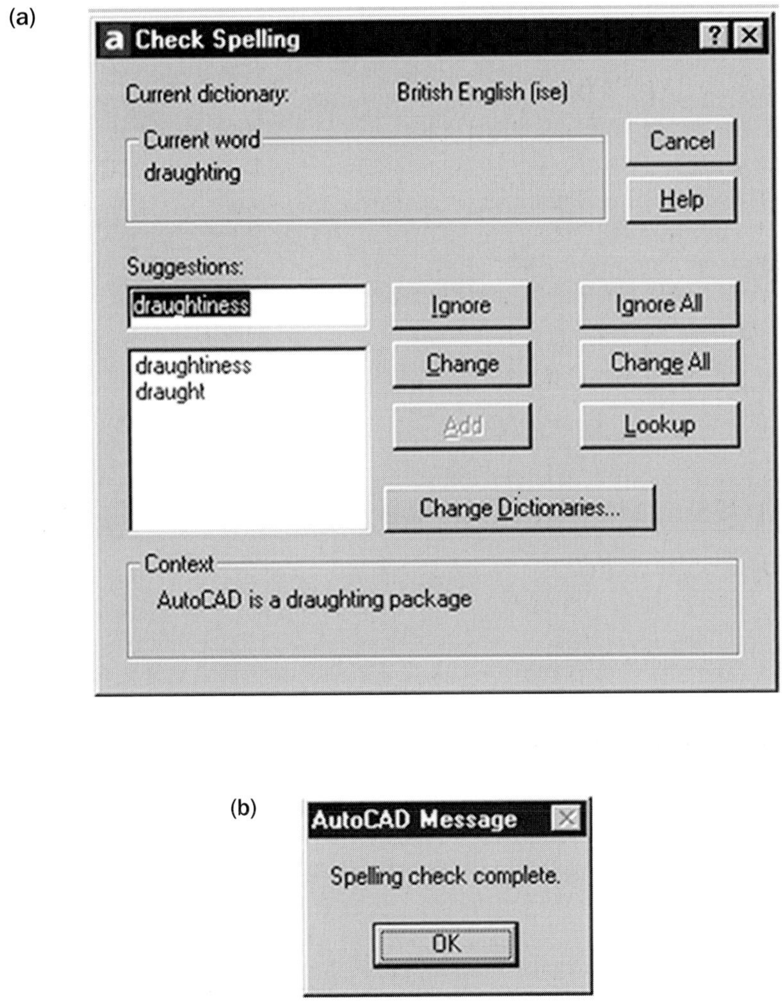

(a)

Figure 18.3 The Check Spelling and Message dialogue boxes.

Text justification

Text items added to a drawing can be 'justified' (i.e. positioned) in different ways, and AutoCAD 2002 has several justification positions, these being:
a) six basic: left, aligned, fitted, centred, middled, right
b) nine additional: TL, TC, TR, ML, MC, MR, BL, BC, BR.

1 Open your A3PAPER standard sheet and refer to Fig. 18.4.

2 With layer OUT current, draw the following objects:
   a) a 100 sided square, the lower left point at (50,50)
   b) a circle of radius 50, centred at (270,150)
   c) five lines:
     1. start point at 220,20 and end point at 320,20
     2. start point at 220,45 and end point at 320,45
     3. start point at 120,190 and end point at 200,190
     4. start point at 225,190 and end point at 275,250
     5. start point at 305,250 and end point at 355,100.

3 Make layer TEXT (blue) current and display the Draw, Modify and Object Snap toolbars.

4 At the command line enter **TEXT <R>** and:
prompt	Specify start point of text or [Justify/Style]
enter	**25,240 <R>**
prompt	Specify height and enter: **6 <R>**
prompt	Specify rotation angle of text and enter: **0 <R>**
prompt	Enter text and enter: **This is NORMAL <R>**
prompt	Enter text and enter:, **i.e. LEFT justified text. <R>**
prompt	Enter text and enter: **It is the default <R>**
prompt	Enter text and enter: **<R>**

5 Menu bar with **Draw-Text-Single Line Text** and:
prompt	Specify start point of text or [Justify/Style]
enter	**J <R>** – the justify option
prompt	Enter an option [Align/Fit/Center/Middle/Right/TL/TC ...
enter	**R <R>** – the right justify option
prompt	Specify right endpoint of text baseline
respond	**Snap to Midpoint icon and pick right vertical line of the square**
prompt	Specify height and enter: **8 <R>**
prompt	Specify rotation angle of text and enter: **0 <R>**
prompt	Enter text and enter: **This is RIGHT <R>**
prompt	Enter text and enter: **justified text <R> and <R>**

6 Repeat the single line text command and:
prompt	Specify start point of text or [Justify/Style]
enter	**J <R>**
prompt	Enter an option [Align/Fit..
enter	**C <R>** – center option
prompt	Specify center point of text
respond	**Snap to Midpoint icon and pick line 1**
prompt	Specify height and enter: **10 <R>**
prompt	Specify rotation angle of text and enter: **0 <R>**
prompt	Enter text and enter: **CENTERED <R> and <R>**

7 Enter **TEXT <R>** at the command line then:
   a) enter **J <R>** for justify
   b) enter **M <R>** for middle option
   c) Middle point: Snap to Midpoint icon and pick line 2
   d) Height: 10 and Rotation: 0
   e) Text: MIDDLE **<R><R>**

8 Activate the single line text command with the Fit justify option and:
   prompt `Specify first endpoint of text baseline`
   respond **Snap to Endpoint icon and pick left end of line 3**
   prompt `Specify second endpoint of text baseline`
   respond **Snap to Endpoint icon and pick right end of line 3**
   prompt `Specify height` and enter: **15 <R>**
   prompt `Enter text` and enter: **FITTED TEXT <R><R>**
   NB: this option has no rotation prompt!

9 Using the TEXT command with the Fit justify option, add the text item FITTED TEXT
   with a height of 40 to line 4.

10 With TEXT again:
    a) select the Align justify option
    b) pick 'top' of line 5 as first endpoint of text baseline
    c) pick 'bottom' of line 5 as second endpoint of text baseline
    d) text item: ALIGNED TEXT
    NB: this option has no height or rotation prompt!

11 Activate the single line text command and:
    a) justify with TL option
    b) Top/Left point: Intersection icon – pick top left of square
    c) Height of 8 and rotation of 0
    d) text item: TOP-LEFT.

12 Finally with the TEXT command and:
    a) justify with MC option
    b) Middle point: Snap to center of circle
    c) height of 8 and rotation of 0
    d) text item: MIDDLE-CENTER.

13 Your drawing should now resemble Fig. 18.4. It can be saved, but we will not use it again.

14 The text justification options are easy to use. Simply enter the appropriate letter for the
    justification option at the command line. The entered letters are:

A: Align	F: Fit	C: Center
M: Middle	R: Right	TL: Top left
MC: Middle center	BR: Bottom right	etc., etc.

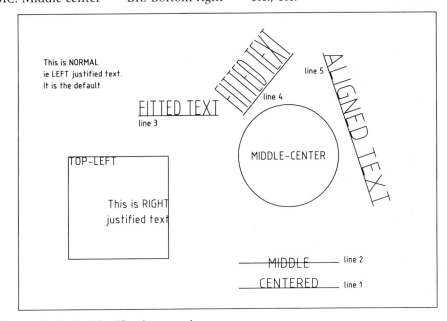

Figure 18.4 Text justification exercise.

Text style

When the single line text command is activated, one of the options available is Style. This option allows the user to select any one of several previously created text styles. This will be discussed in a later chapter which will investigate text styles and fonts.

Assignment

Attempt activity 6 which has four simple components for you to complete. Text should be added, but not dimensions.

The procedure is:

1. Open A3PAPER standard sheet.

2. Complete the drawings using layers correctly.

3. Save as C:\BEGIN\ACT6.

4. Read the summary then progress to the next chapter.

Summary

1. Text is a draw command which allows single/several lines of text to be entered from the keyboard.

2. The command is activated:
   a) from the keyboard with TEXT <R>
   b) from the menu bar with Draw-Text-Single line text.

3. The entered text is dynamic, i.e. the user 'sees' the text on the screen as it is entered from the keyboard.

4. Text can be entered with varying height and rotation angle.

5. Text can be justified to user specifications, and there are several justification options.

6. Screen text can be modified with:
   a) menu bar Modify-Objects-Text-Edit
   b) keyboard entry: DDEDIT.

7. Fitted and aligned text are similar, the user selecting both the start and end points of the text item, but:
   a) fitted: allows the user to enter a text height
   b) aligned: text is automatically 'adjusted' to suit the selected points
   c) neither has a rotation option.

8. Centered and middled text are similar, the user selecting the center/middle point, but:
   a) centered: is about the text baseline
   b) middled: is about the text middle point.

9. *a*) Text styles can be set by the user
   b) Multiple line (or paragraph) text can be set by the user
   c) These two topics will be investigated in later chapters.

Chapter 19

Dimensioning

AutoCAD has both automatic and associative dimensioning, the terms meaning:

Automatic: when an object to be dimensioned is selected, the actual dimension text, arrows, extension lines, etc. are all added in one operation

Associative: the arrows, extension lines, dimension text, etc. which 'make up' a dimension are treated as a single object.

AutoCAD allows different 'types' of dimensions to be added to drawings, these being displayed in Fig. 19.1. They are:

1   Linear: horizontal, vertical and aligned.

2   Baseline and Continue.

3   Ordinate: both X-datum and Y-datum.

4   Angular.

5   Radial: diameter and radius.

6   Leader: taking the dimension text 'outside' the object.

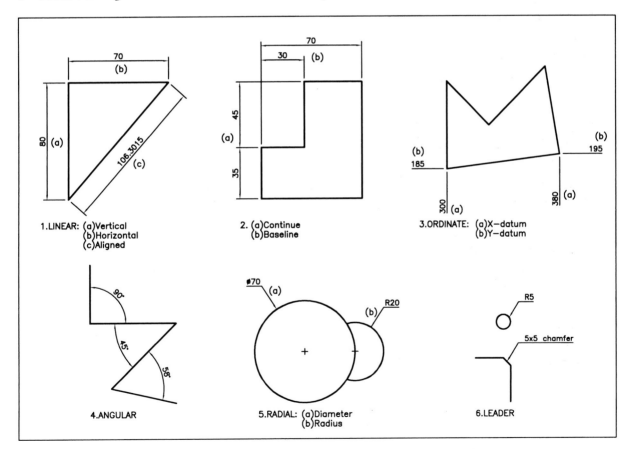

Figure 19.1   Dimension types.

Dimension exercise

To demonstrate how dimensions are added to a drawing, we will use one of our working drawings, so:

1   Open C:\BEGIN\USEREX and with layer OUT current add the following three circles:
    a) centre: 220,180, radius: 15
    b) centre: 220,115, radius: 3
    c) centre: 190,75, radius: 30
    d) trim the large circle to the line as Fig. 19.2.

2   Make layer DIMS current and activate the Object Snap and Dimension toolbars.

3   In the exercises which follow, the appearance of the dimensions in Fig. 19.2 may differ from those on your system. This is to be expected at this stage.

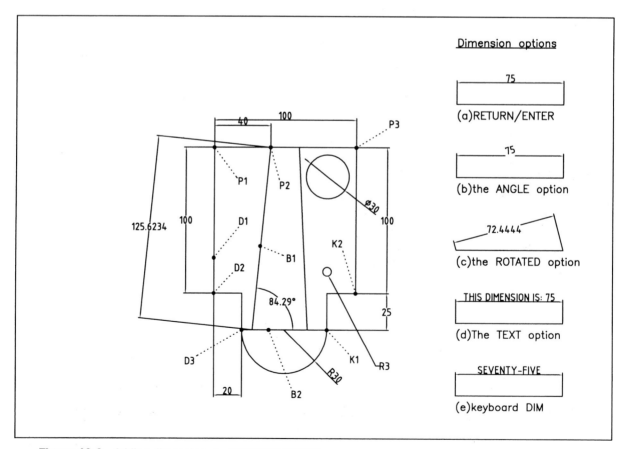

Figure 19.2   Adding dimensions to C:\BEGIN\USEREX.

Linear dimensioning

1   Select the LINEAR DIMENSION icon from the Dimension toolbar and:
   prompt   Specify first extension line origin or <select object>
   respond   **Endpoint icon and pick line D1**
   prompt   Specify second extension line origin
   respond   **Endpoint icon and pick the other end of line D1**
   prompt   Specify dimension line location or [Mtext/Text..
   respond   **pick any point to the left of line D1**

2   From the menu bar select Dimension-Linear and:
   prompt   Specify first extension line origin or <select object>
   respond   **Intersection icon and pick point D2**
   prompt   Specify second extension line origin
   respond   **Intersection icon and pick point D3**
   prompt   Specify dimension line location or..
   respond   **pick any point below object.**

Baseline dimensioning

1   Select the LINEAR DIMENSION icon and:
   prompt   Specify first extension line origin
   respond   **Intersection icon and pick point P1**
   prompt   Specify second extension line origin
   respond   **Intersection icon and pick point P2**
   prompt   Specify dimension line location
   respond   **pick any point above the line.**

2   Select the BASELINE DIMENSION icon from the Dimension toolbar and:
   prompt   Specify a second extension line origin
   respond   **Intersection icon and pick point P3**
   prompt   Specify a second extension line origin
   respond   **press the ESC key** to end command.

3   The menu bar selection Dimension-Baseline could have been selected for step 2.

Continue dimensioning

1   Select the LINEAR DIMENSION icon and:
   prompt   Specify first extension line origin
   respond   **Intersection icon and pick point K1**
   prompt   Specify second extension line origin
   respond   **Intersection icon and pick point K2**
   prompt   Specify dimension line location
   respond   **pick any point to right of the line.**

2   Menu bar with Dimension-Continue and:
   prompt   Specify a second extension line origin
   respond   **Intersection icon and pick point P3**
   prompt   Specify a second extension line origin
   respond   **press ESC.**

3   The Continue-Dimension icon could have been selected for step 2.

Diameter dimensioning

Select the DIAMETER DIMENSION icon from the Dimension toolbar and:
prompt Select arc or circle
respond **pick the larger circle**
prompt Specify dimension line location
respond **drag out and pick a suitable point.**

Radius dimensioning

Select the RADIUS DIMENSION icon and:
prompt Select arc or circle
respond **pick the arc**
prompt Specify dimension line location
respond **drag out and pick a suitable point.**

Angular dimensioning

Select the ANGULAR DIMENSION icon and:
prompt Select arc, circle, line or..
respond **pick line B1**
prompt Select second line
respond **pick line B2**
prompt Specify dimension arc line location
respond **drag out and pick a point to suit.**

Aligned dimensioning

Select the ALIGNED DIMENSION icon and:
prompt Specify first extension line origin
respond **Endpoint icon and pick line B1**
prompt Specify second extension line origin
respond **Endpoint icon and pick other end of line B1**
prompt Specify dimension line location
respond **pick any point to suit.**

Leader dimensioning

Select the LEADER DIMENSION icon and:
prompt Specify first leader point
respond **Nearest icon and pick any point on smaller circle**
prompt Specify next point
respond **drag to a suitable point and pick**
prompt Specify next point
respond **right-click**
prompt Enter text width<0>
respond **right-click**
prompt Enter first line of annotation text
enter **R3 <R>**
prompt Enter next line of annotation text
respond **right-click** to end leader dimension command.

Dimension options

When using the dimension commands, the user may be aware of various options when the prompts are displayed. To investigate these options:

1   Make layer OUT current and draw five horizontal lines of length 75 at the right-side of the screen then make layer DIMS current.

2   *The RETURN option*
   Select the LINEAR DIMENSION icon and:
   prompt   Specify first extension line origin or <select object>
   respond   **press the RETURN/ENTER key**
   prompt   Select object to dimension
   respond   **pick the top line**
   prompt   Specify dimension line location
   respond   **pick above the line** – fig. (a).

3   *The ANGLE option*
   Select the LINEAR DIMENSION icon, press RETURN, pick the second top line and:
   prompt   Specify dimension line location
   and   [Mtext/Text/Angle..
   enter   **A <R>** – the angle option
   prompt   Specify angle of dimension text
   enter   **15 <R>**
   prompt   Specify dimension line location
   respond   **pick above the line** – fig. (b).

4   *The ROTATED option*
   LINEAR dimension icon, right-click, pick third line and:
   prompt   Specify dimension line location
   enter   **R <R>** – the rotated option
   prompt   Specify angle of dimension line<0>
   enter   **15 <R>**
   prompt   Specify dimension line location
   respond   **pick above the line** – fig. (c).

5   *The TEXT option*
   LINEAR dimension icon, right-click, pick the fourth line and:
   prompt   Specify dimension line location
   enter   **T <R>** – the text option
   prompt   Enter dimension text<75>
   enter   **THIS DIMENSION IS: 75 <R>**
   prompt   Specify dimension line location
   respond   **pick above the line** – fig. (d).

6   *Dimensioning with keyboard entry*
   At the command line enter **DIM <R>** and:
   prompt   Dim
   enter   **HOR <R>** – horizontal dimension
   prompt   Specify first extension line origin
   respond   **right-click and pick the fifth line**
   prompt   Specify dimension line location
   respond   **pick above the line**
   prompt   Enter dimension text<75>
   enter   **SEVENTY-FIVE <R>** – fig. (e)
   prompt   Dim and **ESC** to end command.

Note

1   At this stage your drawing should resemble Fig. 19.2.

2   As stated earlier, the dimensions in Fig. 19.2 may differ in appearance from those which have been added to your drawing. This is because I have used the 'default AutoCAD dimension style' and made no attempt to alter it. The object of the exercise was to investigate the dimensioning process and dimension styles will be discussed in the next chapter.

3   The <RETURN> selection is useful if a single object is to be dimensioned. It is generally not suited to baseline or continue dimensions.

4   Object snap is used extensively when dimensioning. This is one time when a running Object Snap (e.g. Endpoint) will assist, but remember to cancel the running object snap!

5   From the menu bar select Format-Layer to display the Layer Properties Manager dialogue box. Note the layer **Defpoints**. We did not create this layer. It is *automatically* made by AutoCAD any time a dimension is added to a drawing. This layer can be turned off or frozen but cannot be deleted. **It is best left untouched**.

Dimension terminology

All dimensions used with AutoCAD objects have a terminology associated with them. It is important that the user has an understanding if this terminology especially when creating dimension styles. It is in the users interest to know the various terms used and Fig. 19.3 explains the basic terminology when dimensions are added to a drawing.

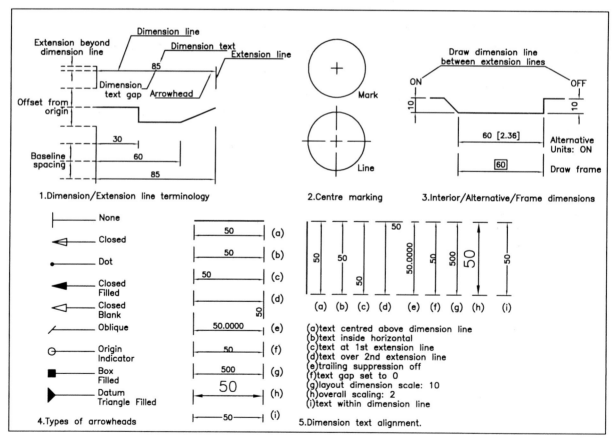

Figure 19.3 Dimension terminology.

1   *The dimension and extension lines.* These are made up of:

   a) dimension line

     – the actual line

     – the dimension text

     – arrowheads

     – extension lines.

   b) extension line

     – an origin offset from the object

     – an extension beyond the line

     – spacing (for baseline)

2   *Centre marking.* This can be:

   a) a mark

   b) a line

   c) nothing.

3   *Dimension text.* It is possible to:

   a) have interior dimension line drawn or not drawn

   b) display alternative units, i.e. [imperial]

   c) draw a frame around the dimension text.

4   *Arrowheads.* AutoCAD 2002 has several arrowheads for the dimension line and has the facility for user defined arrowheads. A selection is displayed.

5   *Dimension text alignment.* It is possible to align the dimension relative to the dimension line by altering certain dimension variables. A selection of dimension text positions is displayed.

Summary

1   AutoCAD 2002 has automatic, associative dimensions.

2   Dimensioning can be linear, radial, angular, ordinate or leader.

3   The diameter and degree symbol are automatically added when using radial or angular dimensions.

4   Object snap modes are useful when dimensioning.

5   A layer DEFPOINTS is created when dimensioning. The user has no control over this layer.

6   Dimensions should be added to a drawing using a Dimension Style.

Chapter 20

Dimension styles 1

Dimension styles allow the user to set dimension variables to individual/company requirements. This permits various styles to be saved for different customers.

To demonstrate how a dimension style is 'set and saved', we will create a new dimension style called **A3DIM**, use it with our WORKDRG drawing and then save it to our standard sheet.

Note

1 The exercise which follows will display several new dialogue boxes and certain settings will be altered within these boxes. It is important for the user to become familiar with these Dimension Style dialogue boxes, as a good knowledge of their use is essential if different dimension styles have to be used.

2 The settings used in the exercise are my own, designed for our A3PAPER standard sheet.

3 You can alter the settings to your own values at this stage.

Getting started

Open your C:\BEGIN\WORKDRG drawing to display the component created from a previous chapter, i.e. red outline with green centre lines. Activate the Draw, Modify, Dimension and Object Snap toolbars.

Setting dimension style A3DIM

1 Either   *a*) menu bar with **Dimension-Style**
    or   *b*) Dimension Style icon from Dimension toolbar
    prompt Dimension Style Manager dialogue box
    with   *a*) Current Dimstyle: ISO-25 or similar
             b) Styles for selection
             c) Preview of current dimstyle
             d) Description of current dimstyle
    respond **pick New**
    prompt Create New Dimension Style dialogue box
    respond 1. alter New Style Name: **A3DIM**
             2. Start with: ISO-25 or similar
             3. Use for: All dimensions – Fig. 20.1
             4. Pick Continue
    prompt New Dimension Style: A3DIM dialogue box
    with   six tab options for selection
    respond continue with the tab selections which follow.

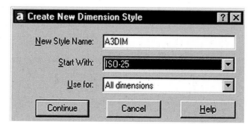

Figure 20.1 Create New Dimension Style dialogue box.

2 **Lines and Arrows**

respond pick Lines and Arrows tab

prompt `Lines and Arrows tab dialogue box` – probably active?

alter 1. Baseline Spacing: 10

2. Extend beyond dim lines: 2.5

3. Offset from origin: 2.5

4. Arrowheads: both Closed Filled

5. Leader: Close Filled

6. Arrow size: 3.5

7. Center mark for Circle; Type: Mark; Size: 2

and dialogue box similar to Fig. 20.2.

3 **Text**

respond pick Text tab

prompt `Text tab dialogue box`

alter 1. Text Style: Standard – this will be altered later

2. Text Color: Bylayer

3. Text height: 4

4. Vertical text placement: Above

5. Horizontal text placement: Centered

6. Offset from dim line: 1.5

7. Text Alignment: ISO Standard

and dialogue box similar to Fig. 20.3.

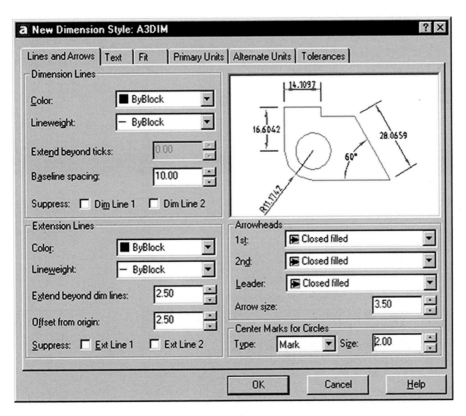

Figure 20.2 Lines and Arrows tab dialogue box.

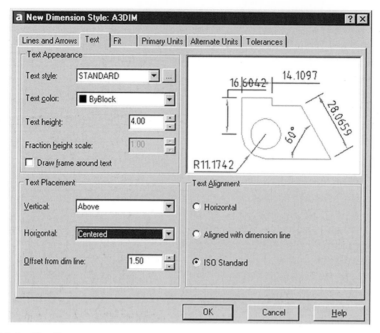

Figure 20.3 The Text tab dialogue box.

4 **Fit**

respond pick Fit tab
prompt Fit tab dialogue box
alter 1. Fit options: Either the text or the arrows, whichever fits best active, i.e. black dot
  2. Text Placement: Beside the dimension line
  3. Scale for Dimension Feature: Overall scale: 1
  4. Fine tuning: both not active, i.e. no black dots
and dialogue box similar to Fig. 20.4.

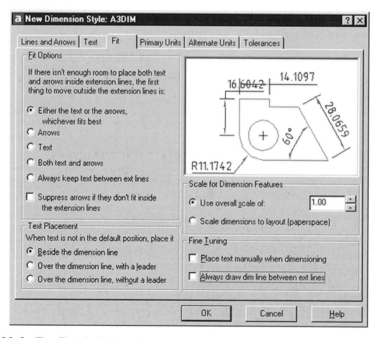

Figure 20.4 The Fit tab dialogue box.

5 **Primary Units**
respond   pick Primary Units tab
prompt   `Primary Units tab dialogue box`
alter   A. Linear Dimensions
             1. Unit Format: Decimal
             2. Precision: 0.00
             3. Decimal separator: '.' Period
             4. Round off: 0
             5. Scale factor: 1
             6. Zero Suppression: Trailing active, i.e. tick in box
        B. Angle Dimensions
             1. Units Format: Decimal Degrees
             2. Precision: 0.00
             3. Zero Suppression: Trailing active
and   dialogue box similar Fig. 20.5.

6 **Alternate Units**
respond   pick Alternate Units tab
prompt   `Alternate Units tab dialogue box`
respond   Display alternate units not active, i.e. no tick
and   dialogue box similar to Fig. 20.6.

7 **Tolerances**
respond   pick Tolerances tab
prompt   `Tolerances tab dialogue box`
respond   Tolerance Format Method: None
and   dialogue box similar to Fig. 20.7.

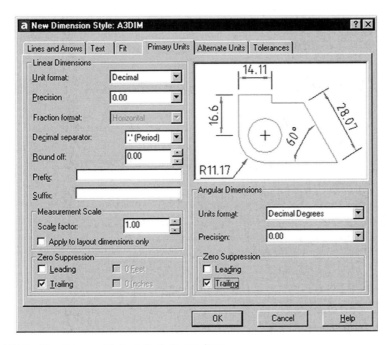

Figure 20.5 The Primary Units tab dialogue box.

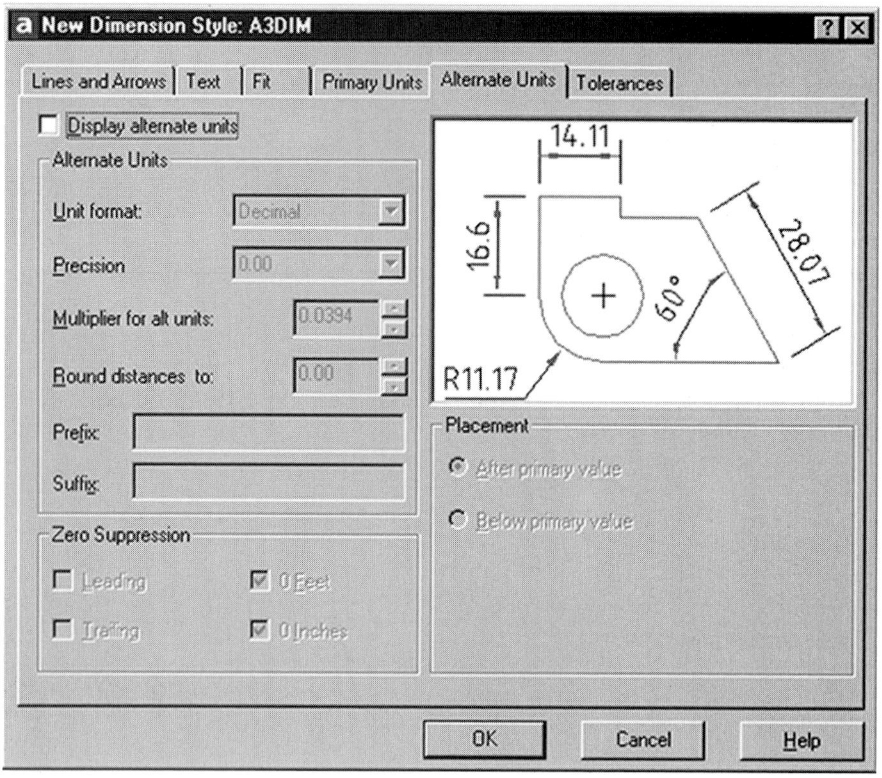

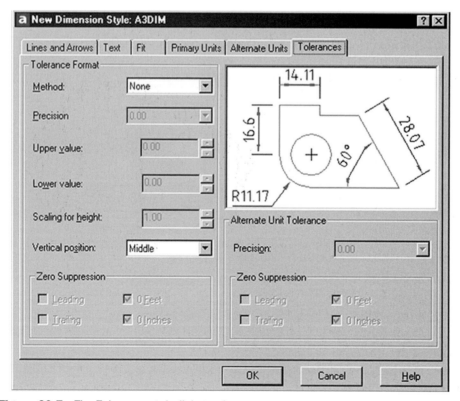

Figure 20.6 The Alternate Units tab dialogue box.

Figure 20.7 The Tolerances tab dialogue box.

8 **Continue:**

respond pick OK from New Dimension Style dialogue box

prompt Dimension Style Manager dialogue box

with
1. A3DIM added to styles list
2. Preview of the A3DIM style
3. Description of the A3DIM style

respond
1. pick A3DIM – becomes highlighted
2. pick Set Current – note description
3. scroll at List and pick: Styles in use
4. dialogue box similar to Fig. 20.8
5. pick Close.

Note if you have been altering other dimension styles during this process, the AutoCAD Alert message about over-rides may be displayed. If it is, respond as you require.

9 *Note*: We have still to 'set' a text style for our A3DIM dimension style. This will be covered in a later chapter and for the present, our A3DIM dimension style is 'complete'. It will be used for all future dimensioning work.

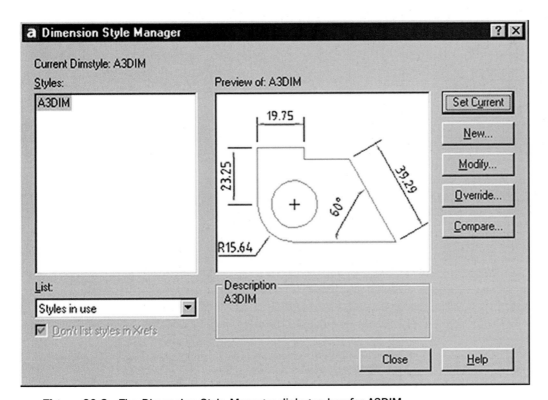

Figure 20.8 The Dimension Style Manager dialogue box for A3DIM.

Using the A3DIM dimension style

1 C:\BEGIN\WORKDRG drawing still on the screen?

2 Make layer DIMS current.

3 Refer to Fig. 20.9 and add the following dimensions as indicated:
   a) linear baseline
   b) linear continue
   c) diameter
   d) radius
   e) angular
   f) leader.

4 With layer TEXT current add suitable text, the height being at your discretion.

5 When all dimensions have been added save the drawing as C:\BEGIN\WORKDRG.

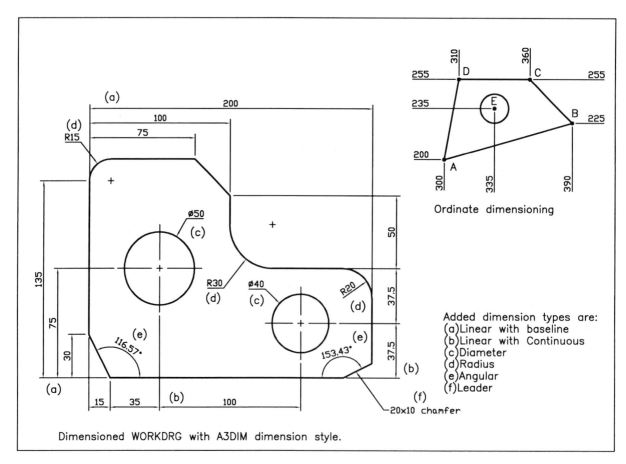

Dimensioned WORKDRG with A3DIM dimension style.

Figure 20.9   Dimensioning C:\BEGIN\WORKDRG with the A3DIM dimension style.

Ordinate dimensioning

This type of dimensioning is very popular with many companies, and although mentioned in the last chapter, was not discussed in detail. We will now investigate how it is used.

1   Continue using WORKDRG and refer to Fig. 20.9.

2   Layer OUT current and toolbars to suit.

3   Draw the following objects:

LINE		*CIRCLE*
First point:	300,200	centre: 335,235
Next point:	@90,25	radius: 10
Next point:	@−30,30	
Next point:	@−50,0	
Next point:	close.	

4   Make the DIMS layer current.

5   Select the ORDINATE dimension icon from the Dimensiontoolbar and:
    prompt   Specify feature location
    respond   **pick point A** – osnap helps
    prompt   Specify leader endpoint or [Xdatum/Ydatum..
    enter   **X <R>** – the Xdatum option
    prompt   Specify leader endpoint or [Xdatum/Ydatum..
    enter   **@0,−10 <R>**
    prompt   Dimension text = 300
    and   command line returned.

6   Menu bar with **Dimension-Ordinate** and:
    prompt   Specify feature location
    respond   **pick point A**
    prompt   Specify leader endpoint or [Xdatum/Ydatum..
    enter   **Y <R>**
    prompt   Specify leader endpoint or [Xdatum/Ydatum..
    enter   **@−10,0 <R>**
    prompt   Dimension text = 200
    and   command line returned.

7   Now add ordinate dimensions to the other named points, using the following X and Y datum leader endpoint values:
    Point B: Xdatum: @0,−35   Point C: Xdatum: @0,10
              Ydatum: @10,0   Ydatum: @40,0
    Point D: Xdatum: @0,10   Point E: Xdatum: @0,−45
              Ydatum: @−20,0   Ydatum: @−45,0.

8   The result should be similar to Fig. 20.9. Save if required but do not use the name WORKDRG.

Saving the A3DIM dimension style to the standard sheet

The dimension style which we have created will be used for all future work when dimensioning is required. This means that we want to have this style incorporated into our A3PAPER standard drawing sheet. We could always open the standard sheet and re-define the A3DIM dimension style, but this seems a waste of time.

We will therefore use the existing layout, so:

1   Make layer OUT current.

2   Erase all objects from the screen **EXCEPT** the black border.

3   Menu bar with **File-Save As** and:
    prompt   Save Drawing As dialogue box
    respond   1. ensure type: AutoCAD 2000 Drawing (*.dwg)
            2. scroll and pick: C:\BEGIN named folder
            3. enter file name: A3PAPER
            4. pick Save
    prompt   AutoCAD message
    with   C:\BEGIN\A3PAPER.dwg already exists
            Do you want to replace it?
    respond   **pick Yes.**

4.   We now have the A3PAPER saved as a drawing file with the A3DIM dimension style defined. This means that we do not have to set layers or dimension styles every time we start a new drawing.

The A3PAPER standard drawing sheet should be used for **all new work**.

Quick dimensioning

AutoCAD 2002 has a quick dimensioning option and we will investigate it with WORKDRG. This topic should really have been investigated in the last chapter, but I decided to leave it until we had discussed dimension styles. To demonstrate the topic:

1   Open the C:\BEGIN\WORKDRG drawing just saved, and erase all dimensions and text.

2   Make layer DIMS current and refer to Fig. 20.10.

3   Menu bar with **Dimension-Quick Dimension**
    prompt   Select geometry to dimension
    respond   **window the shape then right-click**
    prompt   Specify dimension line position or [Continuous/
              Staggered/Baseline..
    enter   **B <R>** – the baseline option
    prompt   Specify dimension line position or..
    respond   **pick any point to the right of the shape.**

4   All vertical dimensions will be displayed from the horizontal baseline of the shape as Fig. 20.10(a). The user can now delete any dimensions which are unwanted.

5   Erase all the dimensions.

6   Select the QUICK DIMENSION icon from the Dimension toolbar and:
    prompt   Select geometry to dimension
    respond   **window the shape then right-click**
    prompt   Specify dimension line position or [Continuous/
              Staggered/Baseline..
    enter   **R <R>** – the radius option
    prompt   Specify dimension line location
    respond   **pick any point to suit.**

7 All radius dimensions for the component will be displayed with the A3DIM dimension style as Fig. 20.10(b).

8 Erase the radius dimensions then use the Quick dimension command to add staggered dimensions to the component as Fig. 20.10(c).

9 The user should realise that it is now possible to dimension a component with the Quick Dimension command. It may be necessary to use the command several times and delete unwanted dimensions every time the command is used before the required dimensions are displayed. This is an alternative to adding dimensions to individual objects. It is the users preference as to whether Quick Dimensions are used.

10 Do not save this part of the dimension exercise.

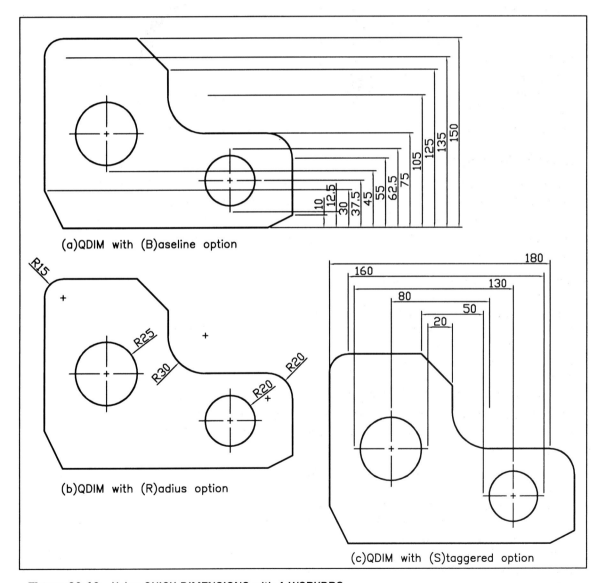

Figure 20.10 Using QUICK DIMENSIONS with A:WORKDRG.

Assignments

As dimensioning is an important concept, I have included four activities which will give you practice with;

a) using the standard sheet with layers
b) using the draw commands with coordinates
c) adding suitable text
d) adding dimensions with the A3DIM dimension style.

In each activity the procedure is the same:

1   Open the A3PAPER standard drawing sheet.

2   Using layers correctly, complete the drawings.

3   Save the completed work as C:\BEGIN\ACT7, etc.

The four activities are:

1   Activity 7: two simple components to be drawn and dimensioned. The sizes are more awkward than usual.

2   Activity 8: two components created mainly from circles and arcs. Use offset as much as possible. The signal arm is interesting.

3   Activity 9: a component which is much easier to complete than it would appear. Offset and fillet will assist.

4   Activity 10: components to be drawn and dimensioned. Remember that practice makes perfect.

Summary

1   Dimension styles are created by the user to 'individual' standards.

2   Different dimension styles can be created and 'made current' at any time.

3   The standard A3PAPER sheet has a 'customised' dimension style named A3DIM which will be used for all future drawing work including dimensioning.

4   Dimension styles will be discussed again in a later chapter.

Chapter 21

Modifying objects

The draw and modify commands are probably the most commonly used of all the AutoCAD commands and we have already used several of each. The modify commands discussed previously have been Erase, Offset, Trim, Extend, Fillet and Chamfer. In this chapter we will use C:\BEGIN\WORKDRG to investigate several other modify commands as well as some additional selection set options.

Getting ready

1 Open C:\BEGINWORKDRG to display the red component with green centre lines. There should be no dimensions displayed.

2 Layer OUT current, with the Draw, Modify and Object Snap toolbars.

3 Freeze layer CL – you will find out why shortly.

4 Menu bar with **View-Zoom-All** and the drawing should 'fill the screen'.

Copy

Allows objects to be copied to other parts of the screen. The command can be used for single or multiple copies. The user specifies a start (base) point and a displacement (or second point).

1 Refer to Fig. 21.1.

2 Select the COPY OBJECTS icon from the Modify toolbar and:
prompt	Select objects
enter	**C <R>** – the crossing selection set option
prompt	Specify first corner and: **pick a point P1**
prompt	Specify opposite corner and: **pick a point P2**
but	**DO NOT RIGHT-CLICK YET**
prompt	**11 found** – note objects not highlighted
then	Select objects
enter	**A <R>** – the add selection set option
prompt	Select objects
respond	**pick objects D1, D2 and D3 then right-click**
and	note command line as each object is selected
prompt	Specify base point or displacement or [Multiple]
enter	**50,50 <R>** – note copy image as mouse moved!
prompt	Specify second point of displacement
enter	**300,300 <R>**

3 The original component will be copied to another part of the screen and **may** not all be visible (if at all?)

4 Don't panic! Select from the menu bar **View-Zoom-All** to 'see' the complete copied effect – Fig. 21.2. You may have to reposition your toolbars?

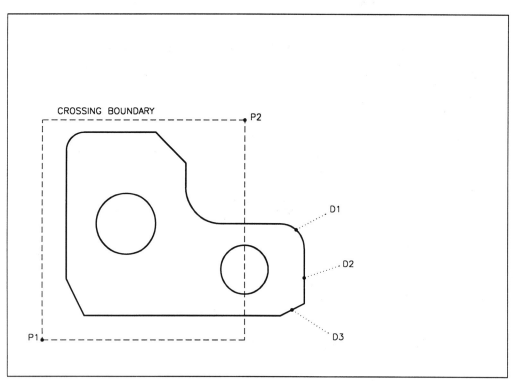

Figure 21.1 WORKDRG with selection points for the COPY command.

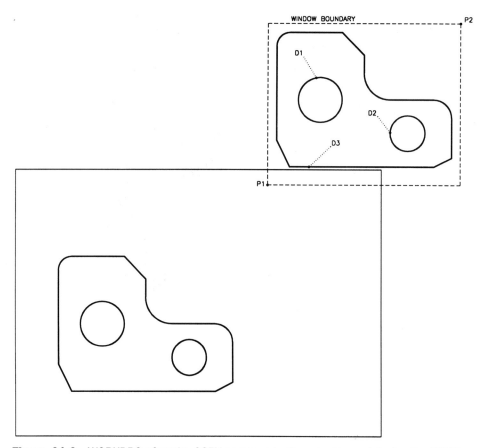

Figure 21.2 WORKDRG after the COPY command with selection points for the MOVE command.

Move

Allows selected objects to be moved to other parts of the screen, the user defining the start and end points of the move by:

a) entering coordinates
b) picking points on the screen
c) referencing existing entities.

1   Refer to Fig. 21.2 and select the MOVE icon from the Modify toolbar:

prompt	Select objects
enter	**W <R>** – the window selection set option
prompt	Specify first corner and: **pick a point P1**
prompt	Specify opposite corner and: **pick a point P2**
but	**DO NOT RIGHT-CLICK YET**
prompt	**14 found**
and	Select objects
enter	**R <R>** – the remove selection set option
prompt	Remove objects
respond	**pick circles D1 and D2 then right-click**
and	note command line as each object is selected
prompt	Specify base point or displacement
respond	**Endpoint icon and pick line D3 at left end**
	(note image as cursor is moved)
prompt	Specify second point of displacement
enter	**@–115,–130 <R>**

2   The result will be Fig. 21.3, i.e. the red outline shape is moved, but the two circles do not move, due to the Remove option.

3   *Task*
Using the Layer Properties Manager dialogue box, THAW layer CL and the centre lines are still in their original positions. They have not been copied or moved. This can be a problem with objects which are on frozen or off layers.
Now: *a*) freeze layer Cl
          b) erase the copied-moved objects to leave the original shape
          c) View-Zoom-All to 'restore' the original screen.

Rotate

Selected objects can be 'turned' about a designated point in either a clockwise (−ve angle) or counter-clockwise (+ve angle) direction. The base point can be selected as a point on the screen, entered as a coordinate or referenced to existing objects.

1   Refer to Fig. 21.3 and draw a circle with:
     centre: 100,220 and radius: 10

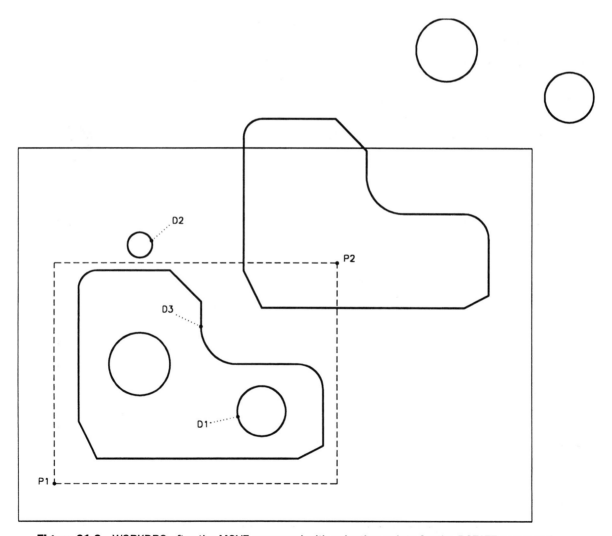

Figure 21.3 WORKDRG after the MOVE command with selection points for the ROTATE command.

2 Menu bar with **Modify-Rotate** and:

prompt   `Current positive angle in UCS: ANGDIR=counterclockwise:`
        `ANGBASE=0.0`

then   `Select objects`

respond **window from P1 to P2 – but NO right-click yet**

prompt   14 found

then   `Select objects`

enter   **R <R>** – the remove option

prompt   `Remove objects`

respond **pick circle D1**

prompt   `1 found, 1 removed, 13 total`

then   `Remove objects`

enter   **A <R>** – the add option

prompt   `Select objects`

respond **pick circle D2**

prompt   `1 found, 14 total`

then   `Select objects`

respond **right-click** to end selection sequence

prompt   `Specify base point`

respond **Endpoint icon and pick line D3** – at 'lower end'

prompt   `Specify rotation angle or [Reference]`

enter   **–90 <R>**

3 The selected objects will be rotated as displayed in Fig. 21.4.

4 Note that two selection set options were used in this single rotate sequence:
a) R – removed a circle from the selection set
b) A – added a circle to the selection set.

5 The ROTATE icon could have been selected for this sequence.

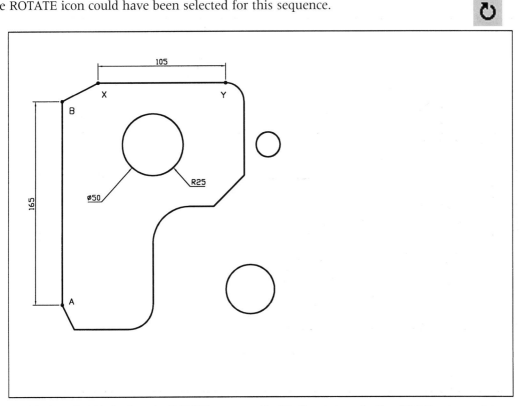

Figure 21.4 WORKDRG after the ROTATE command, with dimensions for the SCALE command.

Scale

The scale command allows selected objects or complete 'shapes' to be increased/decreased in size, the user selecting/entering the base point and the actual scale factor.

1   Refer to Fig. 21.4 and erase the two circles 'outside the shape'.

2   Make layer DIMS current and with the Dimension command (icon or menu bar):
     a) linear dimension line AB
     b) diameter dimension the circle.

3   At the command line enter **DIM <R>** and:

prompt	Dim
enter	**HOR <R>**
prompt	Specify first extension line origin
respond	**Endpoint icon and pick line at X**
prompt	Specify second extension line origin
respond	**Endpoint icon and pick line at Y**
prompt	Specify dimension line location
respond	**pick above line**
prompt	Enter dimension text<105>
enter	**105 <R>**
prompt	Dim
enter	**RAD <R>**
prompt	Select arc or circle
respond	**pick the circle**
prompt	Enter dimension text<25>
enter	**R25 <R>**
prompt	Specify dimension text location
respond	**pick to suit**
prompt	Dim and **ESC** to end sequence.

4   Make layer OUT current.

5   Select the SCALE icon from the Modify toolbar and:

prompt	Select objects
respond	**window the complete shape with dimensions then right-click**
prompt	Specify base point
respond	**Endpoint of line AB at 'B' end**
prompt	Specify scale factor or [Reference]
enter	**0.5 <R>**

6   The complete shape will be scaled as Fig. 21.5.

7   Note the dimensions:
     a) the vertical dimension of 165 is now 82.5
     b) the diameter value of 50 is now 25
     c) the horizontal dimension of 105 is still 105
     d) the radius of 25 is still 25.

8   *Questions:*
     a) Why have two dimensions been scaled by 0.5 and two have not?
       Answer: the scaled dimensions are those which used the icon or menu bar selection and those not scaled had their dimension text values entered from the keyboard.
     b) Which of the resultant dimensions are correct?
       Is it the 82.5 and diameter 25, or the 105 and R25?
       Answer: I will let you reason this one for yourself, but I can assure you that it causes a great deal of debate.

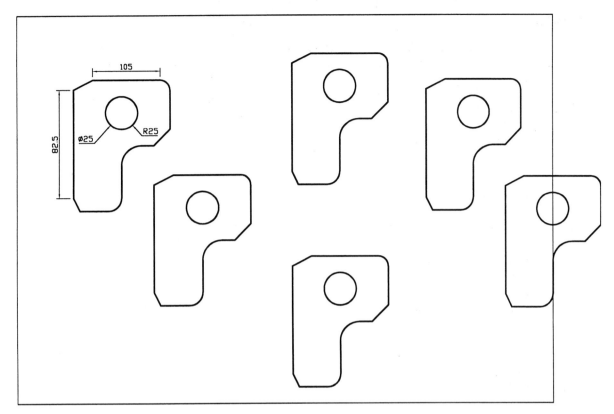

Figure 21.5 WORKDRG after the SCALE and MULTIPLE COPY commands.

Multiple copy

This is an option of the COPY command and does what it says – it produces as many copies of selected objects as you want.

1 Refer to Fig. 21.5.

2 Using the Layer Properties Manager dialogue box, lock layer DIMS.

3 From menu bar select **Modify-Copy** and:
prompt   Select objects
respond   **window the shape and dimensions then right-click**
prompt   17 found and 4 were on a locked layer
then   Specify base point or displacement or [Multiple]
enter   **M <R>** – the multiple copy option
prompt   Specify base point
respond   **Center icon and pick the circle**
prompt   Specify second point of displacement and enter: **145,150 <R>**
prompt   Specify second point and enter: **@170,20 <R>**
prompt   Specify second point and enter: **@170,-135 <R>**
prompt   Specify second point and enter: **@275,0 <R>**
prompt   Specify second point
respond   **Midpoint icon and pick right vertical line of the black 'border'**
prompt   Specify second point
respond   **right-click to end sequence.**

4 Result as Fig. 21.5, the 'shape' being multiple-copied, but the dimensions not being copied, due to the locked layer effect.

Mirror

A command which allows objects to be mirror imaged about a line designated by the user. The command has an option for deleting the original set of objects – the source objects.

1 Erase the five multiple copied shapes to leave the original scaled shape with dimensions. Unlock layer DIMS.

2 Move the shape and dimensions:
Base point: **Centre of circle**
Second point: **@120,20.**

3 Refer to Fig. 21.6 and with layer OUT current, draw the following lines:

AB	MN	PQ	XY
first: 140,270	first: 260,200	first: 240,150	first: 155,150
next: @0,-100	next: @0,70	next: @50<45	next: @60<0.

4 With layer TEXT current, draw the following text items all with 0 rotation and using:
a) centred on 200,257; height: 6; item: AutoCAD
b) centred on 185,180; height: 7, item: R2002.

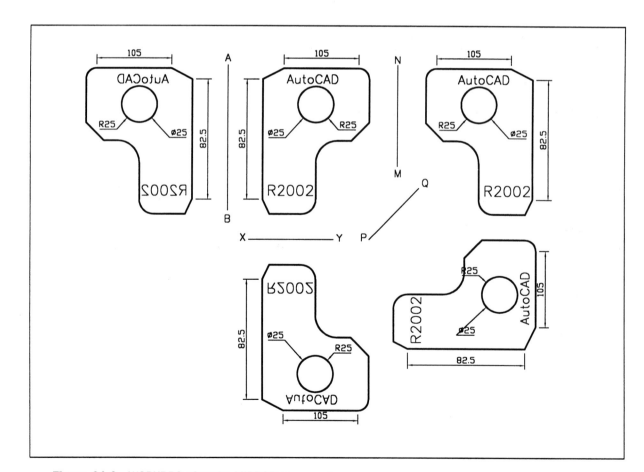

Figure 21.6 WORKDRG after the MIRROR command.

5   Make layer OUT current and select the MIRROR from the Modify toolbar and:
   prompt   `Select objects`
   respond   **window shape and dimensions then right-click**
   prompt   `Specify first point of mirror line`
   respond   **Endpoint icon and pick point A**
   prompt   `Specify second point of mirror line`
   respond   **Endpoint icon and pick point B**
   prompt   `Delete source objects [Yes/No]<N>`
   enter   **N <R>**

6   The selected objects are mirrored about the line AB. Note the text has also been mirrored but the dimension text has not.

7   At the command line enter **MIRRTEXT <R>**
   prompt   `Enter new value for MIRRTEXT<1>`
   enter   **0 <R>**

8   Menu bar with **Modify-Mirror** and:
   prompt   `Select objects`
   enter   **P <R>** – previous election set option
   prompt   19 found
   then   `Select objects`
   respond   **right-click** to end selection
   prompt   `Specify first point of mirror line`
   respond   **pick endpoint of point M**
   prompt   `Specify second point of mirror line`
   respond   **pick endpoint of point N**
   prompt   `Delete source objects` and enter: **N <R>**

9   The shape is mirrored – what about text?

10 At the command line enter **MIRROR <R>** and:
   a) select objects: enter P <R><R> – yes two returns – why?
   b) first point – pick point P
   c) second point – pick point Q
   d) delete – N.

11 Set the MIRRTEXT variable to 1 and mirror the original shape about line XY without deleting the source objects.

12 The final result should be Fig. 21.6.

13 *Task:* Before leaving the exercise, thaw layer CL. The circle centre lines are still in their original positions.

Reference option

The rotate and scale modify commands have a reference option, and we will use the rotate command to demonstrate how this option is used.

1   Close any existing drawing (no save) and open the original WORKDRG. Refer to Fig. 21.7.

2   Draw a circle with centre: 200,180 and radius 30.

3   Activate the rotate command, window the shape, right-click and:
   prompt   Specify base point
   respond **Endpoint icon and pick as indicated**
   prompt   Specify rotation angle or [Reference]
   enter   **R <R>** – the reference option
   prompt   Specify the reference angle and enter **80 <R>**
   prompt   Specify the new angle and enter **90 <R>**

4   The selected shape will be rotated through an angle of 10° which is equivalent to (new–old) – fig. (a).

5   Undo the rotate effect to restore the original WORKDRG.

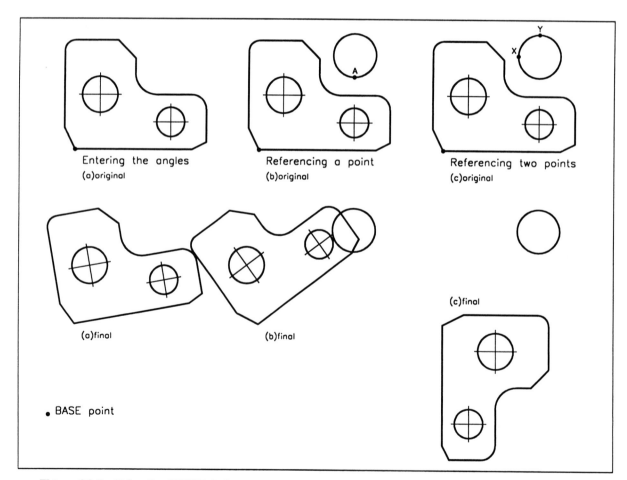

Figure 21.7   Using the ROTATE (reference) command with WORKDRG.

6 With the rotate command:
   a) objects: window the WORKDRG shape then right-click
   b) base point: endpoint icon and pick as indicated
   c) rotation angle: quadrant icon and pick pt A on circle.

7 The selected shape will be rotated about the base point as fig. (b). The actual angle of rotation is from the horizontal to a line from the base point through the selected quadrant of the circle. I dimensioned this angle and it was 37°.

8 Undo the rotate effect.

9 With the rotate command again:
   a) objects: window the shape and right-click
   b) Base point: endpoint icon and pick as indicated
   c) Rotation angle: enter R <R>, the reference option and:
prompt	Specify the reference angle
respond	quadrant icon and pick pt X as indicated
prompt	Specify second point
respond	quadrant icon and pick pt Y as indicated
prompt	Specify new angle
enter	–45 <R>

10 The selected shape will be rotated about the base point as fig. (c). The actual rotation angle is equivalent to –90° and is obtained from the (new – old). I will let you work out how this (new – old) gives –90.

11 This exercise does not need to be saved.

Assignments

Four 'relatively easy' activities have been included for this chapter. Each activity should be completed on the A3PAPER standard drawing sheet and layers should be used correctly. The components do not need to be dimensioned, unless you are feeling adventurous. Remember to save your completed drawings as C:\BEGIN\ACT??, etc. and always use your discretion for sizes which are not given. The four activities are:

1   Activity 11: a vent cover plate.
     The original component has to be drawn and then copied into the rectangular plate. Commands could be rotate and multiple copy, but that is your decision.

2   Activity 12: a well-known symbol.
     Easy to draw and construct. Several ways of using the modify commands, e.g. copy/rotate or mirror.

3   Activity 13: a template.
     This is easier to complete than you may think. Draw the quarter template using the sizes, then MIRROR. The text item is to be placed at your discretion. Multiple copy and scale by half, third and quarter.
     Additional: add the given dimensions to a quarter template.

4   Activity 14: two additional components.
     No help with these two, as one of them is virtually identical to activity 13.

Summary

1   The commands COPY, MOVE, ROTATE, MIRROR and SCALE can all be activated:
     a) from the menu bar with Modify-Copy, etc.
     b) by selecting the icon from the Modify toolbar
     c) by entering the command at the command line, e.g. SCALE <R>

2   The selection set is useful when selecting objects, especially the add/remove options

3   All of the commands require a base point, and this can be:
     a) entered as coordinates
     b) referenced to an existing object
     c) picked on the screen.

4   Objects on layers which are OFF or FROZEN are not affected by the modify commands. This could cause problems!

5   MIRRTEXT is a system variable with a 0 or 1 value and:
     a) value 1: text is mirrored – this is the default
     b) value 0: text is not mirrored.

6   Keyboard entered dimensions are not scaled.

Chapter 22

Grips

Grips are an aid to the draughting process, offering the user a limited number of modify commands. In an earlier chapter we 'turned grips off with GRIPS: 0 at the command line. This was to allow the user to become reasonably proficient with using the draw and modify commands. Now that this has been achieved, we will investigate how grips are used.

Toggling grips on/off

Grips can be toggled on/off using:
a) the Selection tab from the Options dialogue box
b) command line entry.

1   Open your A3PAPER standard sheet with layer OUT current.

2   From the menu bar select **Tools-Options** and:

prompt	Options dialogue box
respond	**pick Selection tab**
prompt	Selection tab dialogue box
with	*a*) Selection modes
	b) Grips
respond	1. Enable grips active, i.e. tick in box
	2. Unselected grip color: Blue
	3. Selected grip color: Red
	4. Grip size: adjust to suit – Fig. 22.1
	5. pick Apply
	6. pick OK.

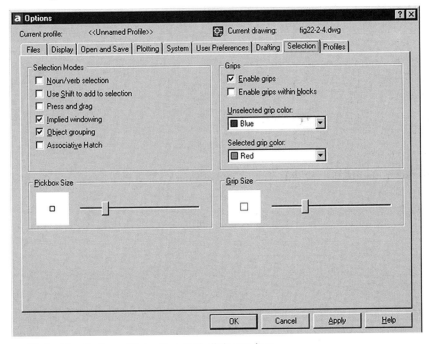

Figure 22.1 The Options (Selection tab) dialogue box.

3   The drawing screen is returned, with the grip box 'attached' to cursor cross-hairs.

4   At the command line enter **GRIPS <R>** and:
    prompt   Enter new value for GRIPS<1>
    respond   **<RETURN>**, i.e. leave the 1 value.

5   The command entry is GRIPS: 1 are ON; GRIPS: 0 are OFF.

6   *Note*
    a) the grip box attached to the cross-hairs should not be confused with the pick box used with modify commands. Although similar in appearance they are entirely different.
    b) when any command is activated (e.g. LINE), the grips box will disappear from the cross-hairs, and re-appear when the command is terminated.

What do grips do and how do they work?

1   Grips provide the user with five modify commands which can also be activated in icon form⁻ or from the menu bar. The five commands are Stretch, Move, Rotate, Scale and Mirror.

2   Grips work in the 'opposite sense' from normal command selection, i.e.
    a) the usual sequence is to activate the command then select the objects, e.g. COPY, then pick object to be copied
    b) with grips, the user selects the objects and then activates on of the five commands.

    To demonstrate using grips:

1   Draw a line, circle, arc and text item anywhere on the screen.

2   Ensure grips are on, i.e. GRIPS: 1 at command line.

3   Refer to Fig. 22.2 and move the cursor to each object and 'pick them' with the grip box and:
    a) blue grip boxes appear at each object 'snap point' and the object is highlighted – fig. (a)
    b) move the grip box to any one of the blue boxes and 'pick it'. The box becomes red solid in appearance and the object is still highlighted. The red box is the **BASE GRIP** – fig. (b)
    c) press the ESC key – blue grips with highlighted object
    d) ESC again to cancel the grips operation.

Figure 22.2 Grip 'states' or types.

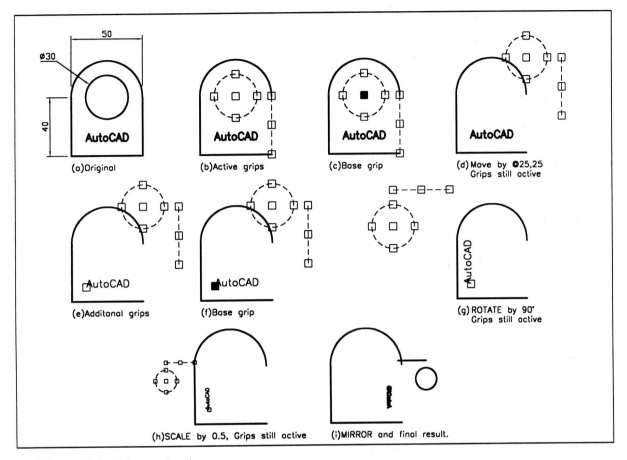

Figure 22.3 Grip exercise 1.

4 *Note*

When grips were first introduced into AutoCAD, the terms warm and hot were used, meaning:

a) Warm grip: appear as blue boxes and the selected objects are highlighted – dashed appearance as fig. (a). The grip options cannot be used in this state.

b) Hot grip: appear as solid red boxes when a cold or warm box is 'picked' as fig. (c). The selected hot grip acts as the base grip and the grip options can be used.

Grip exercise 1

This demonstration is relatively simple but rather long. It is advisable to work through the exercise without missing out any of the steps.

1 Erase all objects from the screen, or re-open A:A3PAPER.

2 Refer to Fig. 22.3 and draw the original shape using the sizes given as fig. (a). Make the lower left corner at the point (100,100) and ensure grips are on.

3 Move the cursor to the circle and pick it, then move to the right vertical line and pick it. Blue grip boxes appear and the two objects are highlighted – fig. (b).

4   Move the cursor grip box to the grip box at the circle centre and left-click, i.e. pick it. The selected box will be displayed in red as it is now the base grip – fig. (c). Observe the command line:

    prompt   ** STRETCH **

               `Specify stretch point or [Base point/Copy/Undo/eXit]`

    respond   **with a <RETURN>**

    prompt   ** MOVE **

               `Specify move point or [Base point/Copy/Undo/eXit]`

    enter   **@25,25 <R>**

5   The following should have happened:

    a) the circle and line are moved

    b) the command prompt line is returned

    c) the grips are still active

    d) there is no base grip – fig. (d).

6   Move the cursor and pick the text item to add it to the grip selection – fig. (e).

7   Make the grip box of the text item the base grip, by moving the cursor pick box onto it, left-clicking as fig. (f), and:

    prompt   ** STRETCH **

               `Specify stretch point or [Base point/Copy..`

    respond   **with a <RETURN>**

    prompt   ** MOVE **

               `Specify move point or [Base point/Copy..`

    respond   **<RETURN>**

    prompt   ** ROTATE **

               `Specify rotation angle or [Base point/Copy..`

    enter   **90 <R>**

8   The circle, line and text item will be rotated and the grips are still active – fig. (g).

9   Make the same text item grip box the base grip (easy!) and:

    prompt   **STRETCH**

               `Specify stretch point or..`

    enter   **SC <R>** – the scale grip option

    prompt   ** SCALE **

               `Specify scale factor or [Base point..`

    enter   **0.5 <R>**

10   The three objects are scaled and the grips are still active as fig. (h).

11   Make the right box on the line hot and:

    prompt   **STRETCH**

    enter   **MI <R>** – the mirror option

    prompt   ** MIRROR **

               `Specify second point or [Base point..`

    enter   **B <R>** – the base point option

    prompt   `Specify base point`

    respond   **Midpoint icon and pick the original horizontal line**

    prompt   ** MIRROR **

               `Specify second point or..`

    respond   **Midpoint icon and pick the arc.**

12   The three objects are mirrored about the selected 'line' and the grips are still active (warm) – fig. (i).

13   Press ESC – removes the grips and ends the sequence

14   The exercise is now complete. Do not exit yet.

Selection with grips

Selecting individual objects for use with grips can be tedious. It is possible to select a window/crossing option when grips are on.

1 Your screen should display the line, circle and text item after the grips exercise has been completed?

2 Refer to Fig. 22.4(a), move the cursor and pick a point 'roughly' where indicated. Move the cursor down and to the right, pick a second point and all complete objects within the window will display grip boxes with highlighted objects, i.e. warm.

3 ESC to cancel the grip effect.

4 Move the cursor to about the same point as step 2, and pick a point – Fig. 22.4(b), then move the cursor upwards and to the left and pick a second point. All objects within or which cross the boundary will display grip boxes with highlighted objects.

5 ESC to cancel grip selection.

6 The effect can be summarised as:
a) window effect to the right of first pick
b) crossing effect to the left of the first pick.

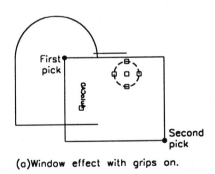

(a)Window effect with grips on.

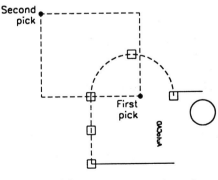

(b)Crossing effect with grips on.

Figure 22.4 Window/crossing selection with grips.

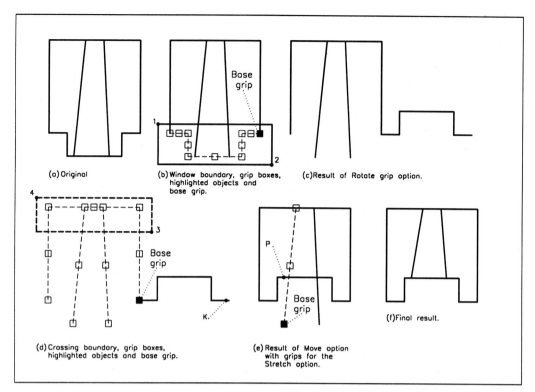

(a) Original

(b) Window boundary, grip boxes, highlighted objects and base grip.

Base grip

(c) Result of Rotate grip option.

(d) Crossing boundary, grip boxes, highlighted objects and base grip.

Base grip

K.

(e) Result of Move option with grips for the Stretch option.

P

Base grip

(f) Final result.

Figure 22.5 Grip exercise 2.

Grips exercise 2

1 Open C:\BEGIN\USEREX to display the component used in the offset and text exercises and refer to Fig. 22.5.

2 Ensure only the red objects displayed – fig. (a).

3 Grips on?

4 Pick a point 1 and drag out a window and pick a point 2. Five lines will display grip boxes and be highlighted as fig. (b).

5 Make the rightmost grip the base grip and:
   prompt   ** STRETCH **..
   respond right-click to display the grip options
   then   **pick Rotate**
   prompt   Specify rotation angle or..
   enter   **180 <R>**

6 Now ESC to cancel the grips – fig. (c).

7 Pick a point 3 and drag out a crossing window and pick a point 4 to display five highlighted lines with grip boxes as fig. (d).

8 Make the lowest grip box on the right vertical line hot and:
   prompt   ** STRETCH **..
   respond **right-click** to display the grip options menu
   then   **pick Move**
   prompt   Specify move point or..
   respond **Endpoint icon and pick line K.**

9 The lines are moved 'onto' the rotated lines – fig. (e).

10 ESC to cancel the grips.

11 Pick one of the inclined lines and:
   a) make the lowest grip box the base grip
   b) activate the STRETCH option
   c) stretch the grip box perpendicular to line P
   d) repeat for the other inclined line
   e) final result – fig. (f).

12 *Note*: this exercise could have been completed using the modify commands, e.g. rotate and trim. It demonstrates that there is no one way to complete a drawing.

13 Save this exercise if required, but not as USEREX.

Assignment

Activity 15 involves the re-positioning of a robotic arm, which has proved relatively successful in my previous 'Beginning' books.

1 Create the robotic arm in the original position using the sizes given – fig. (a). Use your discretion for sizes not given.

2 Upper arm rotate by 45 degrees – fig. (b).
   Two circles and two lines need to be 'picked' with the base grip at the larger circle centre.

3 Both arms mirrored about line through large circle centre.
   Two more lines and a circle added to the grip selection as fig. (c).

4 Both arms rotated to a horizontal position – fig. (d).

5 Finally – fig. (e) – three grip operations:
   a) lower arm stretch by 50
   b) upper arm move
   c) upper arm rotate.

6 Save when completed.

Summary

1 Grips allow the user access to the STRETCH, MOVE, ROTATE, SCALE and MIRROR modify commands without icon or menu bar selection.

2 Grips work in the 'opposite sense' from the normal AutoCAD commands, i.e. select object first then the command

3 Grips **do not have to be used**. They give the user another draughting tool.

4 Grips are toggled on/off using the Grips dialogue box or by keyboard entry. The dialogue box allows the grip box colours and size to be altered.

5 Grip states are warm or hot.

6 If grips are not being used, I would always recommend that they be toggled off, i.e. GRIPS: 0 at the command line.

7 When a grip box is the base grip, the options can be activated by:
   a) return at the keyboard
   b) entering SC, MO, MI, etc.
   c) right-click the mouse to display the grip option menu.

Chapter 23

Drawing assistance

Up until now, all objects have been created by picking points on the screen, entering coordinate values or by referencing existing objects, e.g. midpoint, endpoint, etc.

There are other methods which enable objects to be positioned on the screen, and in this chapter we will investigate three new concepts, these being:

a) point filters
b) construction lines
c) ray lines.

Point filters

This allows objects to be positioned by referencing the X and Y coordinate values of existing objects.

Example 1

1 Open the A3PAPER standard sheet and refer to Fig. 23.1
2 Draw a 50 side square, lower left corner at 20,220. Grips off
3 Multiple copy this square to three other positions
4 A circle of diameter 30 has to be created at the 'centre' of each square and this will be achieved by four different methods:

a) *Coordinates*
Activate the circle command with centre: 45,245; radius: 15

b) *Object snap midpoint*
– draw in a diagonal of the square, then pick the circle icon
– pick the centre icon
– centre point: snap to Midpoint of diagonal
– radius: enter 15.

c) *Object snap from*
– pick the circle icon
– centre: pick the snap from icon
– base point: endpoint icon and pick left end of line AB
– offset: enter @25,25
– radius: enter 15.

d) *Point filters*
Activate the circle command and:

prompt	Specify center point
enter	**X <R>**
prompt	of
respond	**Midpoint icon and pick line PQ**
prompt	(need YZ)
respond	**Midpoint icon and pick line PR**
prompt	Specify radius and enter: **15 <R>**

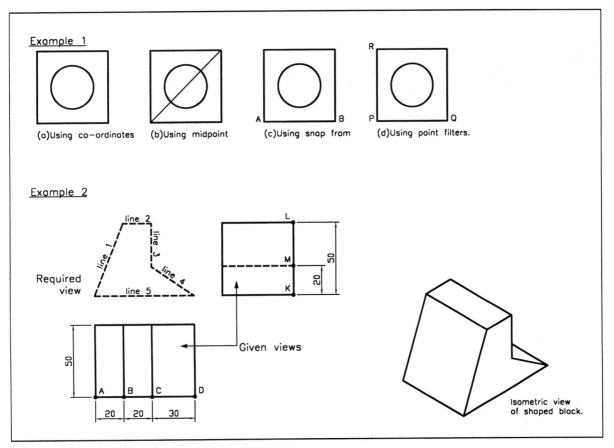

Figure 23.1 Point filter examples.

Example 2

Figure 23.1 displays the top, end and isometric views of a shaped block. The front view has to be created from the two given views, and we will use the point filter technique to achieve this.

1 Draw the top and end views using the sizes given. Use the lower part of the screen. Draw with the snap on.

2 Select the line icon and:

prompt	Specify first point
enter	**.X <R>**
prompt	of
respond	**Intersection icon and pick point A**
prompt	(need YZ)
respond	**Intersection icon and pick point K**
and	cursor 'snaps' to a point on the screen
prompt	Specify next point
enter	**.X <R>**
prompt	of
respond	**Intersection icon and pick point B**
prompt	(need YZ)
respond	**Intersection icon and pick point L** – line 1 is drawn
prompt	Specify next point
enter	**.X <R>** and: **Intersection of point C**
prompt	(need YZ) and: **Intersection of point L** – line 2 is drawn
prompt	Specify next point

enter   **.X <R>** and: **Intersection of point C**
prompt   (need YZ) and: **Intersection of point M** – line 3 is drawn
prompt   Specify next point
enter   **.X <R>** and: **Intersection of point D**
prompt   (need YZ) and: **Intersection of point K** – line 4 is drawn
prompt   To point and: **C <R>** to draw line 5.

3   The front view of the shaped block is now complete.

4   This exercise does not need to be saved.

5   *Note*

   a) The point filter method of creating objects can be rather 'cumbersome' to use. Point filters are another aid to draughting.

   b) I would suggest that Object Snap Tracking is 'easier' than point filters fro creating the front view in this exercise?

Construction lines

Construction lines are lines that extend to infinity in both directions from a selected point on the screen. They can be referenced to assist in the creation of other objects.

1   Open the A3PAPER standard sheet, layer OUT current and display toolbars Draw, Modify and Object Snap. Refer to Fig. 23.2(a).

2   With layer OUT current, draw:
   a) a 100 sided square, lower left corner at 50,50
   b) a circle, centred on 250,220 with radius 50.

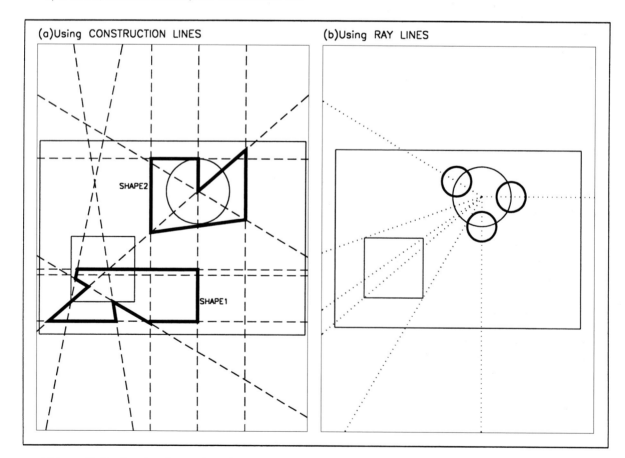

Figure 23.2   Construction and ray lines.

3 Make a new layer (Format-Layer) named CONLINE, colour to suit and with a DASHED linetype. This layer is to be current.

4 Menu bar with **Draw-Construction Line** and:

prompt	`Specify a point or [Hor/Ver/Ang/Bisect/Offset]`
enter	**50,50 <R>**
and	line 'attached' to cursor through the entered point and 'rotates' as the mouse is moved
prompt	`Specify through point`
enter	**80,200 <R>**
prompt	`Specify through point`
respond	**Center icon and pick the circle**
prompt	`Specify through point` and right-click.

5 At the command line enter **XLINE <R>** and:

prompt	`Specify a point or..`
enter	**H <R>** – the horizontal option
prompt	`Specify through point`
enter	**100,20 <R>**
prompt	`Specify through point`
respond	**Midpoint icon and pick a vertical line of square**
prompt	`Specify through point`
respond	**Quadrant icon and pick top of circle**
prompt	`Specify through point` and right-click.

6 Select the CONSTRUCTION LINE icon from the Draw toolbar and:

prompt	`Specify a point or..`
enter	**V <R>** – the vertical option
prompt	`Specify through point`
respond	**Center icon and pick the circle**
prompt	`Specify through point` and right-click.

7 Activate the construction line command and:

prompt	`Specify a point or..`
enter	**O <R>** – the offset option
prompt	`Specify offset distance or [Through]` and enter: **75 <R>**
prompt	`Select a line object`
respond	**pick the vertical line through the circle centre**
prompt	`Select side to offset`
respond	**offset to the right**
prompt	`Select a line object`
respond	**offset the same line to the left**
prompt	`Select a line object` and right-click.

8 Construction line command and at prompt:

enter	**A <R>** – the angle option
prompt	Enter angle of xline (0.0) or [Reference]
enter	**– 30 <R>**
prompt	`Specify through point`
respond	**Center icon and pick the circle**
prompt	`Specify through point`
enter	**135,40 <R>** then right-click.

9 Construction line command for last time and at prompt:

enter **B <R>** – the bisect option
prompt Specify angle vertex point
respond **Midpoint icon and pick top line of square**
prompt Specify angle start point
respond **pick lower left vertex of square**
prompt Specify angle end point
respond **Midpoint icon and pick square right vertical line**
prompt Specify angle end point and right-click.

10 Construction lines can be copied, moved, trimmed, etc. and object snap referenced to create other objects. SHAPE1 and SHAPE2 in Fig. 23.2(a) have been 'drawn' by referencing the existing construction lines. Try some shape creation for yourself.

11 *Note*

   a) Think of how construction lines could have been used to create the front view of the point filters example completed earlier in this chapter.

   b) I would recommend that construction lines are created on their own layer, and that this layer is frozen to avoid 'screen clutter' when not in use. I would also recommend that they are given a colour and linetype not normally used.

12 *Task:* try the following:

   a) At the command line enter **LIMITS <R>** and:
     prompt Specify lower left corner and enter: **0,0 <R>**
     prompt Specify upper right corner and enter: **10000,10000 <R>**

   b) From menu bar select View-Zoom-All and:
     1. drawing appears very small at bottom of screen
     2. the construction lines 'radiate outwards' to the screen edges

   c) Using LIMITS <R> enter the following values:
     lower left corner: –10000,–10000
     upper right corner: 0,0

   d) View-Zoom-All to 'see' the construction lines

   e) Return limits to 0,0 and 420,297 then View-Zoom-All to restore the original drawing screen. A screen regen may be required?

13 The construction line exercise is now complete. The drawing can be saved if required, but we will not use it again.

Rays

Rays are similar to construction lines, but they only extend to infinity in one direction from the selected start point.

1 Make a new layer called RAYLINE, colour to suit and with a dotted linetype. Make this layer current and refer to Fig. 23.2(b).

2 Freeze layer CONLINE and erase any objects to leave the original square and circle.

3 Menu bar with **Draw-Ray** and:

prompt Specify start point
respond **Center icon and pick the circle**
prompt Specify through point and enter: **@100<150 <R>**
prompt Specify through point and enter: **@100<0 <R>**
prompt Specify through point and enter: **@100<–90 <R>**
prompt Specify through point and right-click.

4 With layer OUT current, draw three 25 radius circles at the intersection of the ray lines and circle.

5   Make layer RAYLINE current and at the command line enter **RAY <R>** and:
    prompt   Specify start point
    respond   **pick the circle centre point**
    prompt   Specify through point
    respond   **Intersection icon and pick the four vertices of the square** then right-click.

6   Use the LIMITS command and enter:
    a) lower left: −10000,−10000
    b) upper right: 0,0.

7   Menu bar with View-Zoom-All and note position of 'our drawing'.

8   Return limits to 0,0 and 420,297 the Zoom-Previous to restore the original drawing screen. Is a REGEN needed?

9   *Note*: As with construction lines, it is recommended that ray lines be drawn on 'their own layer' with a colour and linetype to suit. This layer can then be frozen when not in use.

Summary

1   Point filters allow the user a method of creating objects by referencing existing object coordinates.

2   Point filters are activated by keyboard entry.

3   Construction and ray lines are aids to draughting. They allow lines to be created from a selected point and:
    a) construction lines extend to infinity in both directions from the selected start point
    b) ray lines extend to infinity in one direction only from the selected start point.

4   Both construction and ray lines can be activated from the menu bar or by keyboard entry. XLINE is the construction line entry and RAY the ray line entry.

5   The default Draw toolbar only displays the Construction Line icon, although this can be altered to include the ray line icon.

6   Both construction and ray lines are very useful aids for the CAD operator, but it is personal preference if they are used.

Chapter 24

Viewing a drawing

Viewing a drawing is important as the user may want to enlarge a certain part of the screen for more detailed work or return the screen to a previous display. AutoCAD allows several methods of altering the screen display including:

1 The scroll bars.
2 The Pan and Zoom commands.
3 The Aerial view option.
4 Realtime pan and zoom.

In this chapter we will investigate several view options.

Getting ready

1 Open WORKDRG to display the red outline, two red circles, four green centre lines and a black border.

2 With layer TEXT current, menu bar with **Draw-Text-Single Line Text** and:
   a) start point: centred on 90,115
   b) height: 0.1 – yes 0.1
   c) rotation: 0
   d) text item: AutoCAD.

3 With layer OUT current, draw a circle centred on 89.98,115.035 and radius 0.01. The awkward coordinate entries are deliberate.

4 These two objects cannot yet be 'seen' on the screen.

5 Ensure the Zoom toolbar is displayed and positioned to suit.

Pan

Allows the graphics screen to be 'moved', the movement being controlled by the user with coordinate entry or by selecting points on the screen.

1 From the Standard toolbar select the PAN REALTIME icon and:
   prompt Press Esc or Enter to exit, or right-click to display shortcut menu.
   and   cursor changes to a hand
   respond 1. hold down the left button of mouse
             2. move mouse and complete drawing moves
             3. note that scroll bars also move
             4. move image roughly back to original position
             5. right-click and:
                a) pop-up shortcut menu displayed
                b) Pan is active – tick
                c) pick Exit.

2 Menu bar with **View-Pan-Point** and:
   prompt Specify base point or displacement and enter: **0,0 <R>**
   prompt Specify second point and enter: **500,500 <R>**

3 No drawing on screen – don't panic!

4 Use PAN REALTIME to pan down and to the left, and 'restore' the drawing roughly in its original position, then right-click and exit.

Zoom

This is one of the most important and widely used of all the AutoCAD commands. It allows parts of a drawing to be magnified/enlarged on the screen. The command has several options, and it is these options which will now be investigated. To assist us in investigating the zoom options:

a) use the COPY command, window the shape and copy the component from 50,50 to 210,260

b) refer to Fig. 24.1

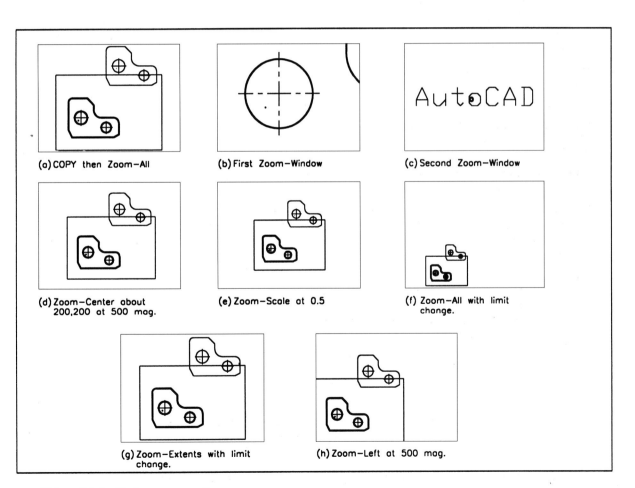

(a) COPY then Zoom—All

(b) First Zoom—Window

(c) Second Zoom—Window

(d) Zoom—Center about 200,200 at 500 mag.

(e) Zoom—Scale at 0.5

(f) Zoom—All with limit change.

(g) Zoom—Extents with limit change.

(h) Zoom—Left at 500 mag.

Figure 24.1 Various zoom options.

Zoom All

Displays a complete drawing including any part of the drawing which is 'off' the current screen.

1 From menu bar select **View-Zoom-All.**
2 The two components and the black border are displayed – fig. (a).

Zoom Window

Perhaps the most useful of the zoom options. It allows areas of a drawing to be 'enlarged' for clarity or more accurate work.

1 Select the ZOOM WINDOW icon from the Zoom toolbar and: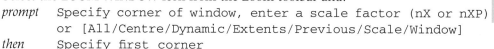
   prompt   Specify corner of window, enter a scale factor (nX or nXP)
              or [All/Centre/Dynamic/Extents/Previous/Scale/Window]
   then   Specify first corner
   respond **window the original left circle** – fig. (b).

2 At the command line enter **ZOOM <R>** and:
   prompt   Specify corner of window, enter a scale factor..
   enter   **W <R>** – the window option
   prompt   Specify first corner and enter: **89.6,114.9 <R>**
   prompt   Specify opposite corner and enter: **90.4,115.2 <R>**

3 The text item and circle will now be displayed – fig. (c).

Zoom Previous

Restores the drawing screen to the display before the last view command.

1 Menu bar with **View-Zoom-Previous** – restores fig. (b).
2 At command line enter **ZOOM <R>** then **P <R>** – restores fig. (a).

Zoom center

Allows a drawing to be centred about a user-defined centre point.

1 Select the ZOOM CENTER icon and:
   prompt   Specify center point and enter: **200,200 <R>**
   prompt   Enter magnification or height<?> and enter **500 <R>**

2 The complete drawing is centred on the drawing screen about the entered point – fig. (d).

3 *Note:*
   a) the 'size' of the displayed drawing depends on the magnification/height value entered by the user and is relative to the displayed default <?> value and:
      1. a value less than the default – magnifies drawing on screen
      2. a value greater than the default – reduces the size of the drawing on the screen
   b) toggle the grid ON and note the effect with the centre option, then toggle the grid off

4 Menu bar with **View-Zoom-Center** and enter the following centre points, all with 1000 magnification:

   a) 200,200   *b*) 0,0   *c*) 500,500.

5 Now Zoom-Previous four times to restore fig. (a).

Zoom Scale

Centres a drawing on the screen at a scale factor entered by the user. It is similar (and easier?) than the Zoom-Center option.

1 Select the ZOOM SCALE icon and:

prompt   Enter a scale factor (nX or nXP) and enter: **0.5 <R>**

2 The drawing is displayed centred and scaled – fig. (e).

3 Menu bar with **View-Zoom-Scale** and enter a scale factor of 0.25.

4 Zoom to a scale factor of 1.5.

5 Zoom-Previous three times to restore fig. (a).

Zoom Extents

Zooms the drawing to extent of the current limits.

1 At command line enter **LIMITS <R>** and:
prompt   Specify lower left corner and enter: **0,0 <R>**
prompt   Specify upper right corner and enter: **1000,1000 <R>**

2 Menu bar with **View-Zoom-All** – fig. (f).

3 Select the ZOOM EXTENTS icon – fig. (g).

4 Menu bar with **Format-Drawing Limits** set the limits back to 0,0 and 420,297 then Zoom-All to restore fig. (a).

Zoom Dynamic

This option will not be discussed. The other zoom options and the realtime pan and zoom should be sufficient for all users needs?

Zoom Realtime

1 Menu bar with **View-Zoom-Realtime** and:
prompt   Press ESC or ENTER to exit, or right click to display
          shortcut menu
and   cursor changes to a magnifying glass with a + and −
respond a) hold down left button on mouse and move upwards to give a magnification
          effect
          b) move downwards to give a decrease in size effect
          c) left-right movement – no effect
          d) right-click to display pop-up menu with Zoom active
          e) pick Exit.

2 View-Zoom-All to display fig. (a).

3 The Zoom Realtime icon is in the Standard Toolbar.

Zoom Left

This option is not available as an icon and does not appear as a command line option. It is an old zoom option that can still be activated.

1 At the command line enter **ZOOM <R>** and:
prompt   Specify corner of window or..
enter   **L <R>** – the left option
prompt   Lower left corner point and enter: **0,0 <R>**
prompt   Enter magnification or height and enter: **500 <R>**

2 Complete drawing is moved to left of drawing screen and displayed at the entered magnification – fig. (h).

3 Zoom-Previous to restore fig. (a).

Aerial View

The aerial view is a navigation tool (AutoCAD expression) and allows the user to pan and zoom a drawing interactively. This view option will be investigated when we construct a large scale drawing.

Transparent Zoom

A **transparent** command is one which can be activated while using another command and zoom has this facility.

1 Restore the original Fig. 23.2(a) – zoom all?

2 Select the LINE icon and:
prompt   Specify first point
enter   **0,0 <R>**
prompt   Specify next point
enter   **'ZOOM <R>** – the transparent zoom command
prompt   All/Center/Dynamic.. – the zoom options
enter   **W <R>** – the window zoom option
prompt   Specify first corner and enter: **89.6,114.9 <R>**
prompt   Specify opposite corner and enter: **90.4,115.2 <R>**
prompt   Specify next point
respond   centre icon and pick the circle created at the start of the chapter, which should now be displayed
prompt   Specify next point
enter   **'Z <R>** then **A <R>** – the zoom all option
prompt   Specify next point
respond   right-click-enter to end line sequence.

3 The transparent command was activated by entering **'ZOOM <R>** or **'Z <R>** at the command line.

4 Only certain commands have transparency, generally those which alter a drawing display, e.g. GRID, SNAP and ZOOM.

5 Do not save this drawing modification.

6 *Note*
In this exercise we enter **Z <R>** to activate the zoom command and this is an example of using an AutoCAD **Alias or Abbreviation**. Many of the AutoCAD commands have an alias, which allows the user to activate the command from the keyboard by entering one or two letters. Typical aliases are:
L line   C circle   E erase   M move   CP Copy

Summary

1 Pan and Zoom are VIEW commands usually activated by icon.

2 Pan does 'not move a drawing' – it 'moves' the complete drawing screen.

3 The zoom command has several options, the most common being:
   All: displays the complete drawing
   Window: allows parts of a drawing to be displayed in greater detail
   Center: centres a drawing about a user-defined point at a user specified
              magnification
   Previous: displays the drawing screen prior to the last pan/zoom command
   Extents: zooms the drawing to the existing limits.

4 Both the pan and zoom commands have a REALTIME option.

Chapter 25

Hatching

AutoCAD 2002 has associated boundary hatching. The hatching (or more correctly section detail) must be added by the user, and there are three types:
a) predefined, i.e. AutoCAD's stored hatch patterns
b) user defined
c) custom – not considered in this book.

When applying hatching, the user has two methods of defining the hatch pattern boundary:
a) by selecting objects which make the boundary
b) by picking a point within the boundary.

There are two hatch commands available:
a) HATCH: command line entry only
b) BHATCH: activated by icon, menu bar or command line entry. This command displays a dialogue box.

In this chapter we will consider all of the above options, with the exception of custom hatch patterns.

Getting ready

1   Open the A3PAPER standard sheet and refer to Fig. 25.1.

2   Draw a 50 unit square on layer OUT and multiple copy it to ten other parts of the screen. Add the other lines within the required squares (any size). Note that I have included additional squares to indicate appropriate object selection and to demonstrate the before and after effect.

3   Make layer SECT current.

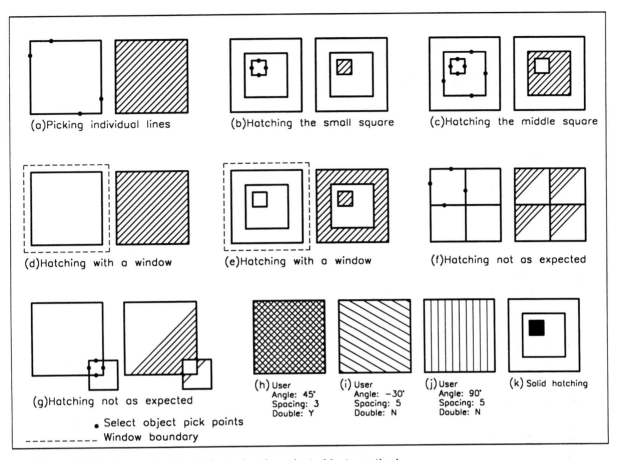

Figure 25.1 User-defined hatching using the select objects method.

User-defined hatch patterns – Select objects method

User-defined hatching consists of straight line patterns, the user specifying:
a) the hatch angle relative to the horizontal
b) the distance between the hatch lines – spacing
c) whether single or double (cross) hatching.

1 At the command line enter **HATCH <R>** and:

prompt   `Enter a pattern name or [?/Solid/User defined]<ANGLE>`
enter   **U <R>** – user-defined option
prompt   `Specify angle for crosshatch lines<0.0>`
enter   **45 <R>**
prompt   `Specify spacing between lines<1.0000>`
enter   **3 <R>**
prompt   `Double hatch area [Yes/No]<N>`
enter   **N <R>**
prompt   `Select objects`
respond   **pick the four lines of first square and right-click**
and   hatching added to the square – fig. (a).

2   Immediately after the hatching has been added, **right-click** and:

    prompt   `Shortcut pop-up menu`
    respond   **pick Repeat HATCH**
    prompt   `Enter a pattern name and enter:` **U <R>**
    prompt   `Specify an angle and enter:` **45 <R>**
    prompt   `Specify spacing and enter:` **3 <R>**
    prompt   `Double hatch and enter:` **N <R>**
    prompt   `Select objects`
    respond   **pick four lines indicated in second square then right-click**
    and   hatching added to the small square – fig. (b).

3   Select the HATCH icon from the Draw toolbar and:

    prompt   `Boundary Hatch dialogue box as Fig. 25.2`
    with   *a*) Type: User-defined
             b) Angle: 45
             c) Spacing: 3, i.e. our previous entries!
    respond   **pick Select Objects**
    and   dialogue box disappears
    prompt   `Select objects` at command line
    respond   **pick eight lines in third square then right-click and pick Enter**
    prompt   `Boundary Hatch dialogue box returned`
    respond   **pick Preview**
    and   1. drawing screen displayed with hatching added – check that it is correct
             2. `<Hit enter or right-click to return to dialog>` at command line
    respond   **right-click**
    prompt   `Boundary Hatch dialogue box`
    respond   **pick OK**
    and   hatching added as fig. (c).

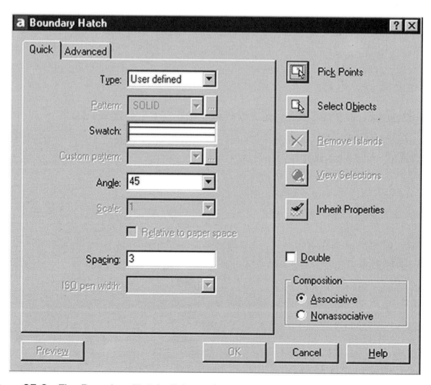

Figure 25.2 The Boundary Hatch dialogue box.

4 Menu bar with **Draw-Hatch** and:

prompt	`Boundary Hatch dialogue box` – settings as before?
respond	**pick Select Objects**
prompt	`Select objects`
enter	**W <R>** – the window selection option
then	**window the fourth square, right-click and Enter**
prompt	`Boundary Hatch dialogue box`
respond	**Preview-right click-OK**
and	square hatched – fig. (d).

5 At the command line enter **HATCH <R>** then:

a) pattern name: U
b) angle: 45
c) spacing: 3
d) double: N
e) Select objects: window the fifth square then right-click
f) hatching as fig. (e)

6 Using HATCH from the keyboard, accept the four defaults of U,45,3,N and pick the four lines of the sixth square to give hatching as fig. (f). Not as expected?

7 Repeat the hatch command and select the four lines indicated in the seventh square – fig. (g).

8 With HATCH <R> at the command line, add hatching to the next three squares using the following entries:

	I	*II*	*III*
Pattern:	U	U	U
Angle:	45	−30	90
Spacing:	3	5	5
Double:	Y	N	N
fig.	(h)	(i)	(j)

9 Finally enter **HATCH <R>** and:

prompt	`Enter a pattern name ........`
enter	**S <R>** – the solid option
prompt	`Select objects`
respond	**pick four lines of small square then right-click**
and	a solid hatch pattern is added – fig. (k).

10 This drawing is complete but does not need to be saved.

User-defined hatch patterns – pick points method

1   Refer to Fig. 25.3 and:
    a) erase all hatching
    b) erase squares not required
    c) add squares and circles as shown – size not important.

2   With layer Sect current, pick the HATCH icon and:
    prompt   Boundary Hatch dialogue box
    respond   1. check/alter:
                    a) Type: User-defined
                    b) Angle: 45
                    c) Spacing: 3
                2. **pick Pick Points**
    prompt   Select internal point
    respond   **pick any point within first square**
    prompt   Various prompts at command line
    then   Select internal point
    respond   **right-click and Enter**
    prompt   Boundary Hatch dialogue box
    respond   **Preview-right click-OK**
    and   hatching added to square – fig. (a).

3   Using the HATCH icon with the Pick Points option from the Boundary Hatch dialogue
    box, add hatching using the points indicated in Fig. 25.3. The only 'problem' is hatching
    the outer square in fig. (d) and fig. (e) – two pick points are required.

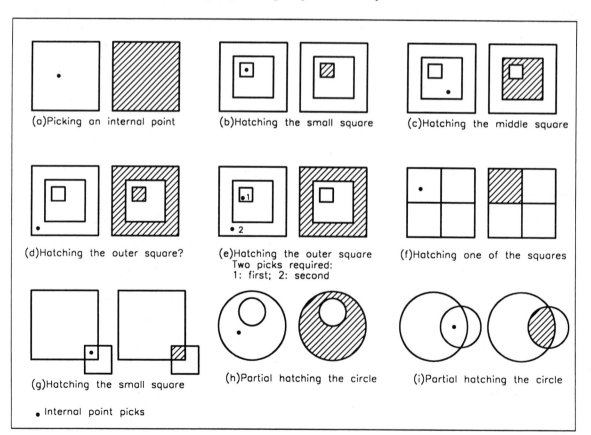

Figure 25.3 User-defined hatching using the pick points method.

Select objects vs pick points

With two options available for hatching, new users to this type of boundary hatch command may be confused as to whether they should Select Objects or Pick Points. In general the Pick Points option is the simpler to use, and will allow complex shapes to be hatched with a single pick within the area to be hatched. To demonstrate the effect:

1   Erase all objects from the screen and refer to Fig. 25.4.

2   Draw two sets of three intersecting circles – any size. We want to hatch the intersection area of the three circles.

3   Select the HATCH icon and from the Boundary Hatch dialogue box
    a) set: User-defined, Angle of 45, Spacing of 3, then:
    b) pick the Select Objects option
    c) pick the three circles
    d) preview-right click-OK and hatching as fig. (a).

4   Select the HATCH icon again and:
    a) pick Pick Points option
    b) pick any point within the area to be hatched
    c) preview-right click-OK and hatching as fig. (b).

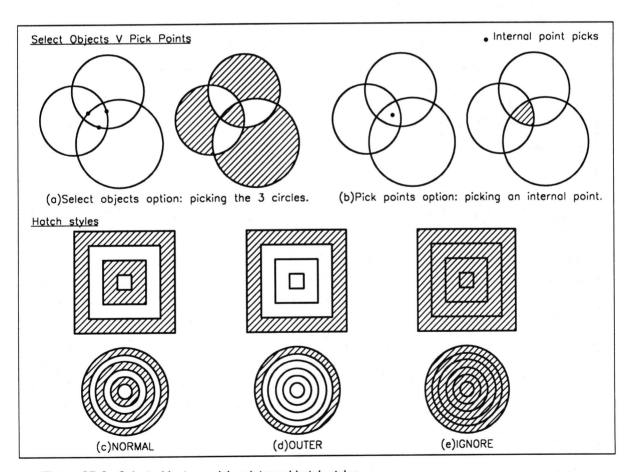

Figure 25.4   Select objects vs pick points and hatch styles.

Hatch style

AutoCAD has a hatch style (called **Island** detection) option which allows the user to control three 'variants' of the hatch command. To demonstrate the hatch style, refer to Fig. 25.4 and draw:

a) a 70 sided square with four smaller squares 'inside it'
b) six concentric circles, smallest radius being 5
c) copy the squares and circles to two other areas of the screen.

1   Select the HATCH icon and:
    prompt   Boundary Hatch dialogue box
    ensure   User-defined, Angle: 45, Spacing: 3
    then   **pick Advanced tab**
    prompt   Boundary Hatch dialogue box with Advanced tab displayed
    note   *a*) Island detection style 'picture'
           b) Style: Normal active – Fig. 25.5
    respond   **pick Select Objects**
    prompt   Select objects at the command line
    respond   **window the first square and circle then right-click and Enter**
    prompt   Advanced boundary hatch dialogue box
    respond   **pick Preview-right click-OK**
    and   Hatching added as fig. (c).

2   Repeat the HATCH icon and with the Advanced tab:
    a) alter Island detection style to Outer
    b) pick Select Objects and window the second square and circle
    c) preview then OK – hatching as fig. (d)

3   Using the Advanced tab of the Boundary Hatch dialogue box, alter the island detection style to Ignore, and add hatching to the third square and circle – fig. (e).

4   *Note*: the user can leave the style at 'Normal' and then control the hatch area by picking the relevant points within the area to be hatched.

5   Save if required, but we will not use this drawing again.

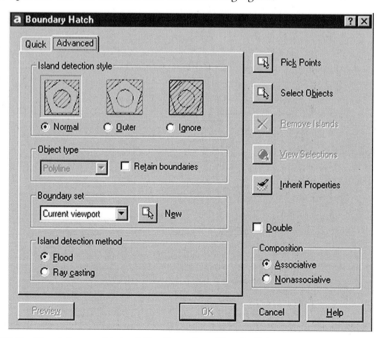

Figure 25.5   The Advanced tab of the Boundary Hatch dialogue box.

Predefined hatch patterns

AutoCAD 2002 has several stored hatch patterns which can be accessed using the command line HATCH entry or from the Boundary Hatch dialogue box, which is slightly easier. With predefined hatch patterns, the user specifies:

a) the scale of the patterns
b) the angle of the pattern.

1   Open your standard sheet, refer to Fig. 25.6(A) and:
   a) draw a 50 unit square
   b) multiple copy the square to eleven other places on the screen
   c) add other lines and circles as displayed.

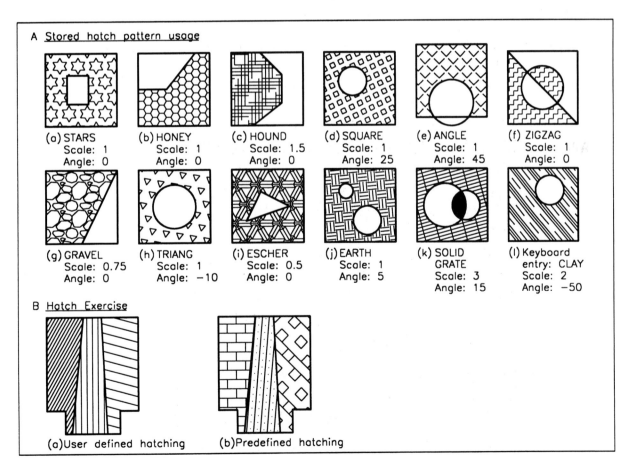

Figure 25.6   Using predefined (stored) hatch patterns.

2   Select the HATCH icon and:

    prompt   `Boundary Hatch dialogue box`

    respond   1. Advanced tab and set:

               a) Island detection style: Normal

           2. Quick tab and:

               a) Type: Predefined

               b) Pattern: **pick** …

    prompt   `Hatch Pattern Palette dialogue box` with four tab selections: ANSI, ISO, Other Predefined, Custom

    respond   1. ensure Other Predefined tab active

           2. scroll until INSUL to ZIGZAG displayed – Fig. 25.7

           3. pick STARS then OK

    prompt   `Boundary Hatch dialogue box`

    with   1. Pattern: STARS

           2. Swatch: display of stars hatch pattern

    respond   1. Angle: 0

           2. Scale: 1 – dialogue box as Fig. 25.8

           3. pick Pick Points

           4. select any internal point in first square

           5. right-click and Enter

           6. Preview-right click-OK

    and   Hatching added as fig. (a).

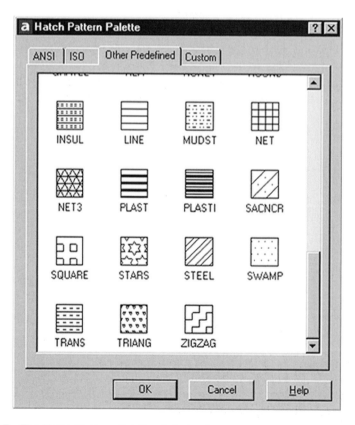

Figure 25.7 The Hatch Pattern Palette dialogue box.

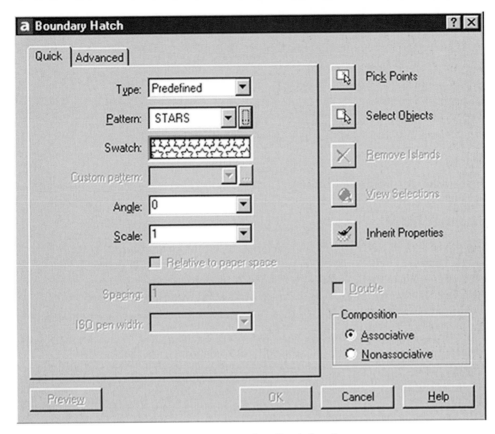

Figure 25.8 The Boundary Hatch dialogue box (Predefined).

3 Repeat the HATCH icon selection and using the following hatch pattern names, scales and angles, add hatching to the other squares using the pick points option:

fig.	pattern	scale	angle
b	HONEY	1	0
c	HOUND	1.5	0
d	SQUARE	1	25
e	ANGLE	1	45
f	ZIGZAG	1	0
g	GRAVEL	0.75	0
h	TRIANG	1	–10
I	ESCHER	0.5	0
j	EARTH	1	5
k	SOLID		
	GRATE	3	15

4 Finally enter **HATCH <R>** and the command line and:

prompt   Enter a pattern name or [?/Solid/..
enter   **CLAY <R>**
prompt   Specify a scale for the pattern and enter: **2 <R>**
prompt   Specify an angle for the pattern and enter: **–50 <R>**
prompt   Select objects
respond **pick lines and circle then right-click** – fig. (l).

5 Exercise is complete, so save if required.

Hatch exercise

1 Open the USEREX and copy the component to another part of the screen. Refer to Fig. 25.6(B). Layer SECT current.

2 Add the following hatching using the Pick Points option:

User-defined
1. angle: 60, spacing: 2
2. angle: 90, spacing: 4
3. angle: −15, spacing: 6

Predefined
1. BRICK, scale: 1, angle: 0
2. SACNCR, scale: 2, angle: 40
3. BOX, scale: 1, angle: 45

3 Save this exercise if required, but not as USEREX.

Associative hatching

AutoCAD 2002 has associative hatching, i.e. of the hatch boundary is altered, the hatching within the boundary will be 'regenerated' to fill the new boundary limits. Associative hatching is applicable to both user-defined and predefined hatch patterns, irrespective of whether the select objects or pick points option was used. We will demonstrate the effect by example so:

1 Open A3PAPER standard sheet, refer to Fig. 25.9 and draw squares and circles as displayed. Note that I have drawn two sets of each squares to demonstrate the 'before and after' effect.

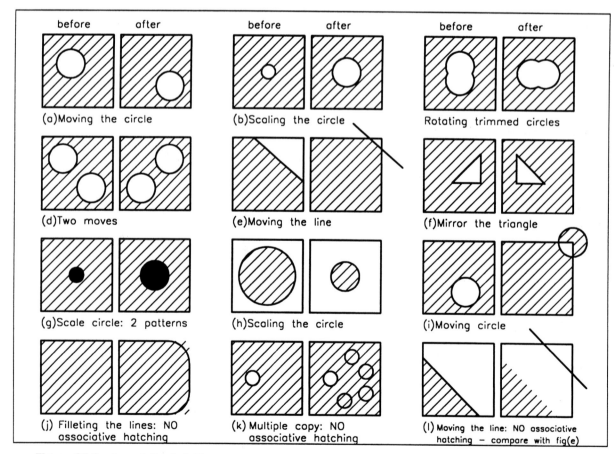

Figure 25.9 Associative hatching.

2   Make layer SECT current.

3   With the HATCH icon:
    a) Type: User defined; angle: 45; spacing: 5
    b) Composition: Associative ON, i.e. tick in box
    c) Pick Points and hatch the first square.

4   Select the MOVE icon and:
    a) pick the circle in the first square
    b) move it to another position in the square
    c) hatching 'changes' – fig. (a), i.e. it is associative.

5   Figure 25.9 displays some associative hatching effects, which you should try for yourself. The effects are:

fig.	effect
b	scaling the circle
c	rotating two trimmed circles
d	two circle moves
e	moving the line
f	mirror the triangle and deleting source objects
g	scaling the circle with two hatch areas
h	scaling the circle
i	moving the circle.

7   *Associative hatch 'quirks'*
    Associative hatching does not occur with every modify operation as is displayed in Fig. 25.9 with:

fig.	operation
j	filleting two corners of the square
k	multiple copy of the circle
l	moving the line – compare with fig. (e) – why compare?

Modifying hatching

Hatching which has been added to an object can be modified, i.e. the angle, spacing or scale can be altered, or the actual pattern changed.

1   Open A3PAPER and refer to Fig. 25.10(A). Display the MODIFY II toolbar.

2   Draw a square with two trimmed circles inside it – any size.

3   *a*) Use the HATCH icon to add hatching to the square with:
        User defined pattern; angle: 45; spacing: 8 – fig. (a)
    b) Ensure that Associative hatching is on.

4   Menu bar with **Modify-Object-Hatch** and:
    prompt   Select associative hatch object
    respond   **pick the added hatching**
    prompt   Hatch Edit dialogue box
    respond   1. alter angle to –45
                2. alter spacing to 4
                3. pick OK – fig. (b).

5 Select the EDIT HATCH icon from the Modify II toolbar and:
  prompt   Select associative hatch object
  respond **pick the altered hatching**
  prompt   Hatch Edit dialogue box
  respond 1. Type: Predefined
            2. Pattern: pick ...
            3. scroll and pick TRIANG then OK
            4. scale: 1 and angle: 0
            5. pick OK – fig. (c).

6 Rotate the trimmed circles by 45 degrees – fig. (d).

7 Using the Edit Hatch icon:
  a) Type: Predefined ESCHER
  b) scale: 0.6 and angle: 15 – fig. (e).

8 Edit the hatching to:
  User-defined, angle: 0, spacing: 2 – fig. (f).

9 Scale the circles by 1.5 – fig. (g). Zoom needed to pick?

10 Move the circles – fig. (h).

11 Finally: *a*)scale the circles by 0.667
            b) move the circles into the square
            c) change hatch pattern to SOLID
            d) fig. (i).

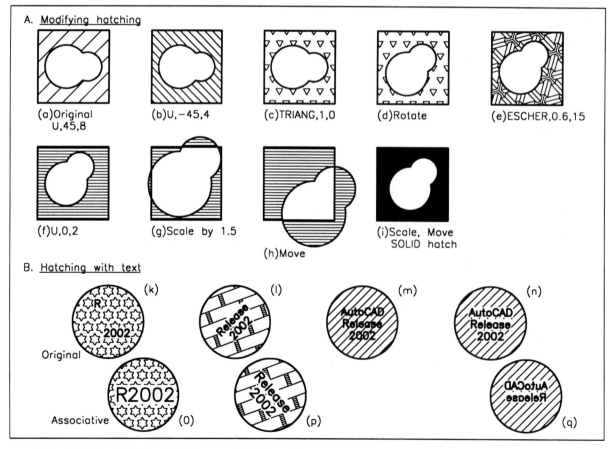

Figure 25.10   Modifying hatching and hatching with text.

Text and hatching

Text which is placed in an area to be hatched can be displayed with a 'clear border' around it.

1 Refer to Fig. 25.10(B) and draw four 50 diameter circles.

2 Add any suitable text as shown. This text can have any height, rotation angle, position, etc.

3 With the HATCH icon:
   a) Type: Predefined STARS with scale: 0.75 and angle: 0
   b) Pick Points and pick any point in first circle
   c) Preview-right click-OK and result as fig. (k).

4 Use the HATCH icon with:
   a) Type: Predefined BRSTONE, scale: 1, angle: 20
   b) Pick Points and pick any point inside second circle
   c) Preview-right click-OK and result as fig. (l).

5 HATCH icon using:
   a) User defined at 45 with 4 spacing
   b) Select Objects and pick third circle only
   c) Preview, etc. – fig. (m).

6 Final HATCH with:
   a) User defined, angle: 45, spacing: 4
   b) Select Objects and pick fourth circle **AND** text items
   c) Preview, etc. – fig. (n).

7 Associative hatching works with text items:
   fig. (o): moving and scaling the two items of text
   fig. (p): rotating the text item
   fig. (q): mirroring the text item, then erase part of it.

8 Save if required – the exercise is complete and will not be used again.

Summary

1 Hatching is a draw command, activated:
   a) from the menu bar
   b) by icon selection
   c) with keyboard entry – HATCH or BHATCH.

2 AutoCAD has three types of hatch pattern:
   a) User-defined: this is line only hatching. The user specifies the angle for the hatch lines, the spacing between these lines and whether the hatching is to be single or double, i.e. cross-hatching.
   b) Predefined: these are stored hatch patterns. The user specifies the pattern scale and angle.
   c) Custom: are patterns designed by the user, but are outside the scope of this book.

3 The hatch boundary can be determined by:
   a) selecting the objects which make the boundary
   b) picking an internal point within the hatch area.

4 Hatching is a single object but can be exploded.

5 AutoCAD 2002 has associative hatching which allows added hatching to alter when the hatch boundary changes.

6 Added hatching can be edited.

Assignments

Some AutoCAD users may not use the hatching in their draughting work, but they should still be familiar with the process. The pick points option makes hatching fairly easy and for this reason I have included five interesting exercises for you to attempt. These should test all your existing CAD draughting skills.

1 Activity 16:
   Three different types of component to be drawn and hatched:
   a) a small engineering component. The hatching is user defined and use your own angle and spacing values
   b) a pie chart. The text is to have a height of 8 and the hatch names and variables are given
   c) a model airport runway system.

2 Activity 17: Cover plate.
   A relatively simple drawing to complete. The MIRROR command is useful and the hatching should give no problems. Add text and the dimensions.

3 Activity 18: Protected bearing housing.
   Four views to complete, two with hatching. A fairly easy drawing to complete.

4 Activity 19: Steam expansion box.
   This activity has proved very popular in my previous books and is easier to complete than it would appear. Create the outline from lines and circles, trimming the circles as required. The complete component uses many commands, e.g. offset, fillet, mirror, etc. The hatching should not be mirrored – why?

5 Activity 20: Gasket cover.
   An interesting exercise to complete. Draw the 'left view' which consists only of circles. The right view can be completed using offset, trim and mirror. Do not dimension.

Chapter 26

Point, polygon and solid

These are three useful draw commands which will be demonstrated by example, so:

1 Open A3PAPER standard sheet with layer OUT current.
2 Activate the Draw, Modify and Object Snap toolbars
3 Refer to Fig. 26.1.

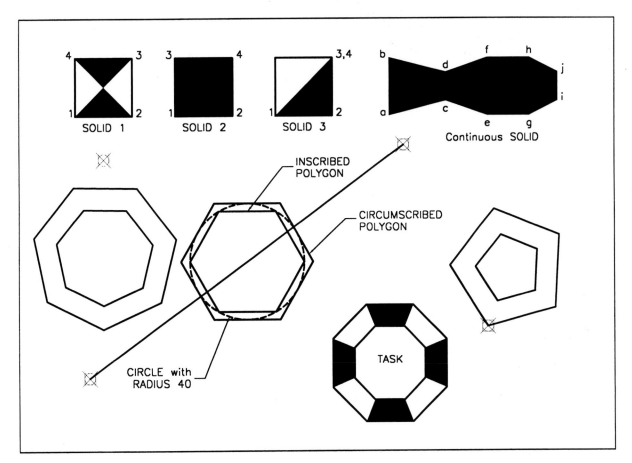

Figure 26.1 The POINT, POLYGON and SOLID draw commands.

Point

A point is an object whose size and appearance is controlled by the user.

1 From the Draw toolbar select the POINT icon and:

prompt Current point modes: PDMODE=0 PDSIZE=0.00 Specify a point

enter **50,50 <R>**

prompt Specify a point and enter: **60,200 <R>**

prompt Specify a point and **ESC** to end the command.

2 Two point objects will be displayed in red on the screen. You may have to toggle the grid off to 'see' these points.

3 From the menu bar select Format-Point Style and:

prompt Point Style dialogue box as Fig. 26.2

respond 1. pick point style indicated
2. set Point Size to 5%
3. Set Size Relative to Screen active
4. pick OK.

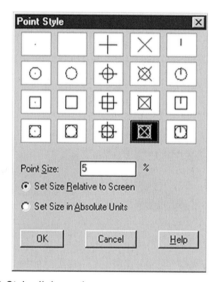

Figure 26.2 The Point Style dialogue box.

4 The screen will be displayed with the two points regenerated to this new point style.

5 Menu bar with **Draw-Point-Multiple Point** and:

prompt Specify a point and enter: **330,85 <R>**

prompt Specify a point and enter: **270,210 <R>**

prompt Specify a point and **ESC.**

6 The screen will now display two additional points in the selected style.

7 Select the LINE icon and:

prompt Specify first point

respond **Snap to Node icon and pick lower left point**

prompt Specify next point

respond **Snap to Node icon and pick upper right point**

prompt Specify next point and right-click then Enter.

8 Task

a) set a new point style and size and note screen display

b) restore the previous point style and size as step 3.

Polygon

A polygon is a multi-sided figure, each side having the same length and can be drawn by the user specifying:
a) a centre point and an inscribed/circumscribed radius
b) the endpoints of an edge of the polygon.

1 Select the POLYGON icon from the Draw toolbar and:

prompt	Enter number of sides<4> and enter: **6 <R>**
prompt	Specify center of polygon or [Edge]
respond	**Snap to Midpoint icon and pick the line**
prompt	Enter an option [Inscribed in circle/Circumscribed about circle]<I>
enter	**I <R>** – the inscribed (default) option
prompt	Specify radius of circle and enter: **40 <R>**

2 Repeat the POLYGON icon selection and:

prompt	Enter number of sides<6> and enter: **6 <R>**
prompt	Specify center of polygon or..
respond	**Snap to Midpoint icon and pick the line**
prompt	Enter an option [Inscribed/Circumscribed..
enter	**C <R>** – the circumscribed option
prompt	Specify radius of circle and enter: **40 <R>**

3 The screen will display an inscribed and circumscribed circle drawn relative to a 40 radius circle as shown in Fig. 26.1.
These hexagonal polygons can be considered as equivalent to:
a) inscribed: ACROSS CORNERS (A/C)
b) circumscribed: ACROSS FLATS (A/F).

4 From the menu bar select **Draw-Polygon** and:

prompt	Enter number of sides<6> and enter: **5 <R>**
prompt	Specify center of polygon or [Edge]
enter	**E <R>** – the edge option
prompt	Specify first endpoint of edge
respond	**Snap to Node icon and pick lower right point**
prompt	Specify second endpoint of edge
enter	**@50<15 <R>**

5 At the command line enter **POLYGON <R>** and:
a) Number of sides: 7
b) Edge/Center: E
c) First point: 60,100
d) Second point: @30<25.

6 Using the OFFSET icon, set an offset distance of 15 and:
a) offset the five-sided polygon 'inwards'
b) offset the seven-sided polygon 'outwards'.

7 *Note*:
A polygon is a POLYLINE type object and has the 'properties' of polylines – next chapter.

Solid (or more correctly 2D Solid)

A command which 'fills-in' lined shapes, the appearance of the final shape being determined by the pick point order.

1   With the snap on, draw three squares of side 40, towards the top part of the screen, as Fig. 26.1.

2   Menu bar with **Draw-Surfaces-2D Solid** and:
    prompt   Specify first point and pick point 1 of SOLID 1
    prompt   Specify second point and pick point 2
    prompt   Specify third point and pick point 3
    prompt   Specify fourth point and pick point 4
    prompt   Specify third point and right-click to end command.

3   At the command line enter **SOLID <R>** and:
    prompt   Specify first point and pick point 1 of SOLID 2
    prompt   Specify second point and pick point 2
    prompt   and pick pints 3 and 4 in order displayed.

4   Activate the SOLID command and with SOLID 3, pick points 1–4 in the order given, i.e. 3 and 4 are the same points.

5   The three squares demonstrate how three- and four-sided shapes can be solid filled.

6   Using the SOLID command and Snap on:
    First point: pick a point a
    Second point: pick a point b
    Third point: pick point c
    Fourth point: pick point d
    Third point: pick point e
    Fourth point: pick point f
    Third point: pick point g
    Fourth point: pick point h
    Third point: pick point i
    Fourth point: pick point j
    Third point: right-click

7   These last entries demonstrate how continuous filled 2D shapes can be created.

8   *Task 1*
    a) Draw an right-sided polygon, centre at 260,60 and circumscribed in a 40 radius circle
    b) Offset the polygon inwards by a distance of 15
    c) Use the SOLID command to produce the effect in Fig. 26.1.
    d) A running object snap to Endpoint will help.

9   *Task 2*
    What are the minimum and maximum number of sides allowed with the POLYGON command?

10 *Task 3*
    a) At the command line enter **FILL <R>** and:
        prompt   Enter mode [ON/OFF]<ON>
        enter   **OFF <R>**
    b) At command line enter **REGEN <R>**
    c) The solid fill effect is not displayed
    d) Turn FILL back on then REGEN the screen.

11 This exercise is complete and can be saved if required.

Assignment

Activity 21 requires a backgammon board to be drawn. This is a nice simple drawing to complete. The filled triangles can be created with the 2D SOLID command or by hatching with the predefined SOLID pattern. The multiple copy and mirror commands are useful.

Summary

1 A point is an object whose appearance depends on the selection made from the Point Style dialogue box.

2 Only ONE point style can be displayed on the screen at any time.

3 The Snap to Node icon is used with points.

4 A polygon is a multi-sided figure having equal sides.

5 A polygon is a polyline type object.

6 Line shapes can be 'solid filled', the appearance of the solid fill being dependent on the order of the pick points.

7 Only three- and four-sided shapes can be solid filled.

Chapter 27

Polylines and splines

A polyline is a single object which can consist of line and arc segments and can be drawn with varying widths. It has its own editing facility and can be activated by icon selection, from the menu bar or by keyboard entry.

A polyline is a very useful and powerful object yet it is probably one of the most under-used draw commands. The demonstration which follows is quite long and several keyboard options are required.

1 Open the A3PAPER standard drawing sheet with layer OUT current and refer to Fig. 27.1. Display the Draw, Modify, Object Snap and Modify II toolbars.

2 Select the POLYLINE icon from the Draw toolbar and:

prompt	`Specify start point`
enter	**15,220 <R>**
prompt	`Specify next point or [Arc//Halfwidth/Length/Undo/Width]`
enter	**@50,0 <R>**
prompt	`Specify next point or [Arc/Close/Halfwidth/Length/Undo/` `Width]`
enter	**@0,50 <R>**
prompt	`Specify next point or` .. and enter **@–50,0 <R>**
prompt	`Specify next point or`.. and enter **@0,–50 <R>**
prompt	`Specify next point or`..
respond	right-click and enter to end command.

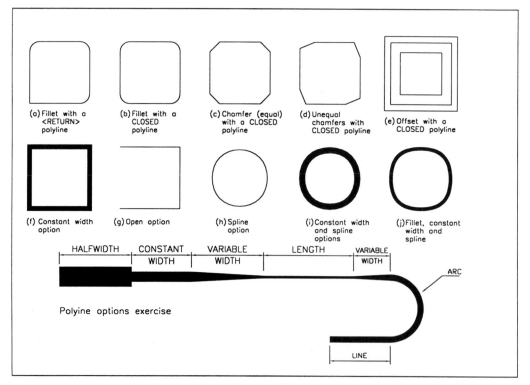

Figure 27.1 Polyline demonstration exercise.

3   From menu bar select **Draw-Polyline** and:
   prompt   Specify start point and enter **90,220 <R>**
   prompt   Specify next point and enter **@50<0 <R>** – polar
   prompt   Specify next point and enter **140,270 <R>** – absolute
   prompt   Specify next point and enter **@–50,0 <R>** – relative
   prompt   Specify next point and enter **C <R>** – close option.

4   Select the COPY icon and:
   prompt   Select objects
   respond pick any point on SECOND square
   and   all 4 lines are highlighted with one pick
   then   right-click
   prompt   Specify base point
   and   **multiple copy the square** to 8 other parts of the screen.

5   Select the FILLET icon and:
   prompt   Select first object or [Polyline/Radius/Trim]
   enter   **R <R>** – the radius option
   prompt   Specify fillet radius
   enter   **8 <R>**
   prompt   Select first object or [Polyline/Radius/Trim]
   enter   **P <R>** – polyline option
   prompt   Select 2D polyline
   respond **pick any point on first square**
   prompt   3 lines were filleted – fig. (a)
   and   command line returned.

6   Repeat the fillet icon selection and:
   a) enter   **P <R>** – the polyline option, radius set to 8
   b) pick any point on second square
   c) prompt   4 lines were filleted – fig. (b).

7   *Note*
   a) The difference between the two fillet operations is:
       1. fig. (a): not a 'closed' polyline, so only 3 corners filleted
       2. fig. (b): a 'closed' polyline, so all 4 corners filleted
   b) The corner 'not filleted' in fig. (a) is the start point.

8   Select the CHAMFER icon and:
   prompt   Select first line or [Polyline/Distance/Angle/Trim/Method]
   enter   **D <R>** – the distance option
   prompt   Specify first chamfer distance and enter: 8
   prompt   Specify second chamfer distance and enter: 8
   prompt   Select first line or [Polyline/Distance/Angle/Trim/Method]
   enter   **P <R>** – the polyline option
   prompt   Select 2D polyline
   respond **pick any point on third square**
   prompt   4 lines were chamfered – fig. (c)
   and   command line returned.

9   *Task 1*
   a) set chamfer distances to 12 and 5
   b) chamfer the fourth square remembering to enter **P <R>** to activate the polyline
       option
   c) result is fig. (d)
   d) note the orientation of the 12 and 5 chamfer distances.

10 *Task 2*

 a) set an offset distance to 5 and offset the fifth square 'outwards'

 b) set an offset distance to 8 and offset the fifth square 'inwards'

 c) The complete square is offset with a single pick – fig. (e).

11 Select the EDIT POLYLINE icon from the Modify II toolbar and:

prompt	Select polyline or [Multiple]
respond	**pick the sixth square**
prompt	Enter an option [Open/Join/Width/Edit vertex/Fit/Spline/ Decurve..
enter	**W <R>** – the width option
prompt	Specify new width for all segments
enter	**4 <R>**
prompt	Enter an option [Open/Join/Width/Edit vertex/Fit/Spline/ Decurve..
respond	right-click-Enter to end command – fig. (f).

12 Menu bar with **Modify-Object-Polyline** and:

prompt	Select polyline or [Multiple]
respond	**pick the seventh square**
prompt	Enter an option [Open/Join/Width/Edit vertex/Fit/Spline/ Decurve..
enter	**O <R>** – the open option
prompt	Enter an option [Close/Join/Width/Edit vertex/Fit/ Spline/Decurve..
enter	right-click-Enter
and	square displayed with 'last segment' removed – fig. (g).

13 At the command line enter **PEDIT <R>** and:

prompt	Select polyline and pick the eighth square
prompt	options and enter **S <R>** – the spline option
prompt	options and enter **X <R>** – exit command
and	square displayed as a splined curve, in this case a circle – fig. (h).

14 Activate the polyline edit command, pick the ninth square then:

 a) enter **W <R>** then 5 <R>

 b) enter **S <R>**

 c) enter **X <R>** – fig. (i).

15 *Task*

 a) set a fillet radius to 12

 b) fillet the tenth square – remember P

 c) use the polyline edit command with options:

  1. width of 3

  2. spline

  3. exit – fig. (j).

16 *Note*

 a) If a polyline is drawn with the close option, then when the edit polyline command is used, the option is: Open

 b) If the polyline was not closed, then the option is: Close

 c) This was demonstrated in step 12, fig. (g).

Polyline options

The polyline command has several options displayed at the prompt line when the start point has been selected. These options can be activated by entering the capital letter corresponding to the option. The options are:

Arc: draws an arc segment
Close: closes a polyline shape to the start point
Halfwidth: user enters start and end halfwidths
Length: length of line segment entered
Undo: undoes the last option entered
Width: user enters the start and end widths.

To demonstrate these options, at the command line enter **PLINE <R>** and enter the options at the command prompt:

prompt	*enter*	*comment*
Start point	**40,80 <R>**	or pick any point
Next point/options	**H <R>**	halfwidth option
Starting half-width<0.00>	**8 <R>**	
Ending half-width<8.00>	**8 <R>**	
Next point/options	**@60,0 <R>**	segment endpoint
Next point/options	**W <R>**	width option
Starting width<16.00>	**8 <R>**	note default <16>
Ending width<8.00>	**8 <R>**	
Next point/options	**@50,0 <R>**	segment endpoint
Next point/options	**W <R>**	width option
Starting width<8.00>	**8 <R>**	
Ending width<8.00>	**2 <R>**	
Next point/options	**@60,0 <R>**	segment endpoint
Next point/options	**L <R>**	length option
Length of line	**75 <R>**	
Next point/options	**W <R>**	width option
Starting width<2.00>	**2 <R>**	
Ending width<2.00>	**5 <R>**	
Next point/options	**@30,0 <R>**	segment endpoint
Next point/options	**A <R>**	arc option
Arc options	**@0,–50 <R>**	arc endpoint
Arc options	**L <R>**	back to line option
Next point/options	**@–50,0 <R>**	segment endpoint
Next point/options		right-click and Enter.

Task

Before leaving this exercise:

1   MOVE the complete polyline shape with a single pick from its start point by **@25,25.**

2   With FILL <R> at the command line, toggle fill off.

3   REGEN the screen.

4   Turn FILL on then REGEN the screen.

5   This exercise is now complete. Save if required, but it will not be used again.

Line and arc segments

A continuous polyline object can be created from a series of line and arc segments of varying width. In the demonstration which follows we will use several of the options and the final shape will be used in the next chapter. The exercise is given as a rather **LONG** list of options and entries, but persevere with it.

1   Open A3PAPER, layer OUT current, toolbars to suit

2   Refer to Fig. 27.2, select the polyline icon and:

prompt	*Enter*	*Ref*	*Comment*
start point	50,50	pt 1	
next pt/options	L		length
length of line	45	pt 2	
next pt/options	W		width
starting width	0		
ending width	10		
next pt/options	@120,0	pt 3	
next pt/options	A		arc
arc end/options	@50,50	pt 4	
arc end/options	L		line
next pt/options	W		width
starting width	10		
ending width	0		
next pt/options	@0,100	pt 5	
next pt/options	220,220	pt 6	
next pt/options	W		width
Starting width	0		
ending width	5		
next pt/options	@30<–90	pt 7	
next pt/options	A		arc
arc end/options	@–40,0	pt 8	
arc end/options	@–50,30	pt 9	
arc end/options	L		line
next pt/options	W		width
starting width	5		
ending width	0		
next pt/options	50,180	pt 10	
next pt/options	C	pt 1 again.	

3   If your entries are correct the polyshape will be the same as that displayed in Fig. 27.2. Mistakes with polylines can be rectified as each segment is being constructed with the **U** (undo) option. Both the line and arc segments have their own command line option entries.

4   Repeat the polyline icon selection and:

prompt	*Enter*	*Ref*	*Comment*
Start point	80,120	pt a	
next pt/options	A		arc
arc end/options	W		width
Starting width	0		
Ending width	5		
arc end/options	@60<0	pt b	
arc end/options	W		width
Starting width	5		
Ending width	15		
arc end/options	@20,20	pt c	
arc end/options	L		line
next pt/options	right-click and Enter.		

5   Save the drawing layout as **C:\BEGIN\POLYEX** for the next chapter, which demonstrates how polylines can be edited.

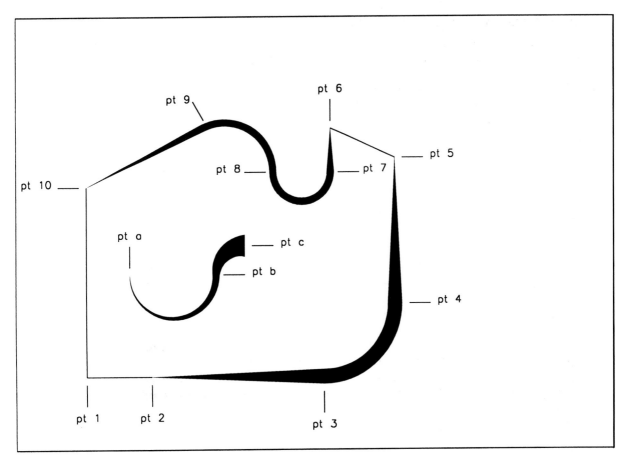

Figure 27.2 Polyline shape exercise.

Polyline tasks

Polyline shapes can be used with the modify commands. To demonstrate their use, open your A3PAPER standard sheet and refer to Fig. 27.3.

Task 1

Draw a 100 closed polyline square and use the sizes given to complete the component in fig. (a). It is easier than you may think. Commands are OFFSET, CHAMFER, FILLET.

Task 2

Draw a 100 closed polyline square of width 7, then use the sizes from task 1 to complete the component – fig. (b).

Task 3

Draw two 50 sided polyline squares each of width 5 but:
a) first square to be drawn as four lines and ended with <RETURN>
b) second square to be closed with close option
c) Note the difference at the polyline start point – fig. (c).

Task 4

Polylines can be trimmed and extended. Try these operations with an 'arrowhead' type polyline with starting width 10 and ending width 0 as fig. (d)

Task 5

a) Draw a polyline arc with following entries:

prompt	enter	comment
start point	pick to suit	
next pt/options	A	arc option
arc end/options	W	width option
starting width	10	
ending width	10	
arc end/options	CE	centre point
center point	@0,–30	relative entry from start point
options	A	angle option
arc end/included angle	270	angle value
arc end/options	<RETURN>.	

b) Draw a donut with ID: 50 and OD: 70, centre: to suit.

c) Draw two lines and trim the donut to these lines.

d) Decide which method is easier – fig. (e).

This exercise is now complete. There is no need to save these tasks.

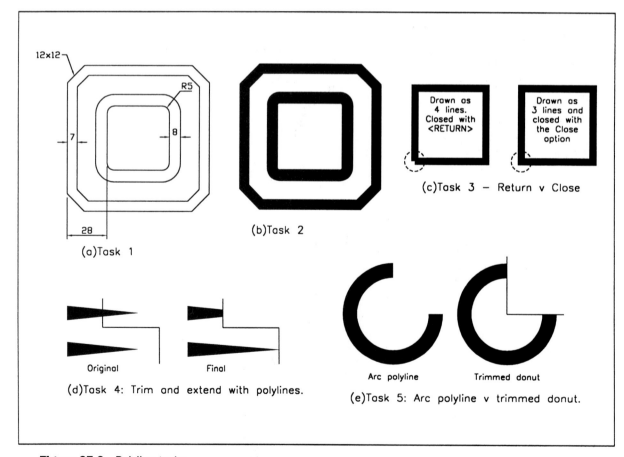

(a)Task 1

(b)Task 2

(c)Task 3 – Return v Close

(d)Task 4: Trim and extend with polylines.

(e)Task 5: Arc polyline v trimmed donut.

Figure 27.3 Polyline tasks.

Splines

A spline is a smooth curve which passes through a given set of points. These points can be picked, entered as coordinates or referenced to existing objects. The spline is drawn as a non-uniform rational B-spline or **NURBS**. Splines have uses in many CAD areas, e.g. car body design, contour mapping, etc. At our level we will only investigate the 2D spline curve.

1   Open your standard sheet or clear all objects from the screen.

2   Refer to Fig. 27.4 and with layer CL current, draw two circles:
a) centre: 80,150 with radius: 50
b) centre: 280,150 with radius: 25.

3   Layer OUT current and select the SPLINE icon from the Draw toolbar and:

prompt	Specify first point or [Object]
respond	**Snap to Center icon and pick larger circle**
prompt	Specify next point and enter **90,220 <R>**
prompt	Specify next point or [Close/Fit tolerance]
enter	**110,80 <R>**
prompt	Specify next point and enter **130,220 <R>**
prompt	Specify next point and enter **150,80 <R>**
prompt	Specify next point and enter **170,220 <R>**
prompt	Specify next point and enter **190,80 <R>**
prompt	Specify next point and enter **210,220 <R>**
prompt	Specify next point and enter **230,80 <R>**
prompt	Specify next point and enter **250,220 <R>**
prompt	Specify next point and enter **270,80 <R>**
prompt	Specify next point and **Center icon and pick smaller circle**
prompt	Specify next point and **right-click-Enter**
prompt	Specify start tangent and enter **80,150 <R>**
prompt	Specify end tangent and enter **280,150 <R>**

4   Menu bar with **Draw-Spline** and:

prompt	Specify first point or [Object]
respond	**Snap to center of larger circle**
prompt	Specify next point and enter **90,190 <R>**
prompt	Specify next point and enter **110,110 <R>**
prompt	Specify next point
respond	˙enter following coordinate pairs:

130,190   150,110   170,190   190,110   210,190
230,110   250,190   270,110   280,150

then	**right-click and Enter**
prompt	Enter start tangent and snap to center of larger circle
prompt	Enter end tangent and snap to center of small circle.

5   At this stage save drawing as **C:\BEGIN\SPLINEX.**

6   The spline options have not been considered in this example.

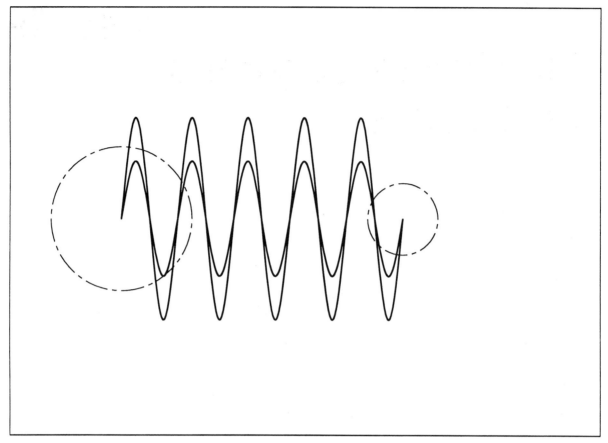

Figure 27.4 Spline curve exercise.

Summary

1 A polyline is a single object which can consist of line and arc segments of varying width.

2 A polyshape which is to be closed should be completed with the Close option.

3 Polyline shapes can be chamfered, filleted, trimmed, extended, copied, scaled, etc.

4 Polylines have their own edit command – next chapter.

5 A spline is NOT a polyline but a NURBS curve (non-uniform rational B-spline, but do not concern yourself about this). Spines have specific properties.

Assignments

Two activities of varying difficulty have been included for you to test your polyline ability.

1 Activity 22: Shapes
Some basic polyline shapes created from line and arc segments. All relevant sizes are given for you. Snap on helps.

2 Activity 23: Printed circuit board.
An interesting application of the polyline command. It is harder to complete than you may think, especially with the dimensions being given in ordinate form. Can you add the dimensions given? All the relevant sizes are given on the drawing. Use your discretion when positioning the 'lines'.

Chapter 28

Modifying polylines and splines

Polylines have their own editing facility which gives the user access to several extra options in addition to the existing modify commands.

1   Open the polyline exercise POLYEX and refer to Fig. 28.1.

2   Select the EDIT POLYLINE icon from the Modify II toolbar and:

prompt   Select polyline

respond   **pick any point on outer polyline shape**

prompt   Enter an option [Open/Join/Width/Edit vertex/Fit/Spline/ Decurve..

respond   enter the following in respond to the prompt.

prompt	enter	ref	comments
Open/Join...	W		constant width
new width	3	fig. (b)	width value
options	D	fig. (c)	decurve
options	S	fig. (d)	spline
options	F	fig. (e)	fit option
options	D	fig. (f)	decurve again
options	O	fig. (g)	open from start point
options	W		constant width
new width	0	fig. (h)	width value
options	S	fig. (i)	spline
options	D	fig. (j)	decurve
options	W		constant width
new width	5	fig. (k)	width value
options	C	fig. (l)	close shape
options	ESC		end command.

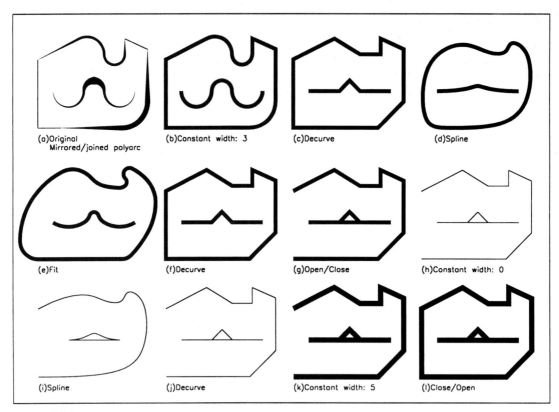

Figure 28.1 Editing polylines with POLYEX.

The join option

This is a very useful option as it allows several individual polylines to be 'joined' into a single polyline object. Refer to Fig. 28.1 and:

1 Use the MIRROR command to mirror the polyarc shape about a vertical line through the right end of the object – ortho on may help, but remember to toggle it off.

2 At the command line enter **PEDIT <R>** and:

prompt   Select polyline
respond **pick any point on original polyarc**
prompt   Enter an option [Close/Join/Width/Edit..
enter   **J <R>**
prompt   Select objects
respond **pick the two polyarcs then right-click**
prompt   2 segments added to polyline
then   Enter an option [Close/Join/Width
enter   **X <R>** to end command.

3 The two arc segments will now be one polyline object

4 Menu bar with **Modify-Polyline** and:
  a) pick the mirrored joined polyshape
  b) enter the same options as step 2 of the first part of the exercise, **EXCEPT** enter C(lose) instead of O(pen), and open instead of close.

5 *Note*: When the Edit Polyline command is activated and a polyline selected, the prompt is:
  a) Open/Join/Width, etc. if the selected polyshape is 'closed'
  b) Close/Join/Width, etc. if the selected polyshape is 'opened'.

Edit vertex option

The options available with the Edit Polyline command usually 'redraws' the selected polyshape after each entry, e.g. if **S** is entered at the options prompt, the polyshape will be redrawn as a splined curve. Using the Edit vertex option is slightly different from this. When **E** is entered as an option, the user has another set of options. Refer to Fig. 28.2 and:

1 Erase all objects from the screen or open A:A3PAPER.

2 Draw an 80 sided **closed** square polyshape and multiple copy it to three other areas of the screen.

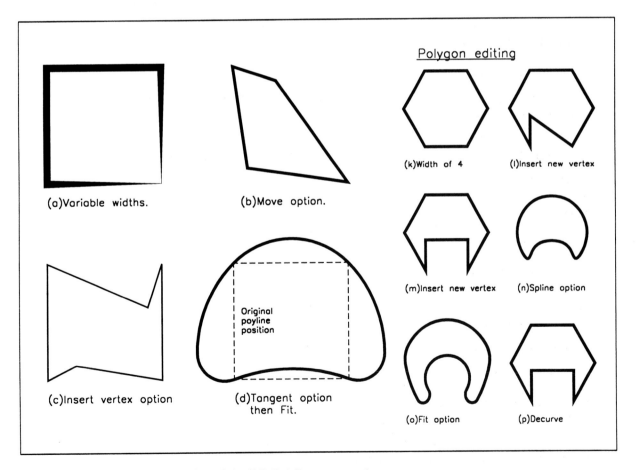

Figure 28.2 Edit vertex option of the Edit Polyline command.

3   Variable widths
     Menu bar with **Modify-Object-Polyline** and:
     prompt   Select polyline or [Multiple]
     respond   **pick the first square**
     prompt   Enter an option [Open/Join/Width..
     enter   **W <R>** and then: **5 <R>**
     prompt   Enter an option [Open/Join/Width..
     enter   **E <R>** – the edit vertex option
     prompt   Enter a vertex editing option [Next/Previous/Break/Insert/..
     and   an X is placed at the start vertex (lower left for me)
     enter   **W <R>**
     prompt   Specify starting width for next segment and enter: **5 <R>**
     prompt   Specify ending width for next segment<5> and enter: **0 <R>**
     prompt   Enter a vertex editing option [Next/Previous/Break..
     enter   **N <R>** and X moves to next vertex (lower right for me)
     prompt   Enter a vertex editing option [Next/Previous/Break..
     enter   **W <R>**
     prompt   Specify starting width for next segment and enter: **0 <R>**
     prompt   Specify ending width for next segment and enter: **5 <R>**
     prompt   Enter a vertex editing option [Next/Previous/Break..
     enter   **X <R>** – to exit the edit vertex option
     then   **X <R>** – to exit the edit polyline command – fig. (a).

4   *Moving a vertex*
     Select the Edit Polyline icon and:
     a) pick the second square
     b) enter a constant width of 2
     c) enter **E <R>** – the edit vertex option
        prompt   Enter a vertex editing option [Next/Previous/Break..
        enter   **N <R>** until X at lower left vertex – probably is?
        then   **M <R>** – the move option
        prompt   Specify new location for marked vertex
        enter   **@10,10 <R>**
        prompt   Enter a vertex editing option [Next/Previous/Break..
        enter   **N <R>** until X at diagonally opposite vertex
        then   **M <R>**
        prompt   Specify new location for marked vertex and enter: **@–50,–10 <R>**
        prompt   Enter a vertex editing option [Next/Previous/Break
        enter   **X <R>** then **X <R>** – fig. (b).

5   *Inserting a new vertex*
     At the command line enter **PEDIT <R>** and:
     a) pick the third square
     b) enter a constant width of 1
     c) pick the edit vertex option
     d) enter N <R> until X at lower left vertex then:
        prompt   Enter a vertex editing option[Next/Previous/Break..
        enter   **I <R>** – the insert new vertex option
        prompt   Specify location of new vertex
        enter   **@20,10 <R>**
        prompt   Enter a vertex editing option [Next/Previous/Break..
        prompt   **N <R>** until X at opposite corner then enter: **I <R>**
        prompt   Specify location of new vertex
        enter   **@–10,–30 <R>**
        then   **X <R>** and **X <R>** – fig. (c).

6   *Tangent-Fit Options*
    Activate the edit polyline command, set a constant width of 2, select the edit vertex
    option with the X at lower left vertex and:

prompt	`Enter a vertex editing option [Next/Previous/Break..`
enter	**T <R>** – the tangent option
prompt	`Specify direction of vertex tangent`
enter	**20 <R>**
and	`note arrowed line direction`
prompt	`Enter a vertex editing option [Next/Previous/Break..`
enter	**N <R>** until X at lower right vertex
then	**T <R>**
prompt	`Specify direction of vertex tangent`
enter	**–20 <R>** – note arrowed line direction
prompt	`Enter a vertex editing option [Next/Previous/Break..`
enter	**X <R>** – to end the edit vertex options
prompt	`Enter an option [Open/Join..`
enter	**F <R>** – the fit option
then	**X <R>** – to end command and give fig. (d).
and	the final fit shape 'passes through' the vertices of the original polyline square.

Editing a polygon

A polygon is a polyline and can therefore be edited with the Edit Polyline command. To
demonstrate the effect:

1   Draw a six-sided polygon inscribed in a 40 radius circle towards the right of the screen

2   Activate the Edit Polyline command, pick the polygon then enter the following option
    sequence:

enter	*ref*	*comment*
W then 4	fig. (k)	constant width
E		edit vertex option
N		until X at lower left vertex
I then @0,30	fig. (l)	new vertex inserted
I then @40,0	fig. (m)	new vertex inserted
X		end edit vertex option
S	fig. (n)	spline curve
F	fig. (o)	fit option
D	fig. (p)	decurve option
X		end command.

Editing a spline curve

Spline curves can be edited in a similar manner to polylines, so open drawing SPLINEX and refer to Fig. 28.3

1   Menu bar with **Modify-Object-Spline** and:

prompt	Select spline
respond	**pick the larger spline curve**
and	blue grip type boxes appear
prompt	Enter an option [Fit Data/Close/Move vertex/Refine/rEverse/Undo]
enter	**F <R>** – the fit data option
and	boxes at each entered vertex of curve
prompt	Enter a fit data option [Add/Close/Delete/Move/Purge/Tangents/toLerance/eXit]
enter	**M <R>** – the move option
prompt	Specify new location or [Next/Previous/Select point/eXit]
and	red box at left end of spline
enter	**N <R>** until red box at second from left lower vertex
then	**@0,–50 <R>**
prompt	Specify new location or [Next/Previous/Select point/eXit]
enter	**N <R>** until red box at third from left lower vertex
then	**@0,–50 <R>**
prompt	Specify new location or [Next/Previous/Select point/eXit]
enter	**X <R>** – to end the move option
prompt	Enter a fit data option [Add/Close/Delete/Move/Purge/Tangents/toLerance/eXit]
enter	**D <R>** the delete option
prompt	Specify control point
respond	pick the sixth from left vertex then right-click
prompt	Enter a fit data option [Add/Close/Delete/Move/Purge/Tangents/toLerance/eXit]
enter	**X <R>**
prompt	Enter an option [Fit Data/Close/Move vertex/Refine/rEverse/Undo]
enter	**X <R>**

2   Select the Edit Spline icon from the Modify II toolbar and:

prompt	Select spline
respond	**pick the smaller spline**
prompt	Enter an option [Fit Data/Close..
enter	**F <R>**
and	blue boxes at entered vertices
prompt	Enter a fit data option [Add/Close/Delete..
enter	**D <R>**
prompt	Specify control point
respond	pick third vertex from left
prompt	Specify control point and pick fourth vertex from left
prompt	Specify control point and pick fifth vertex from left
prompt	Specify control point and pick sixth vertex from left
prompt	Specify control point and pick seventh vertex from left
prompt	Specify control point and right-click
prompt	Enter a fit data option [Add/Close/Delete..
enter	**M <R>** – the move option
prompt	Specify new location or [Next/Previous..

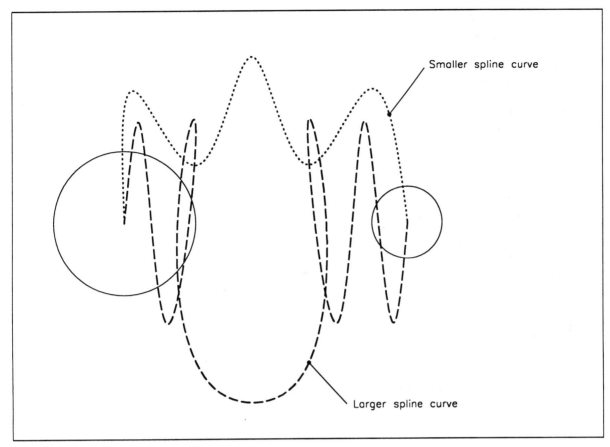

Figure 28.3 shows two spline curves with labels. *Smaller spline curve* points to the dotted curve and *Larger spline curve* points to the dashed curve.

Figure 28.3 Editing spline curves from the SPLINEX exercise.

enter	**N <R>** until second left vertex is red
then	**@0,50 <R>**
then	**N <R>** until fourth left vertex is red
then	**@0,75 <R>**
then	**N <R>** until sixth left vertex is red
then	**@0,50 <R>**
then	**X <R>** and **X <R>** and **X <R>**

3   The two spline curves will be displayed as edited.

4   Save this modified drawing if required.

Summary

1   Polylines and splines can be modified using their own commands.

2   Each edit command has several options.

3   This chapter has been a brief introduction to the edit commands associated with polylines and splines.

Chapter 29

Divide, measure and break

These are three useful commands which will be demonstrated by example, so:

1 Open your standard sheet with later OUT current.

2 Activate the Draw, Modify, Dimension and Object Snap toolbars.

3 Refer to Fig. 29.1 and set the point style and size indicated (Format-Point Style).

4 Draw the following objects:
   a) LINE: from: 20,250 to: @65<15
   b) CIRCLE: centre: 125,250 with radius: 25
   c) POLYLINE: line segment from: 170,270 to: @50,0
                   arc segment to: @0,–50
                   line segment to: @–30,0
   d) SPLINE: draw any spline to suit

5 Copy the four objects below the originals.

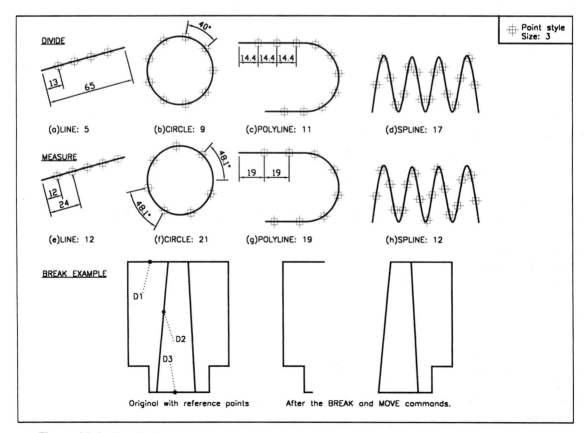

Figure 29.1 The DIVIDE, MEASURE and BREAK commands.

Divide

A selected object is 'divided' into an equal number of segments, the user specifying this number. The current point style is 'placed' at the division points.

1 Menu bar with **Draw-Point-Divide** and:
   prompt Select object to divide
   respond **pick the line**
   prompt Enter the number of segments or [Block]
   enter **5 <R>**
   and the line will be divided into five equal parts, and a point is placed at the end of each segment length as fig. (a).

2 At the command line enter **DIVIDE <R>** and:
   prompt Select object to divide
   respond **pick the circle**
   prompt Enter the number of segments or [Block]
   enter **9 <R>**
   and the circle will be divided into nine equal arc lengths and nine points placed on the circle circumference as fig. (b).

3 Use the DIVIDE command with:
   a) POLYLINE: 11 segments – fig. (c)
   b) SPLINE: 17 segments – fig. (d).

Measure

A selected object is 'divided' into a number of user-specified equal lengths and the current point style is placed at each measured length.

1 Menu bar with **Draw-Point-Measure** and:
   prompt Select object to measure
   respond **pick the line** – the copied one of course!
   prompt Specify length of segment or [Block]
   enter **12 <R>**
   and the line is divided into measured lengths of 12 units from the line start point as fig. (e).

2 At command line enter **MEASURE <R>** and:
   prompt Select object to measure
   respond **pick the circle**
   prompt Specify length of segment or [Block]
   enter **21 <R>**
   and the circle circumference will display points every 21 units from the start point, which is where? – fig. (f).

3 Use the MEASURE command with:
   a) POLYLINE: segment length of 19 – fig. (g)
   b) SPLINE: segment length of 19 – fig. (h).

4 *Task*
   Menu bar with **Dimension-Style** and:
   a) Dimension Style Manager dialogue box with A3DIM current
   b) select Modify then the Primary Units tab
   c) set Linear and Angular Dimensions Precision to 0.0 then OK
   d) pick Close from Dimension Style Manager dialogue box
   e) Add the linear and angular dimensions displayed. The Snap to Node is used for points

5 This completes the exercise – no need to save.

Break

This command allows a selected object to be broken at a specified point – similar to trim, but no erase effect.

1   Open USEREX and refer to Fig. 29.1. We want to split the component into two parts and will use the Break command to achieve the desired effect.

2   Select the BREAK icon from the Modify toolbar and:
    prompt   Select object
    respond   **pick line D1**
    prompt   Specify second point or [First point]
    enter   **F <R>** – the first point option
    prompt   Specify first break point
    respond   **Snap to Endpoint of 'top' of line D2**
    prompt   Specify second break point
    enter   **@ <R>**

3   At command line enter **BREAK <R>** and:
    prompt   Select object and **pick line D3**
    prompt   Specify second point or [First point]
    enter   **F <R>**
    prompt   Specify first break point
    respond   **Snap to Endpoint of 'lower' end of line D2**
    prompt   Specify second break point
    enter   **@ <R>**

4   Now move line D2 and the other lines to the right of D2 (seven lines in total) from any suitable point **by @50,0.**

5   *Note*: *a*)   the @ **entry** at the second point prompt, ensures that the first and second points are the same.

    b)   the BREAK AT POINT icon will allow the above operations (step 3 and 4) to be achieved with a 'single pick'.

6   The exercise is now complete, but do not save.

Summary

1   Objects can be 'divided' into:
    a) an equal number of parts – DIVIDE
    b) equal segment lengths – MEASURE
    c) both commands have a Block option, which will be investigated in a later chapter.

2   The BREAK command allows objects to be 'broken' at specified points.

3   The three commands can be used on line, circle, arc, polyline and spline objects.

Chapter 30

Lengthen, align and stretch

Three useful modify commands which will greatly increase drawing efficiency. To demonstrate the commands, open the A3PAPER drawing standard sheet and refer to Fig. 30.1.

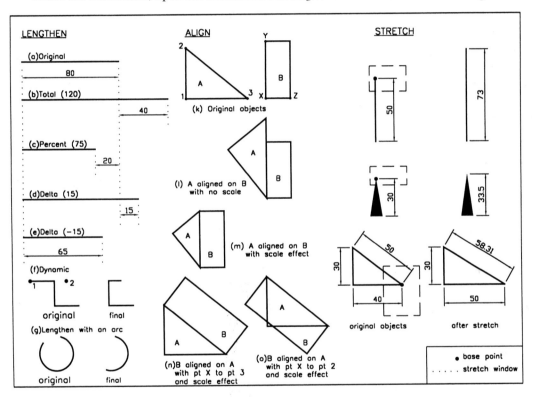

Figure 30.1 The lengthen, align and stretch commands.

Lengthen

This command will alter the length of objects (including arcs) but cannot be used with CLOSED objects.

1 Draw a horizontal line of length 80 and multiple copy it below four times.

2 Select the LENGTHEN icon from the Modify toolbar and:

prompt	`Select an object or [Delta/Percent/Total/Dynamic]`
enter	**T <R>** – the total option
prompt	`Specify total length or [Angle]<1.00>`
enter	**120 <R>**
prompt	`Select an object to change or [Undo]`
respond	**pick the second line**
then	right-click and Enter – fig. (b).

3 Menu bar with **Modify-Lengthen** and:

prompt   Select an object or [Delta/Percent..

enter   **P <R>** – the percent option

prompt   Enter percentage length<100.00>

enter   **75 <R>**

prompt   Select an object to change or [Undo]

respond   **pick the next line**

then   right-click and Enter – fig. (c).

4 At the command line enter **LENGTHEN <R>** and:

enter   DE <R> – the delta option

enter   15 <R> – the delta length

respond   pick the next line – fig. (d).

5 Activate the LENGTHEN command and:

prompt   Select an object or..

respond   **pick the fifth line**

prompt   Current length: 80.00

then   Select an object or..

enter   **DE <R>**

prompt   Enter delta length or [Angle]

enter   **–15 <R>**

prompt   Select an object to change or..

respond   **pick the fifth line** – fig. (e).

6 *a*) Draw a Z shape as fig. (f)

   b) Activate the lengthen command and:

     1. enter DY <R> – the dynamic option

     2. pick point 1 on the line as indicated

     3. move the cursor and pick a point 2

     4. the original shape will be altered as shown.

7 *a*) Draw a three point arc as fig. (g)

   b) Activate the lengthen command and:

     prompt   Select an object or..

     respond   **pick the arc**

     prompt   Current length: 70.87, included angle: 291.0

               (these values depend on the arc you have drawn)

     prompt   Select an object or..

     enter   **T <R>** – the total option

     prompt   Specify total length or [Angle]

     enter   **A <R>** – the angle option

     prompt   Specify total angle

     enter   **180 <R>**

     prompt   Select an object to change or..

     respond   **pick the arc.**

8 *Note:*

   a) An object is 'lengthened' at the end of the line selected

   b) If an object is picked before an option is entered, the length of the object is displayed. With arcs, the arc length and included angle is displayed

   c) When the angle option is used with arcs, the entered total angle is relative to the arc start point.

9 *Task*

   Dimension the five lines to check the various options have been performed correctly.

Align

A very powerful command which combines the move and rotate commands into one operation. The command is mainly used with 3D objects, but can also be used in 2D.

1 Draw a right-angled triangle and a rectangle (own sizes) and copy these two objects to four other parts of the screen – the snap on will help with this operation.

2 We want to align:
a) side 23 of the triangle onto side XY of the rectangle
b) side XZ of the rectangle onto side 23 of the triangle

3 Menu bar with **Modify-3D Operation-Align** and:

prompt	`Select objects`
respond	**pick the three lines of triangle A** then right-click
prompt	`Specify first source point`
respond	**snap to endpoint and pick point 3**
prompt	`Specify first destination point`
respond	**snap to endpoint and pick point X**
and	a line is drawn between points 3 and X
prompt	`Specify second source point`
respond	**snap to endpoint and pick point 2**
prompt	`Specify second destination point`
respond	**snap to endpoint and pick point Y**
and	a line is drawn between points 2 and Y
prompt	`Specify third source point`
respond	right-click as no more selections needed
prompt	`Scale objects based on alignment points? [Yes/No]<N>`
enter	**N <R>**

4 The triangle is moved and rotated onto the rectangle with sides 23 and XY in alignment as fig. (l).

5 Repeat the Align selection and:
a) pick the three lines of the next triangle then right-click
b) pick the same source and destination points as step 3
c) enter **Y <R>** at the Scale prompt
d) the triangle will be aligned onto the rectangle, and side 23 scaled to side XY – fig. (m).

6 At the command line enter **ALIGN <R>** and:

prompt	`Select objects`
respond	**pick the next rectangle** then right-click
prompt	`Specify first source point`
respond	**Snap to Midpoint icon and pick line XY**
prompt	`Specify first destination point`
respond	**Snap to Midpoint icon and pick line 23**
prompt	`Specify second source point` and **pick point X**
prompt	`Specify second destination point` and **pick point 3**
prompt	`Specify third source point` and right-click
prompt	`Scale objects based on alignment points`
enter	**Y <R>**

7 The selected rectangle will be aligned onto the triangle as fig. (n) and scaled to suit.

8 Repeat the align command and:
    a) pick the last rectangle then right-click
    b) Midpoint of side XY as first source point
    c) Midpoint of side 23 as first destination point
    d) point X as second source point
    e) point 2 as second destination point
    f) right click at third source point prompt
    h) Y to scale option – fig. (o).

9 The orientation of the aligned object is thus dependent on the order of selection of the source and destination points.

Stretch

This command does what it says – it 'stretches' objects. If hatching and dimensions have been added to the object to be stretched, they will both be affected by the command – remember that hatch and dimensions are associative.

1 Select a clear area of the drawing screen and draw:
    a) a vertical dimensioned line
    b) a dimensioned variable width polyline
    c) a dimensioned triangle.

2 Select the STRETCH icon from the Modify toolbar and:

prompt	`Select object to stretch by crossing-window or crossing-polygon`
then	`Select objects`
enter	**C <R>** – the crossing option
prompt	`Specify first corner`
respond	**window the top of vertical line and dimension**
prompt	`2 found`
then	`Select objects`
respond	**right-click**
prompt	`Specify base point or displacement`
respond	**pick top end of line** (endpoint)
prompt	`Specify second point of displacement`
enter	**@0,23 <R>**

3 The line and dimension will be stretched by the entered value.

4 Menu bar with **Modify-Stretch** and:

prompt	`Select objects`
enter	**C <R>** – crossing option
prompt	`First corner`
respond	**window the top of polyline and dimension**
then	right-click
prompt	`Specify base point` and **pick top end of polyline**
prompt	`Specify second point` and enter: **@0,3.5 <R>**

5 The polyline and dimension are stretched by the entered value.

6 At the command line enter **STRETCH <R>** and:
    a) enter C <R> for the crossing option
    b) window the vertex of triangle indicated then right-click
    c) pick indicated vertex as the base point
    d) enter @10,0 as the displacement.

7 The triangle is stretched as are the appropriate dimensions.

Stretch example

1   Open your WORKDRG and refer to Fig. 30.2

2   *a*) erase the centre lines
    b) add hatching – own selection
    c) linear dimension the two lines with menu bar selection

3   Activate the STRETCH command and:
    a) enter C <R> – crossing option
    b) first corner: pick a point P1
    c) opposite corner: pick a point P2 then right-click
    d) base point: pick any suitable point
    e) second point: enter @15,0 <R>

4   Repeat the STRETCH command and:
    a) activate the crossing option
    b) pick a point P3 for the first corner
    c) pick a point P4 for the opposite corner
    d) pick a suitable base point
    e) enter @0,–15 as the second point

5   The component and dimensions will be stretched with the entered values.

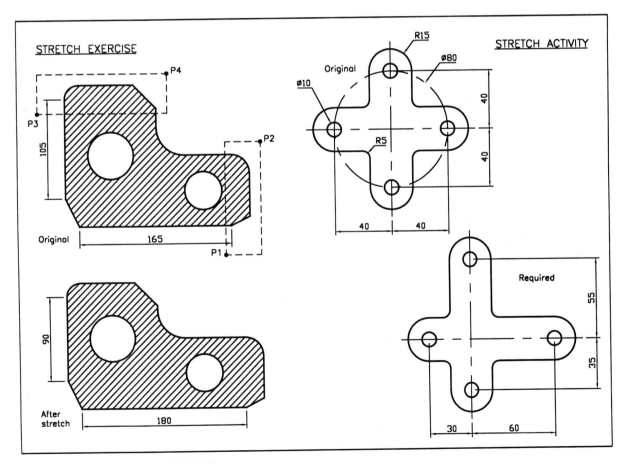

Figure 30.2   Using the STRETCH command.

Stretch activity

Refer to Fig. 30.2 and:

1 Draw the original component as shown adding the four dimensions.

2 Using the STRETCH command only, produce the modified component.

3 There is no need to save this activity.

Summary

1 Lengthen will increase/decrease the length of lines, arcs and polylines. There are several options available.

2 Dimensions are not lengthened with the command.

3 Align is a powerful command which combines move and rotate into one operation. The order of selecting points is important.

4 Stretch can be used with lines, polylines and arcs. It does not affect circles.

5 Dimensions and hatching are stretched due to association.

Chapter 31

Obtaining information from a drawing

Drawings contain information which may be useful to the user, e.g. coordinate data, distances between points, area of shapes, etc. We will investigate how this information can be obtained, so:

1   Open C:\BEGIN\USEREX and refer to Fig. 31.1.

2   Draw a circle, centre at: 190,140 and radius: 30.

3   Activate the Inquiry toolbar.

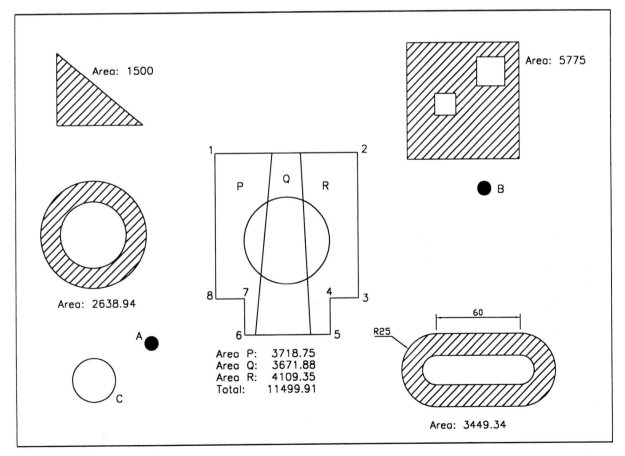

Figure 31.1 Obtaining drawing information.

Point identification

This command displays the coordinates of a selected point.

1   Select the LOCATE POINT icon from the Inquiry toolbar and:
    prompt   Specify point
    respond   **Snap to Midpoint icon and pick line 23.**

2   The command line area displays:

    X = 240.00   Y = 150.00   Z = 0.00.

3   Menu bar with **Tools-Inquiry-ID** Point and:
    prompt   Specify point
    respond   **Snap to Center icon and pick the circle.**

4   Command line display is X = 190.00 Y = 140.00 Z = 0.00.

5   The command can be activated with **ID <R>** at the command line.

Distance

Returns information about a line between two selected points including the distance and the angle to the horizontal.

1   Select the DISTANCE icon from the Inquiry toolbar and:
    prompt   Specify first point
    respond   **pick point 8** (snap to endpoint or intersection)
    prompt   Specify second point
    respond   **pick point 2.**

2   The command prompt area will display:
    Distance=141.42, Angle in XY plane=45.0, Angle from XY plane=0.0
    Delta X=100.00, Delta Y=100.00, Delta Z=0.00.

3   Menu bar with **Tools-Inquiry-Distance** and:
    prompt   Specify first point
    respond   **snap to centre of circle**
    prompt   Specify second point
    respond   **pick midpoint of line 23.**

4   The display at the command prompt is:
    Distance=50.99, Angle in XY plane=11.3, Angle from XY plane=0.0
    Delta X=50.00, Delta Y=10.00, Delta Z=0.00.

5   Using the DISTANCE command, select point 2 as the first point and point 8 as the second point. Is the displayed information any different from step 1?

6   Entering **DIST <R>** at the command line will activate the command.

List

A command which gives useful information about a selected object.

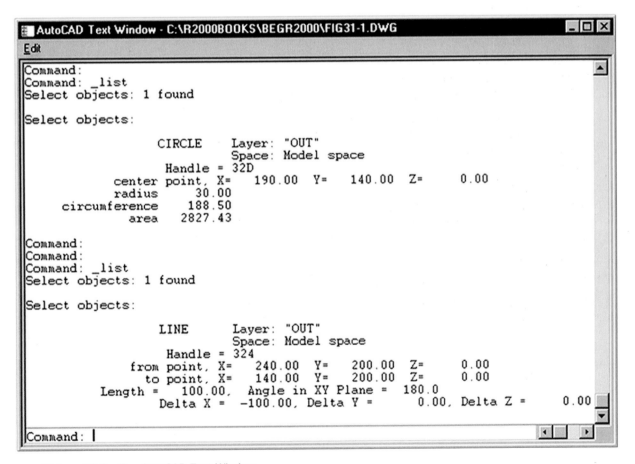

1   Select the LIST icon from the Inquiry toolbar and:
    prompt   Select objects
    respond   **pick the circle** then right-click
    prompt   AutoCAD Text Window with information about the circle.
    respond   F2 to flip back to drawing screen.

2   Menu bar with **Tools-Inquiry-List** and:
    prompt   Select objects
    respond   **pick line 12** then right-click
    prompt   AutoCAD text window
    either   a) F2 to flip back to drawing screen
    or   b) cancel icon from text window title bar.

3   Figure 31.2 is a screen dump of the AutoCAD Text window display for the two selected objects.

4   LIST <R> is the command line entry.

Figure 31.2   The AutoCAD Text Window.

Area

This command will return the area and perimeter for selected shapes or polyline shapes. It has the facility to allow composite shapes to be selected.

1   Select the AREA icon from the Inquiry toolbar and:
    prompt   Specify first corner point or [Object/Add/Subtract]
    enter   **O <R>** – the object option
    prompt   Select objects
    respond   **pick the circle.**

2   The command line will display:
    Area=2827.43, Circumference=188.50.

3   *Question*: are these values the same as from the LIST command?

4   Menu bar with **Tools-Inquiry-Area** and:
    prompt   Specify first corner point or [Object/Add/Subtract]
    respond   **Snap to endpoint and pick point 1**
    prompt   Specify next corner point or press ENTER for total
    respond   **Snap to endpoint and pick point 2**
    prompt   Specify next corner point or..
    respond   **Snap to endpoint and pick points 3,4,5,6,7,8**
    prompt   Specify next corner point or press ENTER for total
    respond   **press <RETURN>.**

5   Command line displays:
    Area=11500.00, Perimeter=450.00.

6   *Question*: Are these figures correct for the shape selected?

7   At the command line enter AREA <R> and:
    prompt   Specify first corner point or [Object/Add/Subtract]
    enter   **A <R>** – the add option
    prompt   Specify first corner point or [Object/Subtract]
    respond   **Endpoint icon and pick point 1**
    prompt   Specify next corner point or press ENTER for total (ADD
                mode)
    respond   **Endpoint icon and pick point 2**
    prompt   Specify next corner point or press ENTER for total (ADD
                mode)
    respond   **Endpoint icon and pick points 3,4,5,6,7,8**
    prompt   Specify next corner point or press ENTER for total (ADD
                mode)
    respond   **press <RETURN>**
    prompt   Area=11500.00, Perimeter: 450.00
                Total area=11500.00
    then   Specify first corner point or [Object/Subtract]
    enter   **S <R>** – the subtract option
    prompt   Specify first corner point or [Object/Add]
    enter   **O <R>** – the object option
    prompt   (SUBTRACT mode) Select objects
    respond   **pick the circle**
    prompt   Area: 2827.43, Circumference: 188.50
                Total area=8672.57
    then   (SUBTRACT mode) Select objects
    respond   **ESC** to end command.

8   *Question*: Is the area value of 8672.57 correct for the outline area less the circle area?

Time

This command displays information in the AutoCAD Text Window about the current drawing, e.g.:

a) when it was originally created
b) when it was last updated
c) the length of time worked on it.

1 The command can be activated:
   a) from the menu bar with **Tools-Inquiry-Time**
   b) by entering **TIME <R>** at the command line.

2 The command has options of Display, On, Off and Reset.

3 A useful command for your boss?

Status

This command gives additional information about the current drawing as well as disk space information. Select the sequence **Tools-Inquiry-Status** to 'see' the status display in the AutoCAD text window.

Calculator

1 AutoCAD 2002 has a built-in calculator which can be used:
   a) to evaluate mathematical expressions
   b) to assist on the calculation of coordinate point data.

2 The mathematical operations obey the usual order of preference with brackets, powers, etc.

3 At the command line enter **CAL <R>** and:
   prompt   Initializing..>> Expression:
   enter   **12.6*(8.2+5.1) <R>**
   prompt   167.58 – is it correct?

4 Enter CAL <R> and:
   prompt   >>Expression:
   enter   **(5*(7-4))^3.5 <R>**
   prompt   13071.3.

5 *Question*:
   What is answer to $((7 - 4)+(2 * (8+1)))$ – a key question?

Transparent calculator

A transparent command is one which can be used 'while in another command' and is activated from the command line by entering the ' symbol. The calculator command has this transparent ability.

1 Activate the DONUT command and set diameters of 0 and 3, then:
   prompt   Specify center of donut
   enter   **'CAL <R>** – the transparent calculator command
   prompt   >>Expression:
   enter   **CEN/2 <R>**
   prompt   >>Select entity for CEN snap
   respond   **pick the circle**
   prompt   (95.0 70.0 0.0)

and	donut at position A
prompt	Specify center of donut
enter	**'CAL <R>**
prompt	>>Expression:
enter	**(MID+INT) <R>**
prompt	>>Select entity for MID snap and: **pick line 65**
prompt	>>Select entity for INT snap and: **pick point 8**
prompt	(330.0 175.0 0.0)
and	donut at position B
prompt	Specify center of donut and right-click.

2 Activate the circle command and:

prompt	Specify center point for circle and enter **55,45 <R>**
prompt	Specify radius of circle
enter	**'CAL <R>**
prompt	>>Expression:
enter	**rad/2 <R>**
prompt	>>Select circle, arc or polyline segment for RAD function
respond	**pick the original circle**
and	circle at position C.

3 Check the donut centre points with the ID command. They should be (a)95,70 and (b)330,175. These values were given at the command prompt line as the donuts were being positioned.

Task

1 Refer to Fig. 31.1 and create the following (anywhere on the screen, but use SNAP ON to help):
   a) right-angled triangle with: vertical side of 50 and horizontal side of 60
   b) square of side 80 and inside this square: two other squares of side 15 and 20
   c) circles: radii 23 and 37 – concentric
   d) polyshape: to size given, offset 15 'inwards'

2 Find the shaded areas using the AREA command.

3 Obtain the areas of the three 'vertical strips' of the original USEREX, i.e. without the circle.

Summary

1 Drawings can be 'interrogated' to obtain information about:
   a) coordinate details
   b) distance between points
   c) area and perimeter of composite shapes
   d) the status and time for the current drawing

2 AutoCAD has a built-in calculator which can be used transparently

Chapter 32

Text fonts and styles

Text has been added to previous drawings without any discussion about the 'appearance' of the text items. In this chapter we will investigate:

a) text fonts and text styles
b) text control codes.

The words 'font' and 'style' are extensively used with text and they can be explained as:

Font: defines the pattern which is used to draw characters, i.e. it is basically an alphabet 'appearance'. AutoCAD 2002 has over 90 fonts available to the user, and Fig. 32.1 displays the text item 'AutoCAD 2002' using 30 of these fonts.

Style: defines the parameters used to draw the actual text characters, i.e. the width of the characters, the obliquing angle, whether the text is upside-down, backwards, etc.

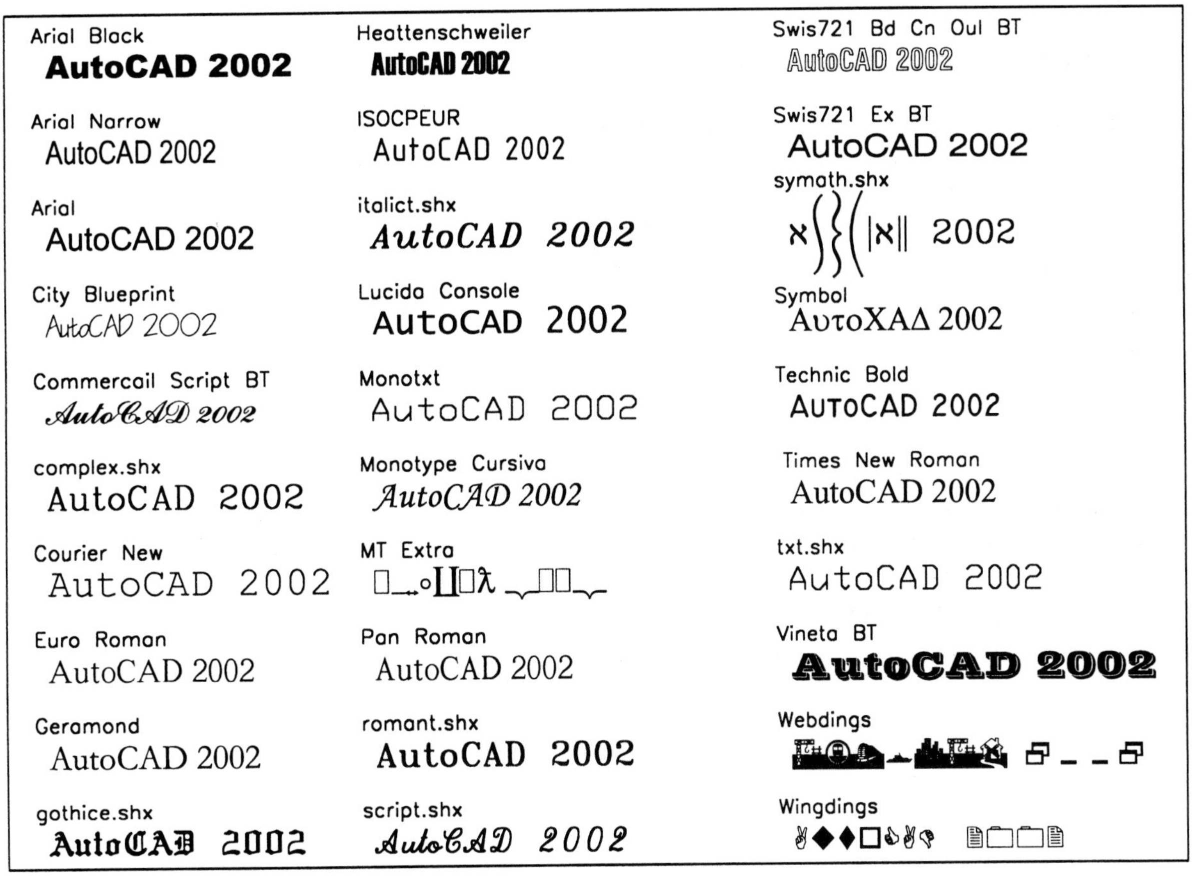

Figure 32.1   Some of AutoCAD 2002's text fonts, all at height 8.

Notes

1 Text fonts are 'part of' the AutoCAD package.
2 Text styles are created by the user.
3 Any text font can be used for many different styles.
4 A text style uses only one font.
5 If text fonts are to be used in a drawing, a text style **must be created** by the user.
6 New text fonts can be created by the user, but this is outside the scope of this book.
7 Text styles can be created
   a) by keyboard entry
   b) via a dialogue box.

Getting started

1 Open A:A3PAPER with layer TEXT current.

2 At the command line enter **-STYLE <R>** and
   prompt   Enter name of text style or [?]<Standard>
   enter   **? <R>** – the 'query' option
   prompt   Enter text style(s) to list<*>
   enter   **<R>**
   prompt   AutoCAD Text Window
   with   Style name: "Standard" Font files: txt
               Height: 0.00 Width Factor: 1.00 Obliquing angle: 0.0
               Generation: Normal.

3 This is AutoCAD 2002's 'default' text style. It has the text style name **STANDARD** and uses the text font **txt.** Realise that your system may have a different text style name and font. If it does, do not worry. It will not affect our exercise.

4 Cancel the text window.

5 With the menu bar sequence **Draw-Text-Single Line Text**, add the text item **AutoCAD 2002** at 110,275 with height 8 and rotation angle 0.

6 The step 2 entry of **–STYLE** was to allow us to use the command line instead of a dialogue box. I thought that it would be easier to understand from the command line.

Creating a text style from the keyboard

1 At the command line enter **–STYLE <R>** and:
   prompt   Enter name of text style or [?]
   enter   **ST1 <R>** – the style name
   prompt   Specify full font name or font filename (TTF or SHX)
   enter   **romans.shx <R>**
   prompt   Specify height of text and enter: **0 <R>**
   prompt   Specify width factor and enter: **1 <R>**
   prompt   Specify obliquing angle and enter: **0 <R>**
   prompt   Display text backwards? and enter: **N <R>**
   prompt   Display text upside-down? And enter: **N <R>**
   prompt   Vertical? and enter: **N <R>**
   prompt   "st1" is now the current style.

2 The above entries of height, width factor, etc. are the parameters which must be defined for every text style created.

Creating a text style from a dialogue box

1 From the menu bar select **Format-Text Style** and

prompt `Text Style dialogue box`
with 1. ST1 as the Style Name
2. romans.shx as the Font Name
3. Height: 0.0
respond **pick New** and:
prompt `New Text Style dialogue box`
respond 1. alter Style Name to **ST2**
2. pick OK
prompt `Text Style dialogue box`
with ST2 as the Style Name
respond 1. pick the scroll arrow at right of romans.shx
2. scroll and pick **italict.shx**
3. ensure that:
Height: 0.00, Width Factor: 1.00, Oblique Angle: 0.0
4. note the Preview box
5. dialogue box as Fig. 32.2
6. pick **Apply** then **Close.**

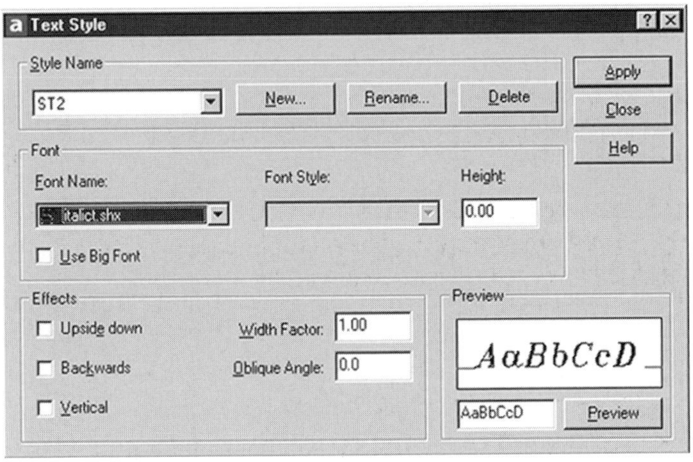

Figure 32.2 The Text Style dialogue box for the ST2 new test style.

2 With the menu bar selection Format-Text Style, use the Text Style dialogue box as step 1 to create the following new text styles:

		Effects					
Style name	*Font name*	*Ht*	*Width factor*	*Obl'g angle*	*Back-wards*	*Upside down*	*Vert'l*
ST3	gothice.shx	12	1	0	OFF	OFF	OFF
ST4	Arial Black	10	1	0	OFF	OFF	OFF
ST5	italict.shx	5	1	30	OFF	OFF	OFF
ST6	Romantic	10	1	0	OFF	ON	—
ST7	scriptc.shx	5	1	–30	OFF	OFF	OFF
ST8	monotxt.shx	6	1	0	OFF	OFF	ON
ST9	Swis721BdOulBT	12	1	0	OFF	OFF	OFF
ST10	complex.shx	5	1	0	ON	OFF	OFF
ST11	isoct.shx	5	1	0	ON	ON	—
ST12	romand.shx	5	1	0	ON	ON	ON

3 *Note*: when using the Text Style dialogue box, a TICK in a box means that the effect is on, a blank box means that the effect is off.

Using created text styles

1   Menu bar with **Draw-Text-Single Line Text** and

    prompt   `Specify start point of text or [Justify/Style]`
    enter   **S <R>** – the style option
    prompt   `Enter style name (or ?)` – ST12 as default name?
    enter   **ST1 <R>**
    prompt   `Specify start point of text or [Justify/Style]`
    enter   **20,240 <R>**
    prompt   `Specify height` and enter: **8 <R>**
    prompt   `Specify rotation angle of text` and enter: **0 <R>**
    prompt   `Enter text` and enter: **AutoCAD 2002 <R>**

2   Using the single line text command, add the text item AutoCAD 2002 using the following information:

Style	Start pt	Ht	Rot
ST1	20,260	8	0 – already entered
ST2	225,265	8	0
ST3	15,145	NA	0
ST4	250,240	NA	0
ST5	25,175	NA	30
ST6	100,220	NA	0
ST7	145,195	NA	–30
ST8	215,235	NA	270 (default angle)
ST9	235,195	NA	0
ST10	365,165	NA	0
ST11	325,145	NA	0
ST12	400,140	NA	270 (default angle)

3   When completed, the screen should display 13 different text styles – the 12 created and the STANDARD default as Fig. 32.3.

4   There is no need to save this drawing but:
    a) erase all the text from the screen
    b) save the 'blank' screen as **C:\BEGIN\STYLEX** – you are really saving the created text styles for future use.

Notes

Text styles and fonts can be confusing to new AutoCAD users due to the terminology, and by referring to Fig. 32.3, the following may be of assistance:

1   *Effects*:
Three text style effects which can be 'set' are upside-down, vertical and backwards. These effects should be obvious to the user, and several of our created styles had these effects toggled on.

2   *Width factor*:
A parameter which 'stretches' the text characters and fig. (a) displays an item of text with six width factors. The default width factor value is 1.

3   *Obliquing angle*:
This parameter 'slopes' the text characters as is apparent in fig. (b) and the default value is 0.

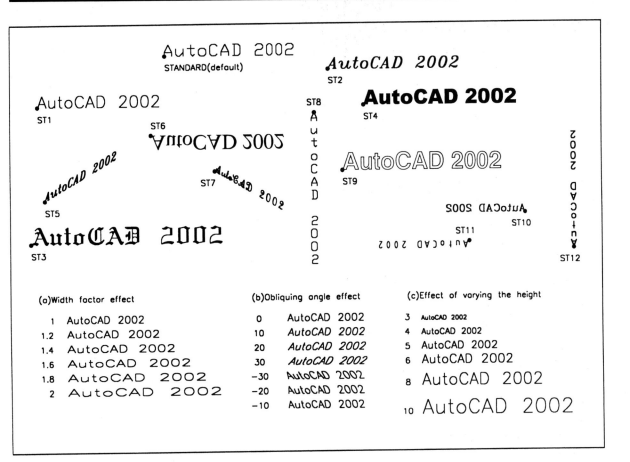

Figure 32.3 Using the created text styles.

4   *Height*:
    When the text command was used with the created text styles, only two styles prompted for a height – ST1 and ST2. The other text styles had a height value entered when the style was created – hence no height prompt. This also means that these text styles cannot be used at varying height values. The effect of differing height values is displayed in fig. (c).

5   *Recommendation*:
    I would strongly recommend that if text styles are being created, the height be left at 0. This will allow you to enter any text height at the prompt when the text command is used.

6   The text items displayed using styles ST5 and ST7 are interesting, these items having:
    ST5   30 obliquing   30 rotation
    ST7   –30 obliquing   –30 rotation

    These styles give an 'isometric text' appearance.

Text control codes

When text is being added to a drawing, it may be necessary to underline the text item, or add a diameter/degree symbol. AutoCAD has several control codes which when used with the **Single Line Text** command will allow underscoring, overscoring and symbol insertion.

The available control codes are:
%%O: toggles the OVERSCORE on/off
%%U: toggles the UNDERSCORE on/off
%%D: draws the DEGREE symbol for angle or temperature (°)
%%C: draws the DIAMETER symbol (∅)
%%P: draws the PLUS/MINUS symbol (±)
%%%:   draws the PERCENTAGE symbol (%)

1   Open C:\BEGIN\STYLEX with the 12 created text styles.

2   Refer to Fig. 32.4, select **Draw-Text-Single Line Text** and:
prompt   Specify start point of text or [Justify/Style]
enter   **S <R>**
prompt   Enter style name or [?]
enter   **ST1 <R>**
prompt   Specify start point of text and enter: **25,250 <R>**
prompt   Specify height and enter: **10 <R>**
prompt   Specify rotation angle of text and enter: **–5 <R>**
prompt   Enter text and enter **%%UAutoCAD 2002%%U <R>**
prompt   Enter text and **<R>**

3   At the command line enter **DTEXT <R>** and:
a) style: ST3
b) start point: 175,255
c) angle: 0
d) text: 123.45%%DF.

4   Activate the single line text command and:
a) style: ST4
b) start point: 35,175
c) angle: 0
d) text: %%UUNDERSCORE%%U and %%OOVERSCORE%%O.

5   With the single line text command
a) style: ST9
b) start point: 285,215
c) angle: 0
d) text: %%C100.

6   Refer to Fig. 32.4 and add the other text items – or text items of your choice. The text style used is at your discretion.

7   Save if required, but we will not use this drawing again.

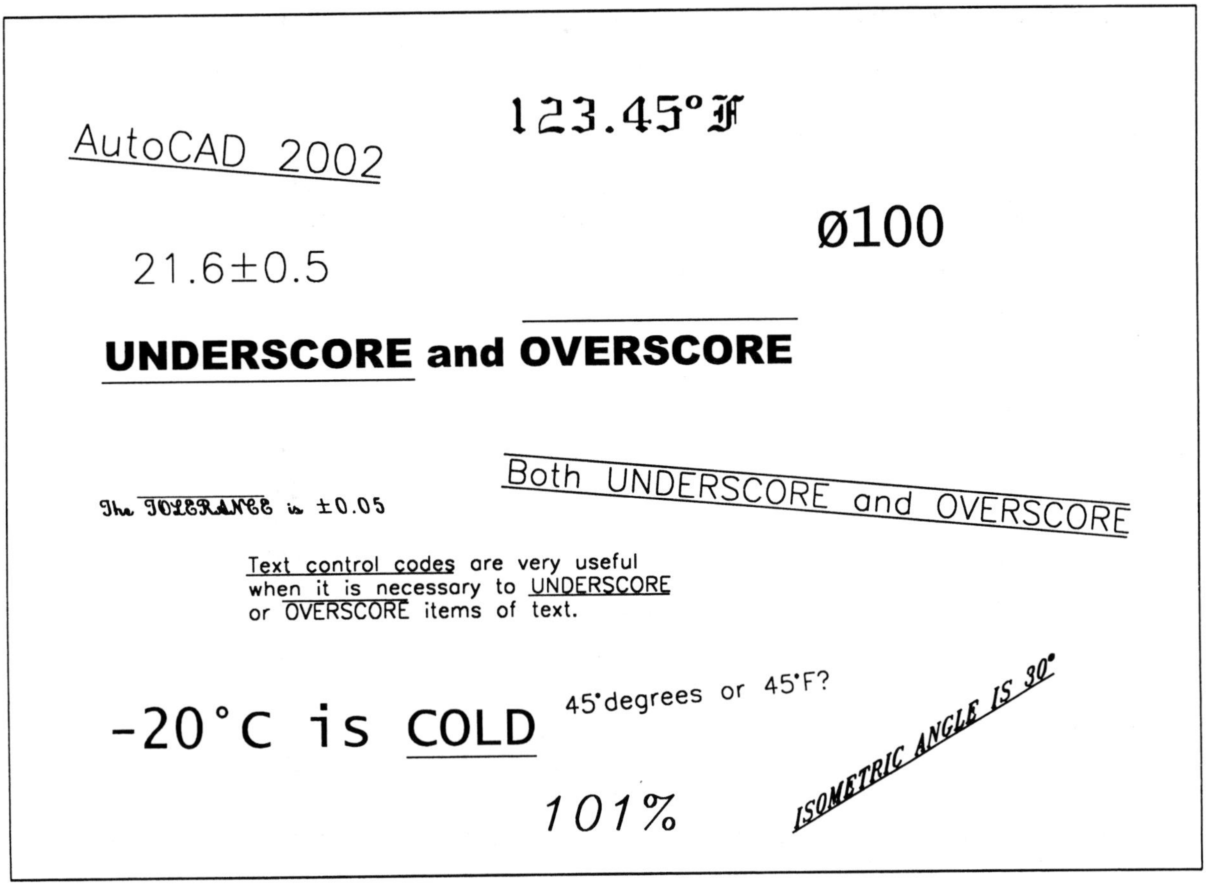

AutoCAD 2002

123.45°F

Ø100

21.6±0.5

UNDERSCORE and OVERSCORE

The TOLERANCE is ±0.05

Both UNDERSCORE and OVERSCORE

Text control codes are very useful
when it is necessary to UNDERSCORE
or OVERSCORE items of text.

-20°C is COLD   45°degrees or 45°F?

ISOMETRIC ANGLE IS 30°

101%

Figure 32.4 Text control codes.

Summary

1   Fonts define the pattern of characters.

2   Styles define the parameters for drawing characters.

3   Text styles must be created by the user.

4   A font can be used for several text styles.

5   Every text style must use a text font.

6   The AutoCAD 2002 default text style is called STANDARD and uses the text font txt.

7   Text control codes allow under/overscoring and symbols to be added to text items.

8   No mention has been made of the different types of font extension used with AutoCAD. This is considered beyond the scope of this book.

Chapter 33

Multiline text

Multiline text was referred to as paragraph text in previous releases. It is a useful draughting tool as it allows large amounts of text to be added to a pre-determined area on the screen. This added text can also be edited.

1 Open the STYLEX with the created text styles.

2 Display toolbars to suit, including the TEXT toolbar.

3 With layer TEXT current, select the TEXT from the Draw toolbar and:

prompt	Specify first corner
enter	**10,275 <R>**
prompt	Specify opposite corner or [Height/Justify/Line spacing/ Rotation/Style/Width]
enter	**S <R>** then **ST1 <R>** – setting the text style
prompt	Specify opposite corner or [Height/Justify/Line spacing/ Rotation/Style/Width]
enter	**H <R>** then **5 <R>** – setting the height
prompt	Specify opposite corner or [Height/Justify/Line spacing/ Rotation/Style/Width]
enter	**125,190 <R>**
prompt	Multiline Text Editor dialogue box
with	flashing cursor in the 'window area'
respond	1. enter the following text **including** the typing errors which have been underlined
	2. **do not try to add the underline effect**
	3. **do not press the return key**
enter	CAD is a draughting <u>tol</u> with many benefits when compared to conventional draughting <u>techniches</u>. Some of these <u>benefitds</u> include <u>incresed</u> productivity, shorter lead <u>tines</u>, standardisation, <u>acuracy</u> <u>amd</u> rapid <u>resonse</u> to change.
and	dialogue box as Fig. 33.1
respond	pick OK.

4 The entered text is displayed as Fig. 33.2(a).

5 *Notes*:
   a) the text 'wraps around' the Text Editor window as it is entered from the keyboard
   b) the text is fitted into the **width** of the selected area of the screen, not the full rectangular area of the dialogue box
   c) the width is determined by the entered coordinates
   d) the entered text in the dialogue box will not appear as the above layout. This is normal, so don't worry about it.

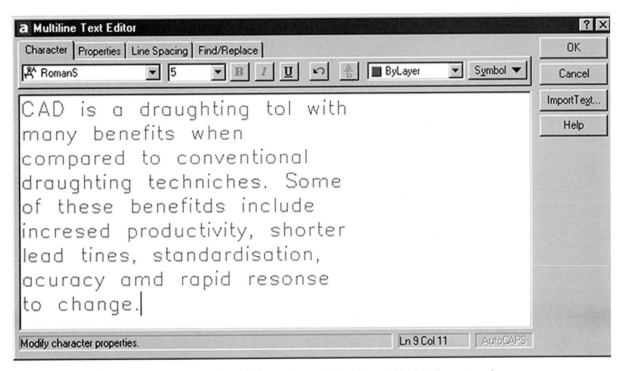

Figure 33.1 The Multiline Text Editor dialogue box with the item of text to be entered.

(a)Original multiline text
CAD is a draughting tol with many benefits when compared to conventional draughting techniches. Some of these benefitds include incresed productivity, shorter lead tines, standardisation, acuracy amd rapid resonse to change.

(b)Text after spell check
CAD is a draughting tool with many benefits when compared to conventional draughting techniques Some of these benefits include increased productivity, shorter lead times, standardisation, accuracy and rapid response to change.

(c)Text after height change
CAD is a draughting tool with many benefits when compared to conventional draughting techniques Some of these benefits include increased productivity, shorter lead times, standardisation, accuracy and rapid response to change.

(d)Style change to ST5
CAD is a draughting tool with many benefits when compared to conventional draughting techniques Some of these benefits include increased productivity, shorter lead times, standardisation, accuracy and rapid response to change.

(e)Justification to TR
CAD is a draughting tool with many benefits when compared to conventional draughting techniques Some of these benefits include increased productivity, shorter lead times, standardisation, accuracy and rapid response to change.

(f)Rotation change to −5
CAD is a draughting tool with many benefits when compared to conventional draughting techniques Some of these benefits include increased productivity, shorter lead times, standardisation, accuracy and rapid response to change.

(g)Width of 120
CAD is a draughting tool with many benefits when compared to conventional draughting techniques Some of these benefits include increased productivity, shorter lead times, standardisation, accuracy and rapid response to change.

(h)Style: ST7, Justication: TC
CAD is a draughting tool with many benefits when compared to conventional draughting techniques Some of these benefits include increased productivity, shorter lead times, standardisation, accuracy and rapid response to change.

(i)Change to lead times
CAD is a draughting tool with many benefits when compared to conventional draughting techniques Some of these benefits include increased productivity, shorter lead times, standardisation, accuracy and rapid response to change.

Figure 33.2 Multiline text exercise.

Spellcheck

AutoCAD 2002 has a built-in spellchecker which has been used in an earlier chapter. It can be activated:
a) from the menu bar with **Tools-Spelling**
b) by entering **SPELL <R>** at the command line

1 Activate the spell check command and:

prompt	Select objects
respond	**pick any part of the entered text then right-click**
prompt	Check Spelling dialogue box
with	1. current dictionary: British English (ise)
	2. current word: probably *draughting*
	3. suggestions: probably draughtiness
	4. context: *CAD is a draughting tol ..*
respond	**pick Ignore All**
note	we have agreed that draughting is correct spelling
prompt	Check Spelling dialogue box
with	1. current word: **tol**
	2. suggestions: toll, tool, to..
respond	1. pick **tool** – becomes highlighted
	2. tool added to Suggestion box
	3. dialogue box as Fig. 33.3
	4. **pick Change**
prompt	Check Spelling dialogue box
with	1. current word: **techniches**
	2. suggestions: tech
respond	1. at suggestions, **alter tech to techniques**
	2. pick **Change**
prompt	Check Spelling dialogue box
respond	change the following as they appear:

original	*alter to*
benefitds	benefits
incresed	increased
acuracy	accuracy
amd	and (manual change required)
resonse	response

then	AutoCAD Message
	Spelling check complete
respond	**pick OK.**

2 The multiline text will be displayed with the correct spelling as Fig. 33.2(b).

3 One of the original spelling mistakes was **'tines'** (times) and this was not highlighted with the spellcheck. This means that the word 'tines' is a 'real word' as far as the dictionary used is concerned although it is wrong to us. This can be a major problem with spellchecks.

4 Enter **DDEDIT <R>** at the command line and:
a) pick any part of the text
b) the Multiline Text Editor dialogue box will be displayed and the word tines can be altered to **times**.
c) when the alteration is complete, pick OK.

5 Multiple copy the corrected text to seven other places on the screen and refer to Fig. 33.2.

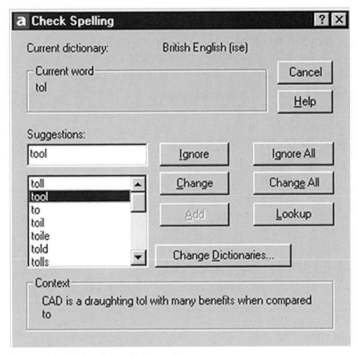

Figure 33.3 The Check Spelling dialogue box.

Editing multiline text

1   Select the EDIT TEXT icon from the Text toolbar and:
    prompt   `Select an annotation object or [Undo]`
    respond   **pick a copied multiline text item**
    prompt   `Multiline Text Editor dialogue box`
    with   text displayed
    respond   1. left-click and drag mouse over CAD
                2. alter height to 7 then pick OK
                3. right-click-Enter to end command
                4. text displayed as fig. (c).

2   Menu bar with **Modify-Object-Text-Edit** and:
    prompt   `Select an annotation object`
    respond   **pick copied text item**
    prompt   `Multiline Text Editor dialogue box`
    respond   1. pick Properties tab
                2. scroll at Style, pick ST5 then pick OK
                3. right-click-Enter – fig. (d).

3   Using the Edit Text command alter the other copied multiline text items using the following information:
    a) Properties: Justification to Top Right – fig. (e)
    b) Properties: Rotation to –5 – fig. (f)
    c) Properties: Width to 120 – fig. (g)
    d) Properties: Style to ST7
                    Justification to Top Centre – fig. (h)
    e) underline lead times and alter height to 8 – fig. (i).

4   This exercise is now complete and can be saved if required.

Text modifications

To investigate some of the other text modifications, erase all the multilane text items except the TR justification – fig. (e). Move this item of text to the top left of the screen and refer to Fig. 33.4.

1 Select the Find and Replace icon from the Text toolbar and:
   prompt   Find and Replace dialogue box
   respond 1. Find text string: enter **CAD**
            2. Replace with: **Computer Aided Draughting**
            3. Search in: Entire drawing
            4. pick Find
   prompt   Context will display text with CAD highlighted
   respond **pick Replace**
   prompt   No more occurrences found – Fig. 33.5
   respond **pick Close.**

3 The text item will be displayed with CAD replaced by Computer Aided Draughting – fig. (b).

4 Multiple copy this replace multiline text item to three other parts of the screen.

5 With the single line text icon from the text toolbar, create an item of text using:
   a) style: ST2
   b) start point: pick to suit
   c) height: 3.75
   d) rotation angle: 5
   e) text: TEST.

6 Select the Scale Text icon from the Text toolbar and:
   prompt   Select objects
   respond **pick a copied multiline text item** then right-click
   prompt   Enter a base point option for scaling
             [Existing/Left/Center/Middle..
   respond **<RETURN>**, i.e. accept the existing default
   prompt   Specify new height or [Match object/Scale factor]
   enter   **M <R>** – the match object option
   prompt   Select a text object with the desired height
   respond **pick TEST item of text**
   and   the selected multiline text will be displayed at a height of 3.75 – fig. (c). This modified text item does not have the style or rotation angle as the TEST item, only the height.

7 Menu bar with **Modify-Object-Text-Scale** and:
   a) object: pick a copied multiline item then right-click
   b) base point for scaling: accept existing
   c) new height: enter 4 – fig. (d).

8 At the command line enter **SCALETEXT <R>** and:
   prompt   Select objects
   respond **pick a copied multiline text item** then right-click
   prompt   Enter a base point option for scaling
             [Existing/Left/Center/Middle..
   respond **<RETURN>**, i.e. accept the existing default
   prompt   Specify new height or [Match object/Scale factor]
   enter   **S <R>** – the scale factor option
   prompt   Specify scale factor
   enter   0.625 <R>
   and   selected text item scaled as fig. (e).

9 As the original text had a height of 5, this scaled text item should have a height of 3.125. Use the list command to check the height of this text. It may be 3.13 – why?

(a)Original TR justified text

CAD is a draughting tool
with many benefits when
compared to conventional
draughting techniques Some
of these benefits include
increased productivity,
shorter lead times,
standardisation, accuracy and
rapid response to change.

(b)Find and Replace CAD

Computer Aided Draughting
is a draughting tool with
many benefits when
compared to conventional
draughting techniques Some
of these benefits include
increased productivity,
shorter lead times,
standardisation, accuracy and
rapid response to change.

(c)Match object scale option

Computer Aided Draughting
is a draughting tool with
many benefits when
compared to conventional
draughting techniques Some
of these benefits include
increased productivity,
shorter lead times,
standardisation, accuracy and
rapid response to change.

(d)Specifying a new height

Computer Aided Draughting
is a draughting tool with
many benefits when
compared to conventional
draughting techniques Some
of these benefits include
increased productivity,
shorter lead times,
standardisation, accuracy and
rapid response to change.

(e)Specifying a scale factor

Computer Aided Draughting
is a draughting tool with
many benefits when
compared to conventional
draughting techniques Some
of these benefits include
increased productivity,
shorter lead times,
standardisation, accuracy and
rapid response to change.

(f)Importing text from Notepad

AutoCAD allows draughting in 2D and 3D and
the following is a brief summary of each:
2D: a)orthograhic layouts in 1st and 3rd
angle can be created
b)detailed working drawings are possible
c)isometric 'views' of 2D objects can be
constructed.
3D: the following models can be created:
- wire-frame
- surface
- solid

Figure 33.4   Text modifications and importing text into AutoCAD.

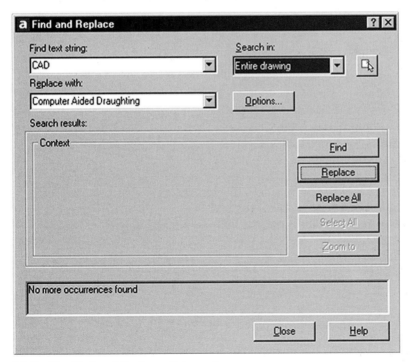

Figure 33.5   The Find and Replace dialogue box.

Importing text files into AutoCAD

Text files can be imported into AutoCAD from other application packages using the Multiline Text Editor dialogue box. To demonstrate the concept we will use a text editor and write a new item of text, save it and then import it into our existing drawing.

1   Save the existing layout as a precaution

2   Select **Start** from the Windows taskbar, then select **Programs-Applications-Notepad** and:
prompt   Blank Notepad screen displayed
enter   the following lines of text with a <R> key press where indicated
        AutoCAD allows draughting in 2D and 3D and the following is a brief summary of each: <R>
        2D: *a*) orthograhic layouts in 1st and 3rd angle can be created <R>
        **   *b*) detailed working drawings are possible <R>**
        **   *c*) isometric 'views' of 2D objects can be constructed <R>**
        3D: the following models can be created: <R>
        **   – wire-frame <R>**
        **   – surface <R>**
        **   – solid <R>**

3   When the text has been entered as above, menu bar with **File-Save As** and:
prompt   Save As dialogue box
respond 1. scroll at Save in and **pick C:\BEGIN**
        2. enter File name as MYTEST
        3. note type: Text Document
        4. pick Save
then   Minimise Notepad (left button from title bar) to return to the AutoCAD screen.

4   Layer TEXT still current?

5   Activate the Multiline Text command with the first corner at 290,140 and the opposite corner at 415,15 (or pick two suitable points of your own) and:
prompt   Multiline Text Editor dialogue box
respond pick Import Text
prompt   Select File dialogue box
respond 1. scroll at Look in and pick C:\BEGIN
        2. pick MYTEST
        3. pick Open
prompt   Multiline Text Editor dialogue box
with   imported text displayed, but it may not appear as you would expect
respond 1. scroll at right until top of text item displayed
        2. pick the Properties tab
        3. hold down the mouse left button and highlight all text by dragging the mouse from the start to the end of the text
        4. scroll at Style and pick ST4
        5. scroll at Justification and pick Top Left
        6. pick the Character tab
        7. alter the height to 3
        8. pick OK.

6   The imported text will be displayed with the ST4 (Arial Black) text style at a height of 3. It will also be top left justified – fig. (e).

7   This exercise is now complete and can be saved, but remember that Notepad may still be open.

Summary

1 Multiline text is also called paragraph text.

2 The text is entered using the Multiline Text Editor dialogue box.

3 The command can be activated by icon, from the menu bar or by entering **MTEXT** at the command line.

4 Multiline text has powerful editing facilities.

5 Text files can be imported into AutoCAD from other application packages with the Multiline Text Editor.

Chapter 34

The ARRAY command

Array is a command which allows multiple copying of objects in either a rectangular or circular (polar) pattern. It is one of the most powerful and useful of the commands available, yet is one of the easiest to use. To demonstrate the command:

1   Open the A3PAPER standard sheet with layer OUT current and the toolbars Draw, Modify and Object snap.

2   Refer to Fig. 34.1 and draw the rectangular shape using the given sizes. Do NOT ADD the dimensions.

3   Multiple copy the rectangular shape from the mid-point indicated to the points A(25,175); B(290,235); C(205,110); D(300,20) and E(35,130). The donuts in Fig. 34.1 are for reference only.

4   Draw two circle, centre at 290,190 with radius 30 and centre at 205,65 with radius 15.

5   Move the original shape to a 'safe place' on the screen.

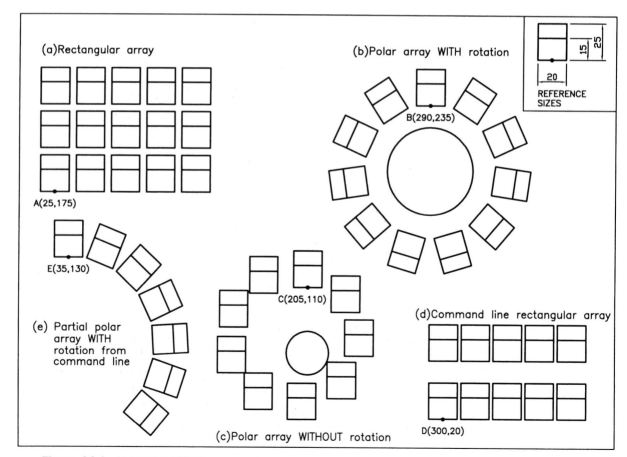

Figure 34.1   Using the ARRAY command.

Rectangular array

1 Select the ARRAY icon from the Draw toolbar and:
   prompt   Array dialogue box
   with   *a*) options: Rectangular Array or Polar Array
              b) rectangular array probably active
              c) information for creating a rectangular array
              d) a preview layout
   respond 1. ensure rectangular array active
              2. enter Rows: 3 and Columns: 5
              3. alter Row offset: 30
              4. alter Column offset: 25
              5. ensure Angle of array: 0
              6. pick Select objects
   prompt   Select objects at the command line
   respond **window the shape at A then right-click**
   prompt   Array dialogue box
   with   5 objects selected and preview – Fig. 34.2
   prompt   pick OK.

2 The shape at A will be copied 14 times into a three row and five column matrix pattern as fig. (a).

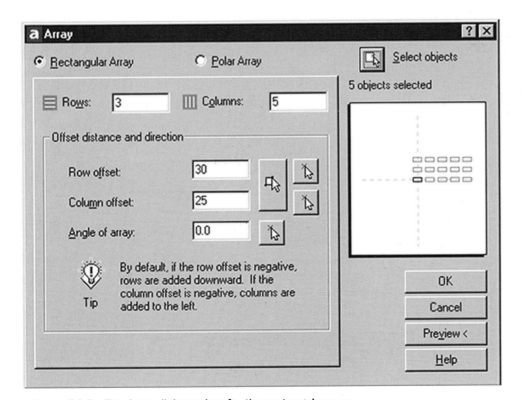

Figure 34.2 The Array dialogue box for the rectangular array.

Polar array with rotation

1 Menu bar with **Modify-Array** and:

    prompt   `Array dialogue box`

    respond   1. Polar Array active

                2. pick **pick Center point**

    prompt   `Select center point of array` at the command line

    respond   **snap to Center icon and pick larger circle**

    prompt   `Array dialogue box`

    with   Selected centre point coordinates

    respond   1. Method: Total number of items & Angle to fill

                2. Total number of items: 11

                3. Angle to Fill: 360

                4. Rotate items as copied: active, i.e. tick

                5. pick Select objects

    prompt   `Select objects` at the command line

    respond   **window the shape at B then right-click**

    prompt   `Array dialogue box` as Fig. 34.3(a)

    respond   **pick Preview<**

    and   1. preview of the entered polar array values

                2. Array message – Fig. 34.3(b)

    respond   **pick Accept** if the array is correct, or Modify to return to the dialogue box for alterations.

2 The shape at B is copied in a circular pattern about the selected centre point. The objects are 'rotated' about this point as they are copied – fig. (b).

(a)

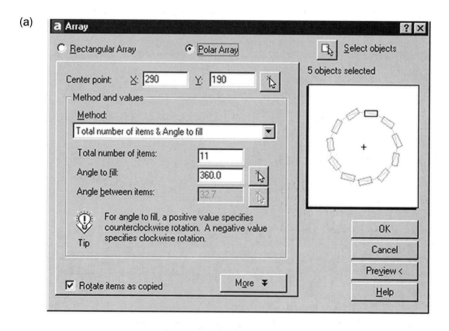

(b)

Figure 34.3 The Array dialogue box for the Polar array and the Array Accept/Cancel message box.

Polar array without rotation

1   At the command line enter **ARRAY <R>** and use the Array dialogue box with the following data:
    a) Polar Array
    b) Center point: pick and snap to center icon of smaller circle
    c) Method: Total number of items & Angle to fill
    d) Total number of items: 9
    e) Angle to fill: 360
    f) Rotate items as copied: Not active, i.e. no tick
    g) Select objects: window shape at C then right-click
    h) preview then accept.

2   The original shape is copied about the selected centre point but is not 'rotated' as it is copied – fig. (c).

Rectangular array from command line

The previous three examples have used the Array dialogue box to enter the details required for the various arrays. The command can also be activated from the command line, so:

1   At the command line enter **–ARRAY <R>** and:
    prompt   Select objects
    respond **Window the shape at D then right-click**
    prompt   Enter the type of array [Rectangular or Polar]
    enter   **R <R>** – the rectangular option
    prompt   Enter the number of rows(---)<1> and enter: **2 <R>**
    prompt   Enter the number of columns(|||)<1> and enter: **5 <R>**
    prompt   Enter the distance between rows or specify unit cell distance (---)
    enter   **40 <R>**
    prompt   Specify the distance between columns(|||)
    enter   **22.5 <R>**

2   The shape at D will be copied into a 2 × 5 column matrix pattern as fig. (d).

Polar array with partial fill angle

Activate the -ARRAY command and:
a) objects: window the shape at D
b) type of array: P
c) center point: enter the point 35,70
d) number of items: 7
e) angle to fill: –130
f) rotate objects: Y.

The result is fig. (e).

Your drawing should resemble Fig. 34.1 and can be saved if required.

Array options

Both the rectangular and polar array commands have variations to those displayed in Fig. 34.1, these being:
a) rectangular: allows the angle of the array to be altered
b) polar: allows for different method items and fill angle.

To demonstrate these options:

1 Erase the array exercises from the screen (save first?) to leave the original, which was moved to a safe place.

2 Refer to Fig. 34.4.

3 Multiple copy the original shape from the point indicated to the points P(45,175), Q(150,175), R(105,115), S(220,115), X(290,275), Y(290,185) and Z(290,25)

4 *a*) rotate the shapes at P and Q by 20 and the shapes at R and S by −10
   b) scale the shapes at X, Y and Z by 0.5
   c) in each case pick the indicated point as the base

5 Draw three circles, radius 20 with centres at the points:
   330,250; 330,160; 330,65

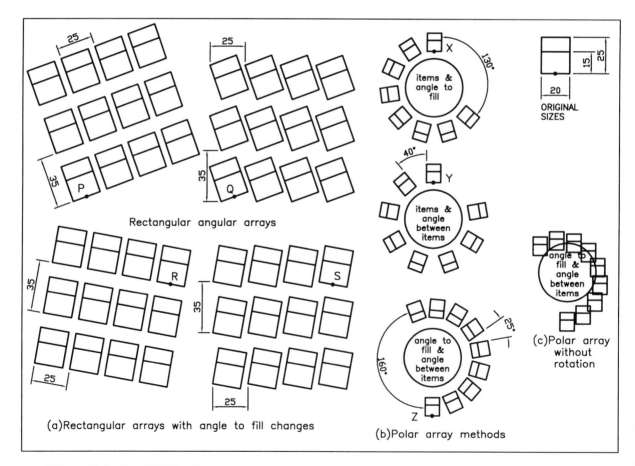

Figure 34.4 The ARRAY options.

6 Using the array dialogue box four times, pick the rectangular type and set:

	1st	*2nd*	*3rd*	*4th*
a) objects	shape P	shape Q	shape R	shape S
b) rows, columns	3,4	3,4	3,4	3,4
c) row offset	35	35	−35	−35
d) column offset	25	25	−25	−25
e) angle of array	20	0	−10	0.

7 The result of the four rectangular operations is fig. (a).

8 Select the ARRAY command and with the polar array type from the dialogue box, enter the following three sets of data:

	1st	*2nd*	*3rd*
a) objects	shape X	shape Y	shape Z
b) method	items & angle to fill	items & angle between	angle to fill & angle between
c) items	8	8	– – –
d) angle to fill	230	– – –	220
e) angle between	– – –	40	25
f) rotate items	yes	yes	yes

9 The results of these three operations is fig. (b).

10 *Task*
   a) linear/align dimension as shown and note the 35 and 25 values for the rectangular array
   b) angular dimension the polar arrays and note the fill angle and angle between the objects values
   c) comment on the 160 degree fill angle. This operation had an angle to fill of 220, and therefore this angle should be 140?
   d) try the polar array operations with no rotate items as copied – fig. (c) displays one operation.

11 This exercise is now complete and can be saved if required.

Summary

1 The ARRAY command allows multiple copying in a rectangular or circular pattern

2 The command can be activated by keyboard entry, from the menu bar or in icon form.

3 The ARRAY dialogue box gives a preview of the entered values

4 Rectangular arrays must have at least one row and one column

5 The rectangular row/column distance can be positive or negative

6 The rectangular angle of array will produce angular rectangular arrays

7 Circular (polar) arrays require a centre point which:
   a) can be entered as coordinates
   b) picked on the screen
   c) reference to existing objects, e.g. Snap to Center icon

8 Polar arrays can be full (360) or partial

9 The polar angle to fill can be positive or negative

10 The polar array has three methods, these being:
   a) number of items and angle to fill
   b) number of items and angle between items
   c) angle to fill and angle between items

11 Polar array objects can be rotated/not rotated about the centre point.

Assignments

It has been some time since any activities have been attempted and seven have been included at this stage. All involve using the array command as well as hatching, adding text, etc. I have tried to make these activities varied and interesting so I hope you enjoy attempting them. As with all activities:

a) start with your A3PAPER drawing standard sheet
b) use layers correctly for outlines, text, hatching, etc.
c) when complete, save as C:\BEGIN\:ACT?? or similar
d) use your discretion for any sizes which have been omitted.

1   Activity 24: Ratchet and Saw Tip Blade.
Two typical examples of how the array command is used.
The ratchet tooth shape is fairly straightforward and the saw blade tooth is more difficult than you would think. Or is it?
Draw each tooth in a clear area of the screen, then copy them to the appropriate circle points. The trim command is extensively used.

2   Activity 25: Fish design.
Refer to Fig. 34.5 and create the fish design using the sizes given. Save this drawing as **C:\BEGIN\:FISH** for future work. When the 'fish' has been completed and saved, create the following array patterns:

a) a 5x3 angular rectangular array at 10 degrees with a 0.2 scale. The distances between the rows and columns are at your discretion
b) three polar arrays with:
    1. scale 0.25 about circle of radius 90 and 9 items
    2. scale 0.2 about circle of radius 70 and 7 items
    3. scale 0.15 about circle of radius 35 and 7 items.

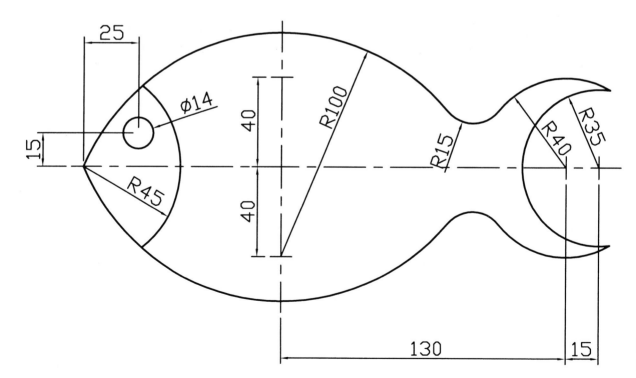

Figure 34.5   C:\BEGIN\FISH information for activity 25.

3   Activity 26: The light bulb
    This was one of my first array exercises and has proved very popular in previous AutoCAD books. It is harder than you would think, especially the R10 arc. The basic bulb is copied and scaled three times. The polar array centre is at your discretion. The position of the smaller polar bulb is relative to the larger polar array, but how is it positioned? – your problem. No hints are given on how to draw the basic bulb.

4   Activity 27: Bracket and Gauge
    The bracket is fairly easy, the hexagonal rectangular array to be positioned to suit yourself. The gauge drawing requires some thought with the polar arrays. I drew a vertical line and arrayed 26 items twice, the fill angles being +150 and –150. Think about this!! The other gauge fill angles are +140 and –140. What about the longer line in the gauge? The pointer position is at your discretion. Another method is to array lines and trim to circles?

5   Activity 28: Pinion gear wheel.
    Another typical engineering application of the array command. The design details are given for you and the basic tooth shape is not too difficult. Copy and rotate the second gear wheel – but at what angle? Think about the number of teeth. I realise that the gears are not 'touching' – that's another problem for you to think about.

6   Activity 29: Propeller blade.
    Draw the outline of the blade using the sizes given. Trimmed circles are useful. When the blade has been drawn, use the COPY, SCALE and polar ARRAY commands to produce the following propeller designs:
    a) 2-bladed at a scale of 0.75
    b) 3-bladed at a scale of 0.65
    c) 4-bladed at a scale of 0.5
    d) 5-bladed at a scale of 0.35
    Note that the four- and five-bladed designs require some 'tidying up'.

7   Activity 30: Leaf design
    Draw the leaf using the reference sizes given. Mirror effect is useful for the outline then 2D solid or hatch. The arrays are with a 1/4 size object. The angular array requires some thought for the angle and the row/column offsets.

Chapter 35

Changing properties

All objects have properties, e.g. linetype, colour, layer, position, etc. Text has also properties such as height, style, width factor, obliquing angle, etc. This chapter will demonstrate how an object's properties can be changed, and this will be achieved by a series of simple exercises.

1   Open A3PAPER with layer OUT current, toolbars to suit and refer to Fig. 35.1.

2   Draw a 40 unit square and copy it to six other parts of the screen.

3   *a*) At the command line enter **LTSCALE <R>** and:
   prompt   Enter new linetype scale factor and enter: **0.6 <R>**
   b) At the command line enter **LWDISPLAY <R>** and:
   prompt   Enter new value for LWDISPLAY and enter: **ON <R>**

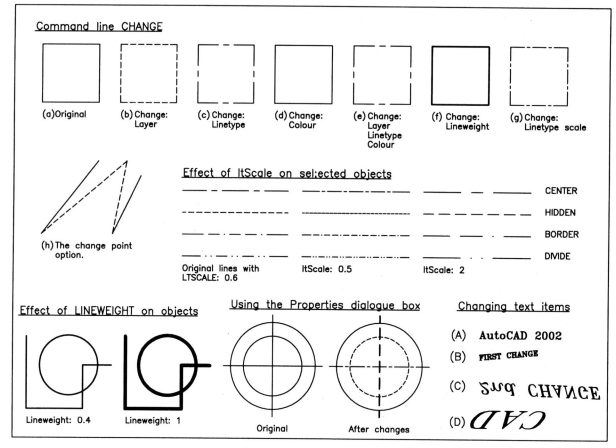

Figure 35.1   Changing properties exercise.

Command line CHANGE

1 At the command line enter **CHANGE <R>** and:
   prompt   `Select objects`
   respond **window the second square then right-click**
   prompt   `Specify change point or [Properties]`
   enter   **P <R>** – the properties option
   prompt   `Enter property to change [Color/Elev/LAyer/LType/ltScale/`
                `LWeight/Thickness]`
   enter   **LA <R>** – the layer option
   prompt   `Enter new layer name<OUT>`
   enter   **HID <R>**
   prompt   `Enter property to change`, i.e. any more property changes
   respond **right-click-Enter.**

2 The square will be displayed as brown hidden lines – fig. (b).

3 Repeat the CHANGE command line entry and:
  a) objects: window the third square then right-click
  b) change point or properties: enter **P <R>** for properties
  c) options: enter **LT <R>** for linetype
  d) new linetype: enter **CENTER <R>** for center linetypes
  e) options: right-click/Enter to end command.

4 The square will be displayed with red centre lines – fig. (c).

5 Use the command line CHANGE with:
  a) objects: window the fourth square then right-click
  b) change point or properties: enter P <R>
  c) options: enter C <R> – the color option
  d) new color: enter BLUE <R>
  e) options: press <R> – fig. (d).

6 Use the CHANGE command with the fifth square and:
  a) enter P <R> then LA <R> – the layer option
  b) new layer: DIMS <R>
  c) options: enter LT <R>
  d) new linetype: CENTER <R>
  e) options: enter: C <R>
  f) new color: enter GREEN <R>
  g) options: right-click/Enter – fig. (e).

7 With the command entry CHANGE:
  a) objects: window the sixth square and right-click
  b) change point or properties: activate properties
  c) options: enter LW <R> – the lineweight option
  d) new lineweight: enter 0.5 <R>
  e) options: press <R> – fig. (f).

8 CHANGE <R> at the command line and:
  a) objects: window the seventh square and right-click
  b) activate properties
  c) options: enter LT <R> then CENTER <R>, i.e. centre linetype
  d) options: enter S <R> – the linetype scale option
  e) new linetype scale: enter 0.5 <R>
  f) options: press <R> – fig. (g).

9   Compare the center linetype appearance of fig. (c) and fig. (g).

10 Menu bar with **Format-Layer** and:
    a) make layer 0 current
    b) freeze layer OUT then OK
    c) only a yellow hidden line square and a green centre line square displayed?
    d) thaw layer OUT and make it current.

11 *Note*
    a) This exercise with the CHANGE command has resulted in:

fig.	square appearance	layer
a	red continuous lines	OUT
b	brown hidden lines	HID
c	red center lines	OUT
d	blue continuous lines	OUT
e	green center lines	DIMS.

    b) I would suggest to you that only fig. (a) and fig. (b) are 'ideally' correct, i.e. the correct colour and linetype for the layer being used. The other squares demonstrate that it is possible to have different colours and linetypes on named layers. This can become confusing.

The change point option

The above exercise has only used the Properties option of the command line CHANGE command, the there is another option – change point.

1   Draw two lines:
    a) start point: 20,140   next point: 60,190
    b) start point: 70,140   next point: 90,180.

2   Activate the CHANGE command and:
    prompt   Select objects
    respond   **pick the two lines then right-click**
    prompt   Specify change point or [Properties]
    respond   **pick about the point 80,190 on the screen.**

3   The two lines are redrawn to this point – fig. (h).

4   This could be a very useful drawing aid?

LTSCALE and ltScale

One of the options available with the CHANGE command is ltScale. This system variable allows individual objects to have their line type appearance changed. The final appearance of the object depends on the value entered and the value assigned to LTSCALE. The LTSCALE system variable is *GLOBAL*, i.e. if it is altered, all objects having center lines, hidden lines, etc. will alter in appearance and this may not be to the user's requirements. Hence the use of the ltScale option of CHANGE. As stated, the final appearance of the object depends on both the LTSCALE value and the entered value for ltScale. Thus if LTSCALE is globally set to 0.6, and the value for ltScale is entered as 0.5, the selected objects will be displayed with an effective value of 0.3. If LTSCALE is 0.6 and 2 is entered for ltScale, the effective value for selected objects is 1.2. This effect is shown in Fig. 35.1 using the Centre, Hidden, Border and Divide linetypes.

Lineweight

LINEWEIGHT allows selected objects to be displayed at varying width. Although this may seem to be similar to the polyline-width type of object, it is not. Objects can be drawn directly with varying lineweight or the LWeight option of the CHANGE command can be used. It is necessary to toggle the LWDISPLAY system variable to ON before lineweight can be used. This was achieved at the start of the chapter.

1 Right-click on **LWT** in the Status bar, pick (left-click)

   Settings and:
   prompt   Lineweight Settings dialogue box
   respond   1. ensure Millimeters active
            2. scroll at Lineweights and pick 0.4mm
            3. dialogue box as Fig. 35.2
            4. note the default value then pick OK.

2 Now draw some lines and circles.

3 From the Object Properties toolbar, scroll at lineweight and set to 1.0 then draw some other lines and circles.

4 Finally set the lineweight back to the default value – 0.25?

5 Figure 35.1 displays some objects at the set lineweights.

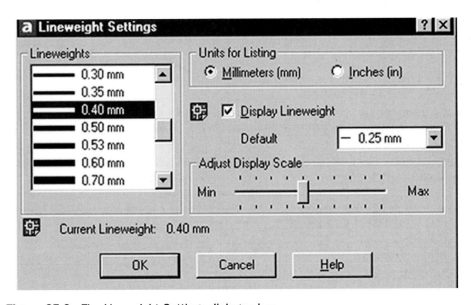

Figure 35.2 The Lineweight Settings dialogue box.

Pickfirst

When objects require to be modified, the normal procedure is to activate the command then select the objects. This procedure can be reversed with the **PICKFIRST** system variable. To demonstrate how this is achieved:

1   Draw a line and circle anywhere on the screen.

2   Menu bar with Modify-Erase and select the two objects then <R>

3   The objects are erased.

4   Draw another line and circle.

5   At the command line enter **PICKFIRST <R>** and:
     prompt   Enter new value for PICKFIRST<0>
     enter   **1 <R>**

6   A pickbox will be displayed on the cursor crosshairs. Do not confuse this pickbox with the grips box. They are different.

7   Pick the line and circle with the pickbox, then menu bar with Modify-Erase. The objects are erased from the screen.

8   Thus PICKFIRST allows the user to alter the selection process and:
     PICKFIRST: 0 – activate the command then select the objects
     PICKFIRST: 1 – select the objects then activate the command.

9   We will use PICKFIRST set to 1 to activate the Properties dialogue box.

Changing Properties using the dialogue box

1   With layer OUT current, draw:
     a) two concentric circles
     b) two lines to represent centre lines.

2   Ensure PICKFIRST is set to 1.

3   With the pick box, select the horizontal line then the PROPERTIES icon from the Standard toolbar and:
     prompt   Properties dialogue box
     with   two tab selections: (*a*) Alphabetical and (*b*)Categorized.

4   The dialogue box gives two types of detail:
     a) General: layer, colour, linetype, etc.
     b) Geometry: start and end point of line, etc.

5   Respond to the dialogue box by:
     a) pick Layer line and scroll arrow appears
     b) scroll and pick layer CL
     c) alter linetype scale value to 0.5 – Fig. 35.3(a)
     d) cancel the dialogue box – top right X box
     e) press ESC.

6   The selected line will be displayed as a green centre line.

7   Pick the vertical line then the Properties icon and from the Properties dialogue box alter:
     a) colour: Magenta
     b) linetype: Hidden
     c) lineweight: 0.5
     d) cancel dialogue box then ESC.

8 Finally pick the smaller circle then the Properties icon and alter:
a) layer: HID
b) linetype scale: 0.75
c) cancel and ESC.

9 Figure 35.1 displays the before and after effects of using the Properties dialogue box to change objects.

10 *Note*
a) the Properties dialogue box gives useful information about the objects which are selected:
1. lines: start and end point; delta values; length; angle
2. circles: centre point; radius; diameter; area; circumference
b) these values can be altered in the dialogue box, and the object will be altered accordingly. You can try this for yourself.

(a)

(b)

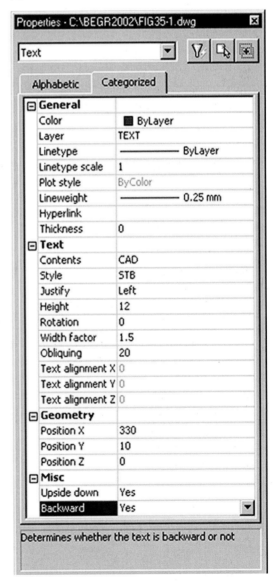

Figure 35.3 The Properties dialogue box for (a) LINE and (b) TEXT.

Changing text

Text has several properties which other objects do not, e.g. style, height, width factor, etc. as well as layer, linetype and colour. These properties can be altered with the command line CHANGE (pickfirst: 0) or the Properties dialogue box (pickfirst:1).

To demonstrate how text can be modified using both methods:

1 Create the following text styles:

name	STA	STB
font	romant	italict
height	0	12
width	1	1
oblique	0	0
backwards	N	N
upside-down	N	Y
vertical	N	N.

2 With layer TEXT current and STA the current style, enter the text item AutoCAD 2002 at height 5 and rotation 0 at a suitable part of the screen – fig. (A).

3 Multiple copy this item of text to three other parts of the screen.

4 At the command line enter **CHANGE <R>** and:

prompt	Select objects
respond	**pick the second text item then right-click**
prompt	Specify change point or [Properties]
respond	**<RETURN>**
prompt	Specify new text insertion point <no change>
respond	**<RETURN>** – no change for text start point
prompt	Enter new text style <STA>
respond	**<RETURN>** – no change to text style
prompt	Specify new height and enter: **3.5 <R>**
prompt	Specify new rotation angle and enter: **5 <R>**
prompt	Enter new text <AutoCAD 2002>
enter	**FIRST CHANGE <R>** – fig. (B).

5 With CHANGE <R> at the command line, pick the third item of text and:
   a) change point: <R>
   b) new text insertion point: <R>
   c) new text style: enter STB <R>
   d) new rotation angle: –2
   e) new text: enter 2nd CHANGE – fig. (C)
   f) Question: why no height prompt?

6 With PICKFIRST on (i.e. set to 1) pick the fourth item of text then the Properties icon.
   a) The Properties dialogue box will display the following details for the text item: General; Text; Geometry and Misc.
   b) Using the Properties dialogue box alter:
       1. text style: STB
       2. contents: CAD
       3. height: 12
       4. width factor: 1.5
       5. obliquing: 20
       6. Backwards: Yes
   c) dialogue box as Fig. 35.3(b)
   d) cancel the dialogue box then ESC
   e) these changes are displayed in fig. (D).

7 This exercise is now complete and can be saved if required.

Combining ARRAY and CHANGE

Combining the array command with the properties command can give interesting results. To demonstrate the effect:

1 Open your standard A3PAPER sheet and refer to Fig. 35.4.

2 Draw the two arc segments as trimmed circles using the information given in fig. (a).

3 Draw the polyline and 0 text item using the reference data.

4 With the ARRAY command, polar array (twice) the polyline and text item using an arc centre as the array centre point:
   a) for 4 items, angle to fill +30° with rotation
   b) for 7 items, angle to fill −60° with rotation – fig. (b).

5 Using (a)the command line CHANGE or (b) the Properties icon, pick each text item and alter:
   a) the text values to 10, 20, 30 to 90
   b) the text height to 8.

6 The final result should be as fig. (c).

7 Save if required as the exercise is complete.

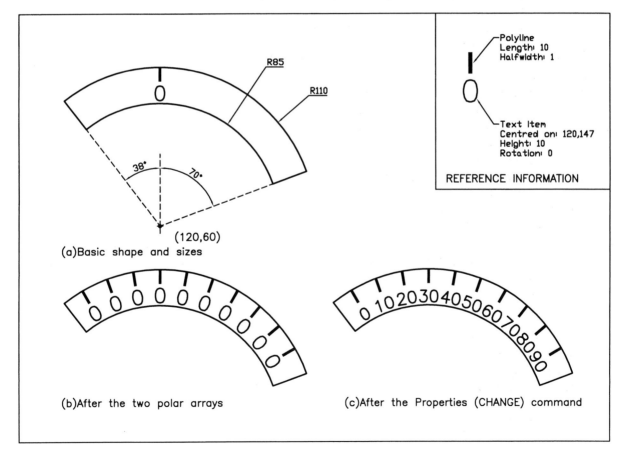

Figure 35.4 The combined ARRAY and CHANGE PROPERTIES command.

Note

The first exercise demonstrated that it was possible to change the properties of objects independent of the current layer. This means that if layer OUT (red, continuous) is current, objects can be created on this layer as green centre lines, blue hidden lines, etc. This is a practice I would **not recommend** until you are proficient at using the AutoCAD draughting package. If green centre lines have to be created, use the correct layer, or make a new layer if required. Try not to 'mix' different types of linetype and different colours on the one layer. Remember that this is only a recommendation – the choice is always left to the user.

Summary

1   Objects have properties such as colour, layer, linetype, etc.

2   An objects properties can be changed with:
    a) the command line CHANGE
    b) the Properties dialogue box.

3   To use the Properties dialogue box, the PICKFIRST variable must be set ON, i.e. value 1

4   With PICKFIRST:
    a) value 0: select the command then the objects
    b) value 1: select the objects then the command.

5   The properties command is very useful when text items are arrayed.

6   Individual objects can have specific linetype scale factors.

Chapter 36

User exercise 3

This exercise will involve creating different arrays with already created text styles.

1   Open C:\BEGIN\STYLEX to display a blank screen but with twelve created and saved text styles (ST1-ST12) from Chapter 32 – I hope? Refer to Fig. 36.1.

2   With layer OUT current, draw a 30 unit square at A(10,10).

3   **Read the note (8)** before the next operation.

4   Multiple copy the square from point A to the points B(85,220), C(200,70), D(280,15) and E(20,255).

5   Array the squares using the following information:
    A: rectangular with 2 rows and 5 columns, both distances 35
    B: polar for 10 items, full circle with rotation, the centre point of the array being 100,170
    C: rectangular with 5 rows and 2 columns and 35 distances
    D: angular rectangular (angle of array –10) with 5 rows and 2 columns, the distances both 35. Remember to rotate first!
    E: rectangular for 1 row and 10 columns, the distance 35.

6   Add/or change the text items using the style names listed.

7   Save the completed layout as A:UEREX3.

8   *Note*
    a) text can be added after the squares are arrayed, or before the first square is multiple copied
    b) after array: the text items are added using single line text, the style, rotation angle and start point being entered by the user. This is fairly straightforward with the exception of the polar and angular arrays – what are the rotation angles?
    c) before the multiple copy: two text items are added to the original square and it is then multiple copied and arrayed. The added text items can then be altered with the change properties command
    d) it is the user's preference as to which method is used. I added the text to the original square, copied, arrayed then changed properties.

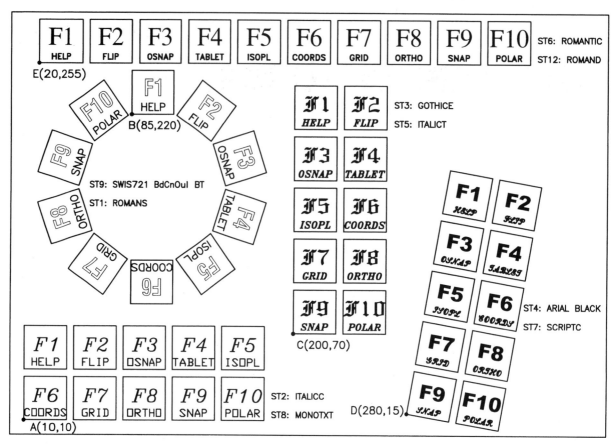

Figure 36.1 User exercise 3.

Assignments

Two activities have been added which require the array and change properties commands to be used. Hopefully these activities will give you some relaxation?

1. Activity 31: Telephone dials
   The 'old-fashioned' type has circles arrayed for a fill angle of ?? Is the text arrayed or just added to the drawing as text?
   You have to decide!
   The modern type has a polyline 'button' with middled text. The array distances are at your discretion as is the outline shape.

2. Activity 32: Flow Gauge and Dartboard
   a) Flow Gauge: A nice simple drawing to complete, but it takes some time! The text is added during the array.
   b) Dartboard: Draw the circles then array the 'spokes'. The filled sections are trimmed donuts. The text is middled, height 10 and ROMANT. Array then change properties?

Chapter 37

Dimension styles 2

In Chapter 20 we investigated how dimensions could be customised to user requirements by setting and saving a dimension style. The dimension style A3DIM was created and saved with our A3PAPER standard sheet.

In this chapter we will:

a) create and use several new dimension styles
b) add tolerance dimensions
c) investigate geometric tolerance.

The process for creating new dimension styles involves altering the values for specific variables which control how the dimensions are displayed on the screen.

Getting started

1   Open your A3PAPER standard drawing sheet.

2   Create the following text styles, all with height 0:

name	*font*
ST1	romans.shx
ST2	scriptc.shx
ST3	italict.shx
ST4	Arial Black.

3   Menu bar with **File-Save** to update the A3PAPER standard sheet with the four created text styles. These are then available for future use if required.

4   Now continue with the exercise.

Creating the new dimension styles

1 The A3PAPER standard sheet still on the screen?
2 Menu bar with **Dimension-Style** and:

prompt	Dimension Style Manager dialogue box with style A3DIM current
respond	**pick New**
prompt	Create New Dimension Style dialogue box
with	*a*) New Style Name: Copy of A3DIM
	b) Start with: A3DIM
	c) Use for: All dimensions
respond	1. alter New Style Name to **DIMST1**
	2. pick **Continue**
prompt	New Dimension Style: DIMST1 dialogue box with tab selections
respond	pick OK at present
prompt	Dimension Style Manager dialogue box
with	Styles: A3DIM and DIMST1
respond	**pick A3DIM then NEW** — step(A)
prompt	Create New Dimension Style dialogue box as before — step(B)
respond	1. alter New Style Name to **DIMST2** — step(C)
	2. pick Continue — step(D)
prompt	New Dimension Style: DIMST2 dialogue box — step(E)
respond	pick OK at present — step(F)
prompt	Dimension Style Manager dialogue box
with	Styles: A3DIM, DIMST1 and DIMST2
respond	using steps A,B,C,D,E and F, pick A3DIM then New and create new dimension styles DIMST3, DIMST4, DIMST5 and DIMST6
then	when DIMST6 has been created
prompt	Dimension Style Manager dialogue box
with	Styles: A3DIM, DIMST1 – DIMST6 as Fig. 37.1
respond	1. pick A3DIM then Set Current
	2. pick Close.

3 We have now created six new dimension styles (DIMST1 – DIMST6) and these all have the same 'settings' as the A3DIM dimension style.
4 These six new dimension styles have now to be individually modified to meet our requirements.

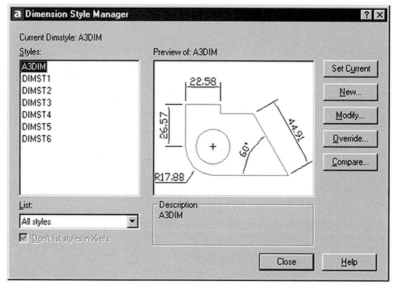

Figure 37.1 The Dimension Style Manager dialogue box with the original A3DIM style and the new DIMST1–DIMST6 styles.

Modifying the new styles

In the exercise which follows, we will only alter certain dimension variables. The rest of the dimension variables will have the same 'settings' as A3DIM.

We will start with DIMST1, so:

1 Menu bar with **Dimension-Style** and:
   prompt   Dimension Style Manager dialogue box
   respond   1. pick DIMST1
                2. pick Modify
   prompt   Modify Dimension Style: DIMST1 dialogue box with tab selections
   respond   1. pick the Text tab
                2. Draw frame around text active, i.e. tick
                3. Text Placement: Vertical: Centered
                                    Horizontal: Centered
                4. Text Alignment: Align with dimension line
                5. Text Appearance: Text style: ST1 with height: 4
                6. pick OK
   prompt   Dimension Style Manager dialogue box
   respond   pick Close

2 The above process is the basic procedure for 'customising' dimension styles, the steps being:
   a) pick Dimension-Style or the Dimension Style icon
   b) pick required style name, e.g. DIMST1
   c) pick Modify
   d) pick required tab and alter as required
   e) pick OK then Close

3 Using the procedure described, alter the named dimension styles to include the following:

Style	*Tab*	*Alteration*
DIMST2	Lines and Arrows	Arrowheads: all Box filled
		Arrow size: 4
		Center Mark Type: Line
		Size: 3
	Text	Appearance Style: ST1 with height: 4
		Alignment: Aligned with dimension line
	Primary Units	Zero suppression
		Trailing off: linear and angular
DIMST3	Lines and Arrows	Arrowheads: all Oblique, size: 4
		Center Mark Type: Mark with size: 4
	Text	Appearance Style: ST2 with height: 4
	Alternative Units	Display alternative units active Unit
		format: Decimal Precision: 0.00
		Multiplier: 0.03937 (default?)
		Round distances to: 0
DIMST4	Lines and Arrows	Arrowheads: all Dot Small, height: 8
	Text	Appearance Style: ST1, height: 4
		Alignment: Horizontal
	Primary Units	Measurement scale factor: 2
DIMST5	Text	Appearance Style: ST3 with height: 5
		Offset from dimension line: 0
DIMST6	Text	Appearance Style: ST4 with height: 6
		Offset from dimension line: 3

4 When the six dimension styles have been altered, return to the Dimension Style Manager dialogue box, pick **A3DIM-Set Current** then Close.

Using the created dimension styles

1   Using Fig. 37.2 for reference, draw seven horizontal and vertical lines, seven circles and seven angled lines – the sizes are not of any significance but try and 'fit' all the objects into your drawing 'frame'.

2   Dimension one set of objects, i.e. a horizontal and vertical line, a circle with diameter and an angled line. The dimensions added will be with the A3DIM style as fig. (a).

3   Menu bar with **Dimension-Style** and:
      prompt   Dimension Style Manager dialogue box
      respond 1. pick DIMST1
              2. pick Set Current
              3. pick Close.

4   Using the set dimension style (DIMST1) dimension another set of four objects.

5   Display the dimension toolbar (View-Toolbars) and note that DIMST1 is displayed, i.e. it is the current dimension style.
      a) pick the scroll arrow to display the other styles – Fig. 37.3
      b) pick DIMST2 to make it current
      c) cancel or reposition the dimension toolbar
      d) dimension a set of objects with the DIMST2 style.

6   At the command line enter **–DIMSTYLE <R>** and:
      prompt Current dimension style: DIMST2
      then   Enter a dimension style option [Save/Restore..
      enter   **R <R>** – the restore option
      prompt Enter a dimension style name, [?] or <select dimension>
      enter   **DIMST3 <R>**
      and   DIMST3 displayed in Dimension toolbar if displayed
      note   if the text window appears, cancel it.

7   Dimension a set of objects with the DIMST3 style.

8   Using the menu bar selection Dimension-Style, the Dimension toolbar or the command line –DIMSTYLE entry:
      a) make each created dimension style current
      b) dimension a set of objects with each style.

9   Figure 37.2 displays the result of the exercise which can be saved if required.

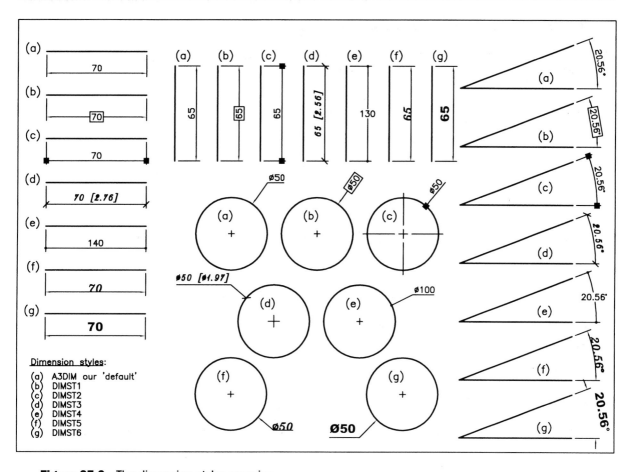

Figure 37.2 The dimension styles exercise.

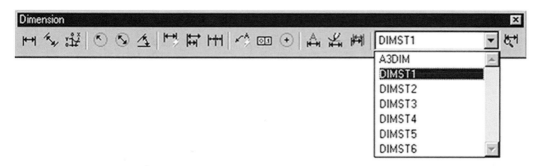

Figure 37.3 The Dimension toolbar with created styles.

Comparing dimension styles

It is possible to compare dimension styles with each other and note any changes between
them.

1   Menu bar with **Dimension Style** and:

prompt   Dimension Style Manager dialogue box

respond **pick DIMST4 then Compare**

prompt   Compare Dimension Styles dialogue box

with   Compare: DIMST4

respond **scroll and alter With name to A3DIM** – Fig. 37.4

prompt   Compare Dimension Styles dialogue box with a comparison between
the different dimension variables of DIMST4 and A3DIM

respond note the differences then Close both dialogue boxes.

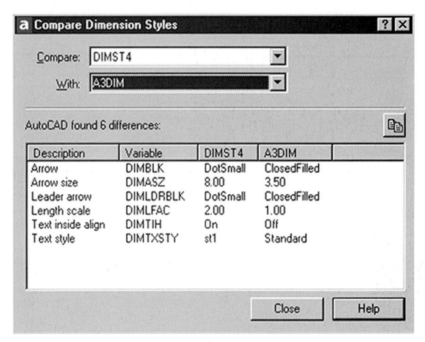

Figure 37.4 The Compare Dimension Styles dialogue box.

Dimension variables

The above comparison between the two dimension styles displays some of the AutoCAD dimension variables. All dimension variables **(dimvars)** can be altered by keyboard entry.

To demonstrate this:

1 Make DIMST1 the current dimension style and note the dimensions which used this style – the second set of dimensioned objects.

2 At the command line enter **DIMASZ <R>** and:
prompt   Enter new value for DIMASZ <3.50>
enter   **8 <R>**

3 Pick the Dimension Style icon from the Dimension toolbar and:
prompt   Dimension Style Manager dialogue box
with   *a*) Current Style: DIMST1
            b) DIMST1
                     └──<style overrides>
            c) Description: DIMST1＋Arrow size＝8.00 – Fig. 38.5
              (i.e. DIMST1 has been altered)
respond   1. **move pointer over <style overrides>**
             2. **right-click the mouse**
prompt   Shortcut menu displayed
respond   1 pick Save to current style
             2 pick Close.

4 The dimensions which used the DIMST1 dimension style will be displayed with an arrow size of 8.

5 Now change the DIMASZ variable for DIMST1 back to 3.5.

6 This exercise is now complete and can be saved if required.

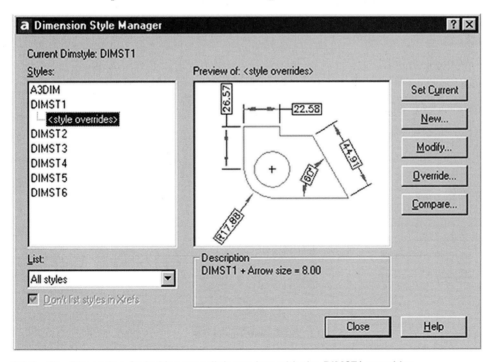

Figure 37.5 The Dimension Style Manager dialogue box with the DIMST1 overrides.

Task

1 Open your A3PAPER standard sheet with the four saved text styles ST1–ST4

2 Using the Dimension Style Manager dialogue box:
 a) check current style is A3DIM
 b) with Modify-Text tab, set the text style to ST1 with height 4
 c) pick OK then close the dialogue box

3 Menu bar with **File-Save** to update the A3PAPER standard sheet with the modified A3DIM dimension style.

Tolerance dimensions

Dimensions can be displayed with tolerances and limits, the 'types' available being symmetrical, deviation, limits and basic. These are displayed in Fig. 37.6(a) with the standard default dimension type, i.e. no tolerances displayed. To use tolerance dimensions correctly, a dimension style should be created for each 'type' which is to be used in the drawing. As usual we will investigate the topic with an exercise so:

1 Open your A3PAPER standard sheet and refer to Fig. 37.6.

2 Draw a line, circle and angled line and copy them to four other areas of the screen.

3 Dimension one set of objects with your standard A3DIM dimension style setting – fig. (p).

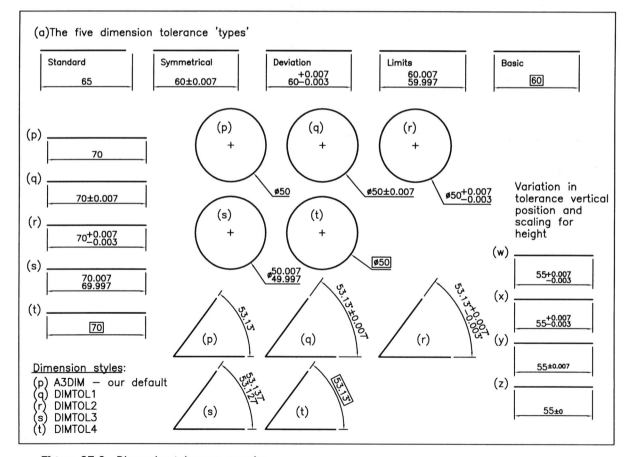

Figure 37.6 Dimension tolerance exercise.

Creating the tolerance dimension styles

1 Using the Dimension Style Manager dialogue box, create four new named dimension styles, all starting with A3DIM. The four new style names are DIMTOL1, DIMTOL2, DIMTOL3 and DIMTOL4.

2 With the Dimension Style Manager dialogue box:
   a) pick each new dimension style
   b) pick Modify then select the:
      1. Primary Units tab and alter both the Linear and Angular precision to 0.000
      2. Tolerances tab and alter as follows:

	DIMTOL1	*DIMTOL2*	*DIMTOL3*	*DIMTOL4*
Method	Symmetrical	Deviation	Limits	Basic
Precision	0.000	0.000	0.000	---
Upper value	0.007	0.007	0.007	---
Lower value	---	0.003	0.003	---
Scaling for height	1	1	1	---
Vertical position	Middle	Middle	Middle	Middle

3 Making each new dimension style current, dimension a set of three objects. The effect is displayed in Fig. 37.6 with:
   fig. (q) : style DIMTOL1   fig. (r) : style DIMTOL2
   fig. (s) : style DIMTOL3   fig. (t) : style DINTOL4

4 *Task*
   Investigate the vertical position and the scaling for height options available in the Tolerances tab dialogue box. Figure 37.6 displays the following:

fig.	*method*	*vertical pos*	*scaling for height*
w	deviation	top	0.8
x	deviation	bottom	0.8
y	symmetrical	top	0.7
z	symmetrical	bottom	0.7

5 This exercise is complete and can now be saved.

Geometric tolerancing

AutoCAD 2002 has geometric tolerancing facilities and to demonstrate this feature:

1   Open your A3PAPER standard sheet.

2   Menu bar with **Dimension-Tolerance** and:
   prompt   Geometric Tolerance dialogue box
   with   Feature Control Frames
   respond move pointer to box under Sym and right-click
   prompt   Symbol dialogue box as Fig. 37.7
   respond 1. press ESC to cancel the Symbol dialogue box
           2. study the Geometric Tolerance dialogue box
           3. pick Cancel at present.

Figure 37.7   The Symbol dialogue box.

Feature control frames

Geometric tolerances define the variations which are permitted to the form or profile of a component, its orientation and location as well as the allowable runout from the exact geometry of a feature. Geometric tolerances are added by the user in **feature control frames**, these being displayed in the Geometric Tolerances dialogue box. A feature control frame consists of two/three sections or compartments. These are displayed in Fig. 37.8(a) and are:

A) *The geometric characteristic symbol*
    This is a symbol for the tolerance which is to be applied. These geometric symbols can represent location, orientation, form, profile and runout. The symbols include symmetry, flatness, straightness, angularity, concentricity, etc.
B) *The tolerance section (1 and 2)*
    This creates the tolerance value in the Feature Control Frame. The tolerance indicates the amount by which the geometric tolerance characteristic can deviate from a perfect form. It consist of three boxes:
    a) first box: inserts an optional diameter symbol
    b) second box: the actual tolerance value
    c) third box: displays the material code which can be:
        1. M : at maximum material condition
        2. L : at least material condition
        3. S : regardless of feature size
C) *The primary, secondary and tertiary datum information*
    This can consist of the reference latter and the material code. Typical types of geometric tolerance 'values' are displayed in Fig. 37.8(b).

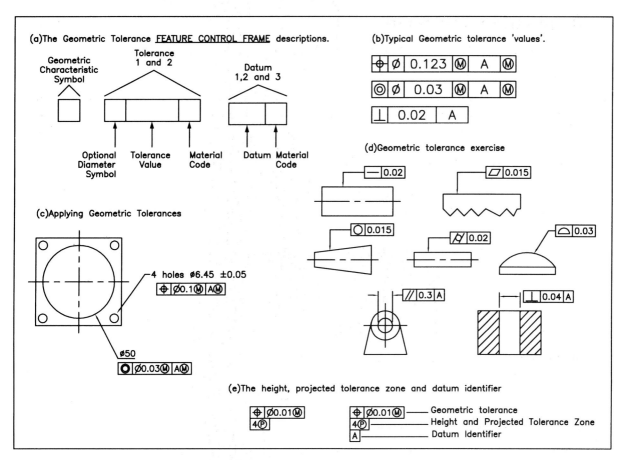

Figure 37.8 Geometric tolerance terminolgy and usage.

Note

a) Feature control frames are added to a dimension by the user.
b) A feature control frame contains all the tolerance information for a single dimension. The tolerance information must be entered by the user.
c) Feature control frames can be copied, moved, erased, stretched, scaled and rotated once 'inserted' into a drawing.
d) Feature control frames can be modified with DDEDIT.

Geometric tolerance example

1   Refer to Fig. 37.8(c) and create a square with circles as shown. The size and layout is of no importance.

2   Diameter dimension the larger circle using the A3DIM dimension style (should be current) and the DIMS layer.

3   Select the TOLERANCE icon from the Dimension toolbar and:

prompt	`Geometric Tolerance dialogue box`
respond	**pick top box under Sym**
prompt	`Symbol dialogue box`
respond	**pick Concentricity symbol** (2nd top left)
prompt	`Geometric Tolerance dialogue box`
with	Concentricity symbol displayed
respond	1. at Tolerance 1, pick left box
	2. at Tolerance 1, enter 0.03
	3. at Tolerance 1, pick MC box and:

    prompt `Material Condition dialogue box`
    respond **pick M**

    4. at Datum 1, enter A
    5. at Datum 1, pick MC box, pick M
    6. Geometric Tolerance dialogue box as Fig. 37.9

then	**pick OK**
prompt	`Enter tolerance location`
respond	**pick 'under' the diameter dimension.**

4   Menu bar with **Dimension-Tolerance** and:

prompt	`Geometric Tolerance dialogue box`
respond	pick top Sym box
prompt	`Symbol dialogue box`
respond	**pick Position symbol** (top left)
prompt	`Geometric Tolerance dialogue box`
respond	1. at Tolerance 1 pick left box (for diameter symbol)
	2. at Tolerance 1 enter value of 0.1
	3. at Tolerance 1 pick MC box and pick M
	4. at Datum 1 enter datum A
	5. at Datum 1 pick MC box and pick M
	6. pick OK
prompt	`Enter tolerance location`
respond	**pick to suit** – refer to Fig. 37.8(c).

5   Select the LEADER icon from the Dimension toolbar and:

prompt	`Specify first leader point` and pick a smaller circle
prompt	`Specify next point` and pick a point to suit then right-click
prompt	`Specify text width` and enter: **0 <R>**
prompt	`Enter first line of annotation text` and right-click
prompt	`Multiline Text Editor dialogue box`
enter	4 holes diameter 6.45 plus/minus 0.05 (use symbols)
then	pick OK.

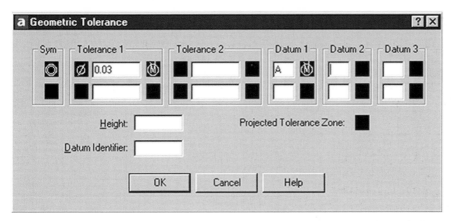

Figure 37.9 The Geometric Tolerance dialogue box.

Geometric tolerance exercise

1 Refer to Fig. 37.8(d) and create 'shapes' as shown – size is not important.

2 Add the geometric tolerance information.

3 Investigate the height, projected tolerance zone and datum identifier options from the Geometric Tolerance dialogue box as fig. (e).

Summary

1 Dimension styles are created by the user and 'customized' for company/customer requirements.

2 Dimension styles are created with the Dimension Style Manager dialogue box using several tab selections.

3 It possible to alter dimension variables (dimvars) from the command line.

4 Tolerance dimensions can be created as dimension styles. The four 'types' are symmetrical, deviation, limits and basic.

5 Geometric tolerances can be added to drawings using feature control frames.

6 There are 14 geometric tolerance 'types' available.

7 All geometric tolerance information must be added by the user in feature control frames.

8 Feature control frames can be copied, moved, scaled, etc.

Assignments

Two assignments on dimension styles, tolerance dimensions and geometric tolerancing. Adding dimensions to a drawing is generally a tedious process and if dimension styles and geometric tolerancing are required, the process does not get any easier.

1 Activity 33: Tolerance dimensions
The two drawings are easy to complete.
Adding the dimensions requires four created styles, all from the A3DIM default. The dimension styles use the symmetrical, limits, deviation and basic tolerance methods. You have the enter the upper and lower values.

2 Activity 34: Geometric tolerances.
A slightly more complex drawing to complete.
Several dimension styles have to be created and two geometric tolerances have to be added.

Chapter 38

Drawing with different sizes

Many new user to the AutoCAD draughting package frequently ask specific questions about drawing, three of the most common being:

1 Can I draw with imperial units?

2 Can you set a scale at the start of a drawing?

3 Can you work with very large/small components?
   The answer to these questions is yes:
   a) you can draw in feet and inches
   b) you can set a scale at the start of a drawing
   c) you can draw large/small components on an A3 sized sheet

The most difficult concept for certain CAD users is that **all work should be completed full size**. This is a concept that certain CAD users cannot accept.

In this chapter we will investigate three different types of exercise – a component drawn and dimensioned in inches, a large scale drawing and a small scale drawing. Each exercise will involve modifying the existing A3PAPER standard sheet and, hopefully, all users will appreciate what it means by drawing everything full size.

Drawing in inches

To draw in inches we require a new standard sheet which could be created:
a) from scratch
b) using the Wizard facility
c) altering the existing A3PAPER standard sheet which is set to metric sizes.

We will alter our existing standard sheet, the only reason being that it has already been 'customised' to our requirements, i.e. it has layers, a text style and a dimension style. This may save us some time? The conversion from a metric standard sheet to one which will allow us to draw in inches is relatively straightforward. We have to alter several parameters, e.g. units, limits, drawing aids and the dimension style variables.

1 Open your A3PAPER standard sheet and erase the black border as it is the wrong size for drawing in inches.

2 Units
   a) Menu bar with **Format-Units** and:
      prompt   `Drawing Units dialogue box`
      respond set the following:
              1. *Length*:
                 Type: Engineering; Precision: 0′–0.00″
              2. *Angle*:
                 Type: Decimal Degrees; Precision: 0.00
              3. *Insert blocks*: Scale to inches
     then   pick OK.

3 Limits
   Menu bar with **Format-Drawing Limits** and:
   prompt Specify lower left corner and enter: **0,0 <R>**
   prompt Specify upper right corner and enter: **16",12" <R>**
   Note: shift 2 for the inches (").

4 Drawing area
   a) make layer 0 current
   b) menu bar with **Draw**-Rectangle and:
      prompt Specify first corner point and enter: **0,0 <R>**
      prompt Specify other corner point and enter: **15.5",11.75"**
   c) menu bar with **View-Zoom-All.**

5 Drawing Aids
   a) set grid spacing to 0.5
   b) set snap spacing to 0.25
   c) set global LTSCALE value to 0.025.

6 Dimension Style
   Menu bar with **Dimension-Style** and:
   prompt Dimension Style Manager dialogue box
   respond create a new dimension style with:
       a) New Style Name: STDIMP
       b) Start with: A3DIM
       c) Tab alterations as follows:

1. *Lines and Arrows*	Baseline spacing: 0.75"
	Extend beyond dim line: 0.175"
	Offset from origin: 0.175"
	Arrowheads: Closed Filled, Size: 0.175"
	Centre: None
2. *Text*	Style: ST1 with height: 0.18"
	Offset from dim line: 0.08"
	Text alignment: with dimension line
3. *Primary Units*	Unit Format: Engineering
	Precision: 0'-0.00"
	Degrees: Decimal
	Precision: 0.00
	Zero supression: both ON, i.e. tick
	Scale factor: 1.

7 *a*) Make layer OUT current
   b) save the drawing as **CL\BEGIN\STDIMP** – it may be useful.

8 Refer to Fig. 38.1 and draw the component as shown. Add all text and dimensions. You should have no problems in completing the drawing or adding the dimensions. What should be the value for text height? Use the grid spacing as a guide.

9 When complete, save the drawing – plot?

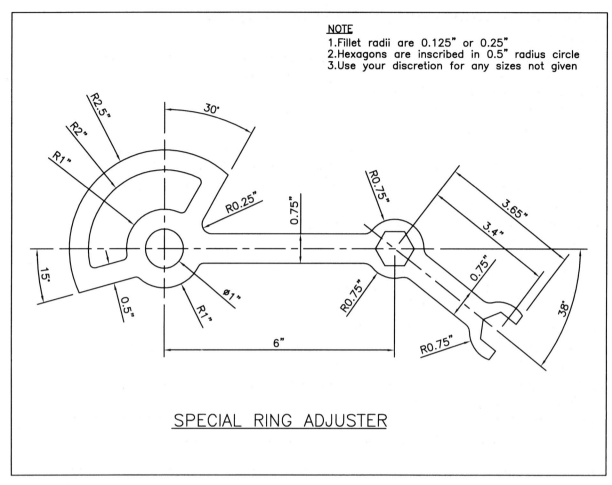

NOTE
1. Fillet radii are 0.125" or 0.25"
2. Hexagons are inscribed in 0.5" radius circle
3. Use your discretion for any sizes not given

SPECIAL RING ADJUSTER

Figure 38.1 Drawing in inches.

Large-scale drawing

This exercise will require the A3PAPER standard sheet to be modified to allow a very large drawing to be drawn (full-size) on A3 paper.

1 Open A3PAPER standard sheet and erase the black border – it is too small for the limits to be used.

2 Make layer 0 current.

3 At the command line enter **MVSETUP <R>** and:

prompt	Enable paper space? (No/Yes)
enter	**N <R>**
prompt	Enter units type [Scientific/Decimal..
enter	**M <R>** – for metric units
prompt	AutoCAD Text Window with Metric scales
and	Enter the scale factor
enter	**50 <R>**, i.e. scale of 1:50
prompt	Enter the paper width and enter: **420 <R>**
prompt	Enter the paper height and enter: **297 <R>**

4 A polyline black border will be displayed. This is our drawing area.

5 Pan the drawing to suit the screen.

6 With the menu bar **Tools-Inquiry-ID Point**, obtain the coordinates of the lower left and upper right corners of the black border. These should be:
   a) lower left: X=0.00 Y=0.00 Z=0.00
   b) upper right: X=21000.00 Y=14850.00 Z=0.00.

7 Set the grid to 40 and the snap to 20.

8 With layer OUT current refer to Fig. 38.2 and complete the factory plan layout using the sizes as given, the start point being at A(2000,2000). The wall thickness is 250.

9 Create the 300 square section column at the point B(16000,4500) then rectangular array it using the information given.

10 With the Dimension Style icon create a new dimension style.
   a) New Style Name: STDLRG
   b) start with: A3DIM
   c) Fit tab: alter **Use overall scale to: 50**
   d) make STDLRG the current dimension style
   e) close the Dimension Style dialogue box.

11 Add appropriate text – but at what height?

12 Save this drawing as **CL\BEGIN\LARGESC** as it will be used to demonstrate the model/paper space concept in a later chapter.

13 If you have access to a plotter/printer, obtain a hard copy on A3 paper – the problem is the plot figures (50 is useful!).

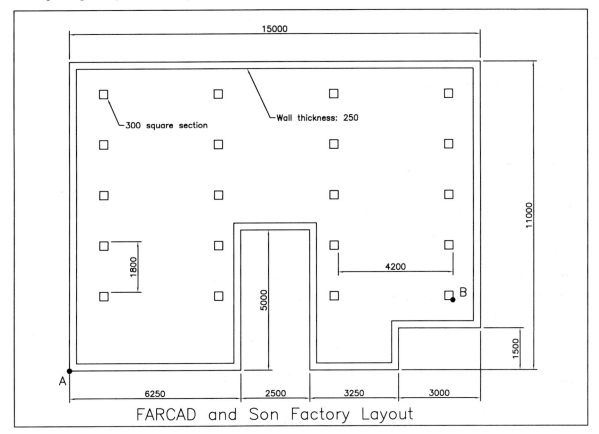

Figure 38.2 Large scale drawing on A3 paper.

Small-scale drawing

This exercise requires the A3 standard sheet to be modified to suit a very small scale drawing, which (as always) will be drawn full size.

1 Open A3PAPER standard sheet and erase the border.

2 Make layer 0 current, enter MVSETUP <R> and:
prompt   Enable paper space? and enter: **N <R>**
prompt   Enter units type and enter: **M <R>** – metric
prompt   Enter the scale factor and enter: **0.001 <R>**
prompt   Enter the paper width and enter: **420 <R>**
prompt   Enter the paper height and enter: **297 <R>**

3 A black border is displayed on the screen – the drawing area.

4 Set UNITS to four decimal places, then ID the lower left and upper right corners
a) lower left: X=0.0000 Y=0.0000 Z=0.0000
b) upper right: X=0.4200 Y=0.2970 Z=0.0000.

5 Set the grid to 0.01 and the snap to 0.005.

6 With layer OUT current, refer to Fig. 38.3 and draw the toggle switch (full size) using the sizes given.

7 Add all dimensions after altering the A3DIM dimension style with:
a) Fit tab: overall scale: 0.001
b) Primary units tab: Linear precision: 0.0000.

8 Add text, but you have to decide on the height.

9 Save as the drawing is complete.

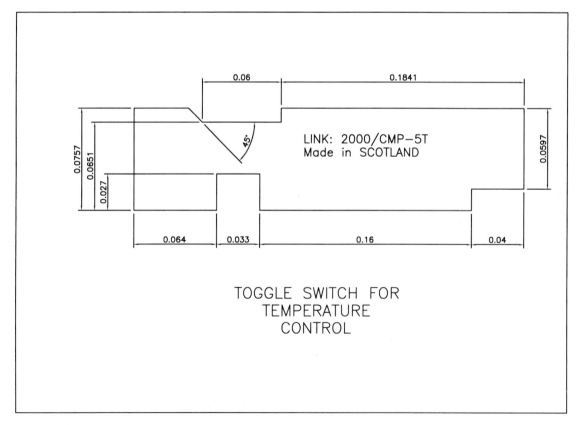

Figure 38.3 Small scale drawing on A3 paper.

Summary

1 Scale drawings require the MVSETUP command to be used.

2 Large and small scale drawings can be 'fitted' onto any size of paper.

3 Dimensions with scaled drawings are controlled with the Overall scale factor in the Dimension Styles dialogue box – Fit tab.
This is equivalent to altering the dimension variable **DIMSCALE.**

4 Text is scaled according to the overall scale factor.

Assignment

A single assignment requiring a large scale drawing to be created on A3 paper. The assignment requires a house plan to be drawn at a scale of 1:50. I would suggest:

1 Open the LARGESC drawing saved earlier in the chapter as it has all the required 'settings'.

2 Erase the factory layout.

3 Complete the house plan using the metric sizes given.

4 Add the text and dimensions, remembering that the drawing is 50 times larger than the A3 paper. This means that the text and dimensions should be 50 times larger than normal. The text height is easy to enter, and the dimension overall scale factor should have been 'set' to 50.

As an extra, modify the dimension style primary tab to Architectural units and note the effect.

Chapter 39

Multilines, complex lines and groups

Layers have allowed us to display continuous, centre and hidden linetypes, but AutoCAD has the facility to display multilines and complex lines, these being defined as:

Multiline: parallel lines which can consist of several line elements of differing linetype. They must be created by the user.

Complex: lines which can be displayed containing text items and shapes. They can be created by the user, although AutoCAD has several 'stored' complex linetypes.

In this chapter we will only investigate a two element multiline and only use the stored AutoCAD complex linetypes. Creating both multilines with several different elements and complex linetypes is outwith the scope of this book.

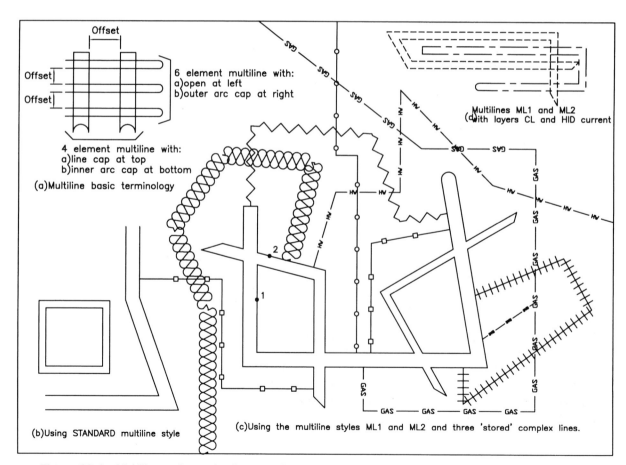

Figure 39.1 Multiline and complex line exercise.

Multilines

Multilines have their own terminology, the basic terms being displayed in Fig. 39.1(a).
To investigate how to use multilines:

1 Open your A3PAPER standard sheet with layer OUT current and refer to Fig. 39.1.

2 Select the MULTILINE icon from the Draw toolbar and:

prompt	`Justification=Top,Scale=??,Style=STANDARD`
then	`Specify start point or [Justification/Scale/STyle]`
enter	**S <R>** – the scale option
prompt	`Enter mline scale<??>`
enter	**10 <R>**
prompt	`Specify start point` and enter: **20,40 <R>**
prompt	`Specify next point` and enter: **@80,0 <R>**
prompt	`Specify next point` and enter: **@70<110 <R>**
prompt	`Specify next point` and enter: **@0,50 <R>**
prompt	`Specify next point` and right-click/Enter.

3 Menu bar with **Draw-Multiline** and:
a) set scale to 5
b) draw a square of side 40 from the point 20,60 using the close option – fig. (b).

4 From the menu bar select **Format-Multiline Style** and:

prompt	`Multiline Styles dialogue box`
respond	1. alter Name to: **ML1**
	2. enter description as: **first attempt**
	3. pick Element Properties
prompt	`Element Properties dialogue box` – Fig. 39.2(a)
respond	note layout then pick OK
prompt	`Multiline Styles dialogue box`
respond	pick Multiline Properties
prompt	`Multiline Properties dialogue box`
respond	1. pick **Line-Start**
	2. pick **Outer arc-End** – Fig. 39.2(b)
	3. pick OK.
prompt	`Multiline Styles dialogue box` – Fig. 39.3
respond	1. pick Add
	2. pick OK.

5 At the command line enter **MLINE <R>** and:

prompt	`Specify start point or [Justification/Scale/STyle]`
enter	**ST <R>** – the style option
prompt	`Enter mline style name` and enter: **ML1 <R>**
prompt	Specify start point or [Justification/Scale/Style]
enter	**S <R>** – the scale option
prompt	`Enter mline scale` and enter: **10 <R>**
prompt	`Specify start point` and enter: **170,170 <R>**
prompt	`Specify next point` and enter: **@0,–100 <R>**
prompt	`Specify next point` and enter: **@150,0 <R>**
prompt	`Specify next point` and enter: **@120<100 <R>**
prompt	`Specify next point` and right-click/Enter.

6 Menu bar with **Format-Multiline Style** and:

prompt Multiline Styles dialogue box
respond 1. alter name to: ML2
  2. enter description as: second attempt
  3. pick Multiline Elements
prompt Element Properties dialogue box
respond note layout then pick OK
prompt Multiline Styles dialogue box
respond pick Multiline Properties
prompt Multiline Properties dialogue box
respond 1. cancel any existing end caps
  2. pick Line-Start
  3. pick Line-End
  4. alter angles to 45 and 30
  5. pick OK
  6. Multiline Styles dialogue returned
  7. pick Add the OK.

(a)

(b)

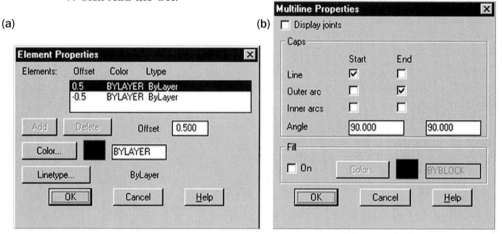

Figure 39.2 The Element and Multiline Properties dialogue boxes.

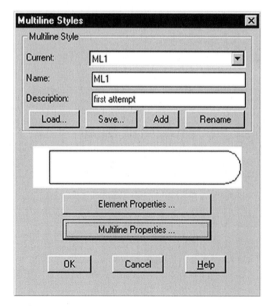

Figure 39.3 The Multiline Styles dialogue box.

7 Activate the MLINE command and:
    prompt   Specify start point or [Justification/Scale/STyle]
    enter   **ST <R>** – the style option
    prompt   Enter mline style name (or ?)
    enter   **? <R>** – the query option
    prompt   AutoCAD Text Window with:

Name	Description
ML1	first attempt
ML2	second attempt
STANDARD	

    respond   **cancel the text window**
    prompt   Enter mline style name and enter: ML2 <R>
    prompt   Specify start point or [Justification/Scale/STyle
    enter   **S <R>** then **8 <R>** – the scale option
    prompt   Specify start point and enter: **210,40 <R>**
    prompt   Specify next point and enter: **@0,80 <R>**
    prompt   Specify next point and enter: **@–80,20 <R>**
    prompt   Specify next point and press <RETURN>.

8 Repeat the MLINE command and:
    a) set the scale to 5
    b) ensure ML2 is current style
    c) draw from: 350,160 to: @100<–150 to: @80<–60 to: <R>

9 Menu bar with **Modify-Object-Multiline** and:
    prompt   Multiline Edit Tools dialogue box – Fig. 39.4
    respond   **pick Open Cross then OK** (middle row left)
    prompt   Select first mline and pick multiline 1
    prompt   Select second mline and pick multiline 2
    prompt   Select first mline
    then   pick as required until all the 'crossed' multilines are opened – fig. (c).

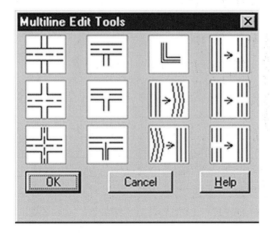

Figure 39.4 The Multiline Edit Tools dialogue box.

10 *Task*
    The multilines created so far have been with layer OUT current. Make layers CL and HID current and draw multilines ML1 and ML2 using your own scale values.

Complex linetypes

1   Continue with the exercise on the screen.

2   Menu bar with **Format-Layer** and create a new layer:
    a) Name: L1
    b) Colour: to suit
    c) Linetype: pick 'Continuous' in L1 layer line and:

prompt	Select Linetype dialogue box
respond	pick Load
prompt	Load or Reload Linetypes dialogue box with **acadiso.lin** file
respond	scroll and pick **GAS_LINE** then OK
prompt	Select Linetype dialogue box
with	GAS_LINE displayed
respond	pick GAS_LINE then OK
prompt	Layer Properties Manager dialogue box
with	layer L1 displayed with GAS_LINE linetype
respond	pick OK.

3   Using step 2 as a guide, create another six layers and load the appropriate linetype using the following information:

Name	Colour	Linetype
L2	to suit	BATTING
L3	to suit	FENCELINE1
L4	to suit	FENCELINE2
L5	to suit	HOT_WATER_SUPPLY
L6	to suit	TRACKS
L7	to suit	ZIGZAG.

4   With each layer current, refer to fig. (c) and draw lines with each of the loaded linetype

5   *Task*
    Using the CHANGE PROPERTIES command (command line CHANGE or icon) use the ltScale option to optimise the appearance of the new added linetypes.

6   The exercise is now complete and can be saved if required.

Groups

A group is a named collection of objects. Groups are stored with a saved drawing and group definitions can be externally referenced. To demonstrate how groups are created and used:

1   Open your standard sheet, layer OUT current and refer to Fig. 39.5.

2   Draw the reference shape using your discretion for any sizes not given. Ensure the lowest vertex is at the point 210,120.

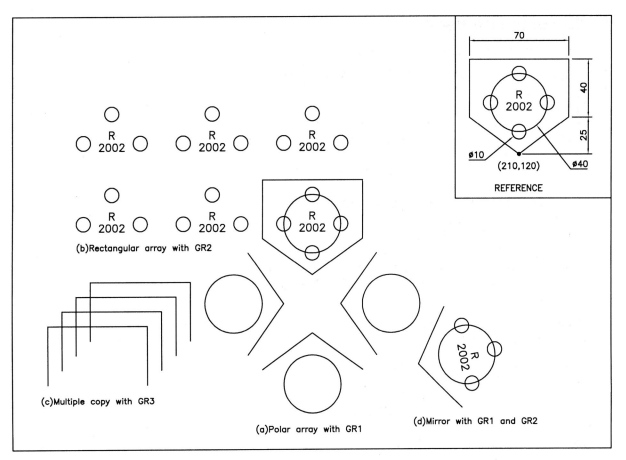

Figure 39.5 Group exercise.

3 At the command line enter **GROUP <R>** and:
prompt Object Grouping dialogue box
respond 1. at Group Name enter: **GR1**
2. at Description enter: **first group**
3. ensure Selectable is ON, i.e. tick in box
4. pick Create Group: **New<**
prompt Select objects for grouping – at the command line
then Select objects
respond pick the two inclined lines and the large circle then right-click
prompt Object Grouping dialogue box
with   *Group Name*   *Selectable*
     GR1   Yes
respond pick OK.

4 At the command line enter **GROUP <R>** and:
prompt Object Grouping dialogue box
respond 1. at Group Name enter: **GR2**
2. at Description enter: **second group**
3. Selectable ON – should be
4. pick Create Group: **New<**
prompt Select objects
respond pick the text and the top three small circles then right-click
prompt Object Grouping dialogue box with GR2 listed?
respond pick OK.

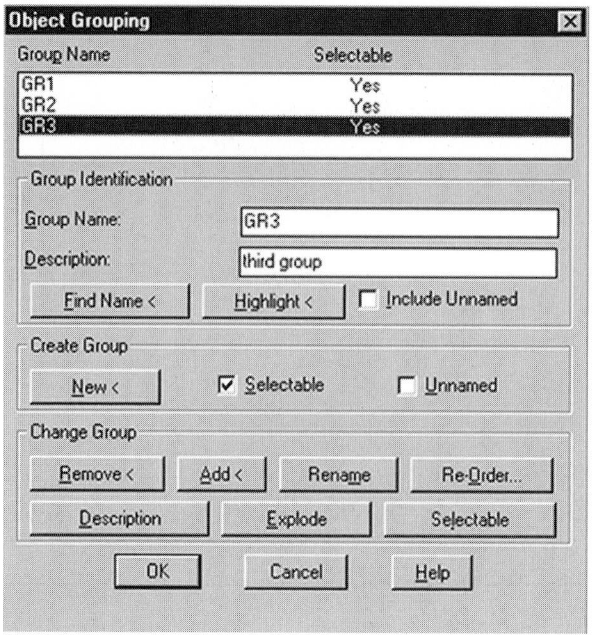

Figure 39.6 The Object Grouping dialogue box.

5 Repeat the GROUP command and:
  a) Name: GR3
  b) Description: third group
  c) Create Group: New<
  d) Objects: pick the two vertical and the horizontal lines
  e) Dialogue box as Fig. 39.6
  f) pick OK.

6 At the command line enter **–ARRAY <R>** and:
prompt	Select objects
enter	**GROUP <R>**
prompt	Enter group name and enter: **GR1 <R>**
prompt	3 found, then Select objects
respond	**right-click**
prompt	Enter type of array and enter: **P <R>**
prompt	Specify center point of array and enter: **210,100 <R>**
prompt	Enter the number of items and enter: **4 <R>**
prompt	Specify the angle to fill and enter: **360 <R>**
prompt	Rotate arrayed objects and enter: **Y <R>**

7 The named group (GR1) will be displayed as fig. (a).

8 Select the ARRAY icon and:
prompt	Array dialogue box
respond	**pick Select objects**
prompt	Select objects at command line
enter	**GROUP <R>** then **GR2 <R><R>** – why two returns?
prompt	Array dialogue box
respond	alter as follows:

    a) Rectangular array
    b) Rows: 2 and Columns: 3
    c) Offsets: Row 55 and Column –70
    d) preview then Accept – fig. (b).

9 With the COPY command:
   a) objects and enter: GROUP <R>
   b) group name and enter: GR3 <R><R>
   c) base point and enter: M <R> – multiple copy option
   d) base point and enter: 175,145 <R> – why these coordinates?
   e) second point and enter: @–150,–100 <R>
   f) second point and enter: @–140,–90 <R>
   g) second point and enter: @–130,–80 <R>
   h) second point and enter: @–120,–70 <R><R> – fig. (c).

10 Select the MIRROR icon and:
   a) select objects and enter: GROUP <R>
   b) group name and enter: GR1 <R>
   c) 3 found then select objects and enter: GROUP <R>
   d) group name and enter: GR2 <R>
   e) 5 found, 8 total then select objects and right-click
   f) first point of mirror line – Center icon and pick bottom circle
   g) second point – Intersection icon and pick top right of border
   h) delete source objects and enter: N <R> – fig. (d).

11 Did your text item mirror? How did mine not?

12 The exercise is complete and can be saved if required.

13 Do not quit the drawing.

Task

1 Erase the original text item – complete group (GR2) erased?

2 Undo this effect with the UNDO icon.

3 Select the explode icon and select any group object and:
   '? were not able to be exploded' displayed at prompt line.

4 At the command line enter GROUP <R> and from the dialogue box:
   a) pick GR2 line
   b) pick Explode
   c) pick OK.

5 Now erase the text item.

Summary

1 Multilines can be created by the user with different linetype and can have several elements in their definition – not considered in this book.

2 Multilines have line or arc end caps.

3 Multilines have their own editing facility.

4 Complex linetypes can be have text or shape items in their definition and can be created by the user – not considered in this book.

5 Groups are collections of objects defined by the user.

6 Once created, a group must be exploded from the Object grouping dialogue box.

7 The GROUP command can only be entered from the keyboard.

Blocks

A block is part of a drawing which is 'stored away' for future recall **within the drawing in which it was created**. The block may be a nut, a diode, a tree, a house or any part of a drawing. Blocks are used when repetitive copying of objects is required, but they have another important feature – text can be attached to them. This text addition to blocks is called **attributes** and will be considered in a later chapter.

Note: remember the statement **within the drawing in which it was created**. We will return to this in a later chapter.

Getting started

1 Open the A3PAPER standard sheet and refer to Fig. 40.1.

2 Draw the house shape using the reference sizes given with:
   a) the outline on layer OUT
   b) the circular windows on layer OUT but green (CHANGE command?)
   c) a text item on layer TEXT
   d) four dimensions on layer DIMS
   e) remember to use your discretion for any size not given.

3 Make layer OUT current

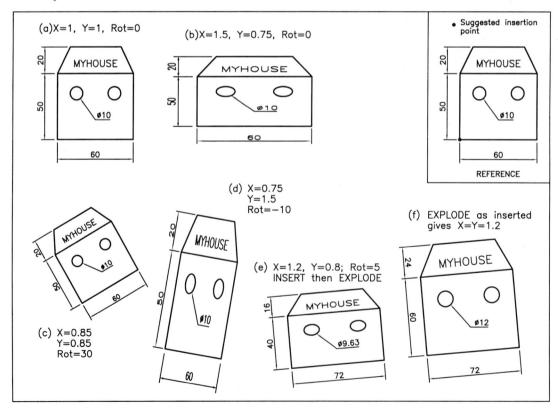

Figure 40.1 First block creation and insertion exercise.

Creating a block

Blocks can be created from both the command line and via a dialogue box. This first exercise will consider the keyboard entry method.

1   At the command line enter **–BLOCK <R>** and:
   prompt   Enter block name or [?]
   enter   **HOUSE <R>**
   prompt   Specify insertion base point
   respond   **Intersection icon and pick lower left corner**
   prompt   Select objects
   respond   **window the house and dimensions**
   prompt   14 found then Select objects
   respond   **right-click.**

2   The house shape may disappear from the screen. If it does, don't panic!

3   *Note:*
   a) the house shape has been 'stored' as a block within the current drawing
   b) this drawing has not yet been saved
   c) the –BLOCK keyboard entry will 'bypass' the dialogue box.

Inserting a block

Created blocks can be inserted into the current drawing by either direct keyboard entry or via a dialogue box. We will investigate both methods with this example.

Keyboard insertion

1   At the command line enter **–INSERT <R>** and:
   prompt   Enter block name or [?]
   enter   **HOUSE <R>**
   prompt   Specify insertion point or [Scale/X/Y/Z/Rotate/PScale..
   and   'ghost' image of block attached to cursor
   enter   **35,200 <R>**
   prompt   Enter X scale factor, specify opposite corner..
   enter   **1 <R>** – an X scale factor of 1
   prompt   Enter Y scale factor <use X scale factor>
   enter   **1 <R>** – a Y scale factor of 1
   prompt   Specify rotation angle<0>
   enter   **0 <R>**

2   The house block is inserted full-size as fig. (a).

3   Repeat the –INSERT command and:
   prompt   Block name and enter: **HOUSE <R>**
   prompt   Insertion point and enter: **145,210 <R>**
   prompt   X scale factor and enter: **1.5 <R>**
   prompt   Y scale factor and enter: **0.75 <R>**
   prompt   Rotation angle and enter: **0 <R>** – fig. (b).

Dialogue box insertion

1 From the Draw toolbar select the INSERT BLOCK icon and:
   prompt   Insert dialogue box
   with   HOUSE as the Name. This is because we entered HOUSE at the command line
               INSERT command.
   respond   ensure that the following **Specify On-screen** prompts are active, i.e. tick in box
               1. Insertion point
               2. Scale
               3. Rotation
               4. pick OK
   prompt   Specify insertion point and enter: **55,75 <R>**
   prompt   Enter X scale factor and enter: **0.85 <R>**
   prompt   Enter Y scale factor and enter: **0.85 <R>**
   prompt   Specify rotation angle and enter: **30 <R>** – fig. (c).

2 Menu bar with **Insert-Block** and:
   prompt   Insert dialogue box
   with   HOUSE as block name
   respond   1. deactivate the three On-screen prompts (no tick)
               2. alter Insertion Point to X: 120, Y: 40, Z: 0
               3. alter Scale to X: 0.75, Y: 1.5, Z: 1
               4. alter Rotation to Angle: –10
               5. dialogue box as Fig. 40.2
               6. pick OK – fig. (d).

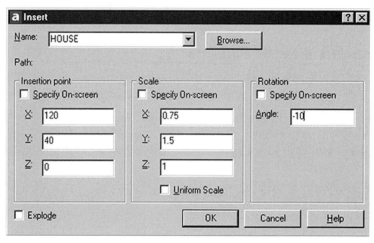

Figure 40.2 The Insert dialogue box.

3 *Notes*
   a) An inserted block is a **single object**. Select the erase icon and pick any point on one of
       the inserted blocks then right-click. The complete block is erased. Undo the erase effect.
   b) Blocks are inserted into a drawing with layers 'as used'. Freeze the DIMS layer and
       turn off the TEXT layer. The four inserted blocks will be displayed without dimensions
       or text. Now thaw the DIMS layer and turn on the TEXT layer.
   c) Blocks can be inserted at varying X and Y scale factors and at any angle of rotation.
       The default scale is X=Y=1, i.e. the block is inserted full size. The default rotation
       angle is 0.
   d) Dimensions which are attached to inserted blocks are not altered if the scale factors
       are changed.
   e) A named block can be redefined and will be discussed later.

Exploding a block

The fact that a block is a single object may not always be suitable to the user, i.e. you may want to copy parts of a block. AutoCAD uses the EXPLODE command to 'convert' an inserted block back to its individual objects.

The explode option can be used:
a) after a block has been inserted
b) during the insertion process.

1   At the command line enter **–INSERT <R>** and:
   prompt   `Block name` and enter: HOUSE
   prompt   `Insertion point` and enter: 220,25
   prompt   `X scale factor` and enter: 1.2
   prompt   `Y scale factor` and enter: 0.8
   prompt   `Rotation angle` and enter: 5

2   Select the EXPLODE icon from the Modify toolbar and:
   prompt   `Select objects`
   respond **pick the last inserted block then right-click.**

3   The exploded block is restored to its individual objects and the dimensions are **scaled to the factors entered** – fig. (e). The individual objects of this exploded block can now be modified if required.

4   Select the INSERT BLOCK icon and using the Insert dialogue box:
   a) Deactivate the three Specify On-screen prompts, i.e. no tick
   b) Insertion Point X: 325; Y: 35; Z: 0
   c) Scale X: 1.2; Y: 0.8; Z: 1
   d) Rotation angle: 5
   e) Explode: ON, i.e. tick and *note the scale factor values*
   f) pick OK.

5   The block is exploded as it is inserted at a scale of X=Y=1.2 and the dimensions display this scale effect – fig. (f).

6   Compare the dimensions of fig. (e) with those of fig. (f) and note the dimension 'orientation' between the two exploded blocks.

7   *Note*
   a) a block exploded after insertion will retain the original X and Y inserted scale factors
   b) a block exploded as it is inserted has X=Y scale factors.

Block exercise

1 Open the A3PAPER standard sheet with layer OUT current.

2 Refer to Fig. 40.3 and draw the four lines using the sizes given. Do not add the dimensions.

Figure 40.3

3 Menu bar with **Draw-Block-Make** and:
   prompt   Block Definition dialogue box
   respond   1. enter Name: **SEAT**
               2. at Base point: **select Pick point**
   prompt   Specify insertion base point at command line
   respond   **midpoint icon and pick line indicated in Fig. 40.3**
   prompt   Block Definition dialogue box
   with   Base point: X and Y coordinates of selected point
   respond   at Objects: **select Select objects**
   prompt   Select objects at command line
   respond   **window the four lines then right-click**
   prompt   Block Definition dialogue box
   with   preview icon displayed
   respond   1. ensure Objects: Delete active
               2. Insert units: scroll and pick Millimeters
               3. at Description enter: **CHAIR block**
               4. pick OK.

4 The four lines should disappear.

5 *Note*
   The Objects part of the Block dialogue box allows the user one of three options, these being:
   a) Retain: will keep the selected objects as individual objects on the screen after the block has been made.
   b) Convert to block: will display the selected objects on the screen as a block after the block has been made.
   c) Delete: will remove the selected objects from the screen after the block has been made – our option.
   d) It is user preference as to what one of these options is used. I generally select delete, although retain is also useful?

6 Draw two 25 radius circles with centres at 100,150 and 250,150. These centre points are important.

7 Select the INSERT BLOCK icon from the Draw toolbar and use the Insert dialogue box to insert the block SEAT twice with:
   a) Insertion point: X=130; Y=150; Z=0
       Scale: X=Y=Z=1
       Rotation: Angle=0
   b) Insertion point: X=280; Y=150; Z=0
       Scale: X=Y=Z=1
       Rotation: Angle=0.

8 Using the ARRAY dialogue box, array the two inserted seat blocks with the following information:

	first array	*second array*
objects:	pick left block	pick right block
type:	Polar	Polar
centre:	100,150	250,150
items:	5	7
angle:	360	360
rotate:	Y	Y.

9 Select the MAKE BLOCK icon from the Draw toolbar and:

prompt	Block Definition dialogue box
respond	1. enter Name: TABLE1
	2. enter Base point as X: 100; Y: 150; Z: 0
	3. at Objects: **pick Select objects**
prompt	Select objects at command line
respond	**window the 5 seat arrangement then right-click**
prompt	Block Definition dialogue box with icon preview display
respond	1. ensure Objects: Delete active
	2. ensure Create icon from block geometry
	3. enter Description: **5 SEATER TABLE**
	4. Insert units: Millimeters
	5. dialogue box as Fig. 40.4
	6. pick OK.

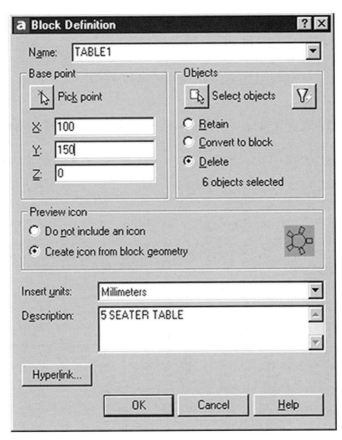

Figure 40.4 The Block Definition dialogue box for the 5 seater.

10 Repeat the Make Block icon selection and:
   a)name: enter TABLE2
   b) base point: enter X: 250; Y: 150; Z: 0
   c) pick select objects
   d) window the seven seat arrangement then right-click
   e) delete objects active
   f) description: enter 7 SEAT TABLE
   g) insert units: Millimeters
   h) pick OK.

11 Now erase any objects from the screen.

12 Menu bar with Insert-Block and:
   prompt   Insert dialogue box
   respond   1. scroll at Name and pick TABLE1
             2. Insertion point: X 60; Y 240; Z 0
             3. Scale: X 0.75; Y 0.75; Z 1
             4. Rotation: Angle 0
             5. pick OK.

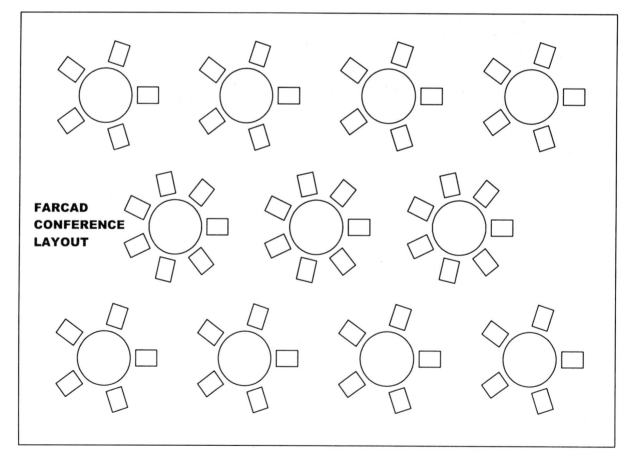

Figure 40.5 Block exercise.

13 Repeat the Insert-Block sequence with:
   a) name: TABLE2
   b) insertion point: X 110; Y 150; Z 0
   c) scale: X 0.75; Y 0.75; Z 1
   d) rotation: 0
   e) pick OK.

14 Menu bar with **Modify-Array** and using the Array dialogue box:
   a) objects: pick any point on the 7 table block and right-click
   b) type: Rectangular array
   c) Rows: 1 and Columns: 3
   d) Column offset: 100 and Row offset: 0 or 1 (which?)
   e) Preview and Accept.

15 Now Rectangular array the 5 table block:
   a) for 2 rows and 4 columns:
   b) row offset: −180
   c) column distance: 100.

16 The final layout should be as Fig. 40.5 and can be saved.

Notes on blocks

Several points are worth discussing about blocks:

1 *The insertion point.*
   When a block is being inserted using the Insert dialogue box, the user has the option to decide whether the parameters are entered via the dialogue box or at the command line. The Specify On-screen selection allows this option.

2 *Exploding a block.*
   I would recommend that blocks are inserted before they are exploded. This maintains the original X and Y scale factors. Remember that blocks do not need to be exploded.

3 *Command line.*
   Entering −BLOCK and −INSERT will 'byepass' the dialogue boxes and allow all parameters to be entered from the keyboard.

4 *The ? option.*
   Both the command line −BLOCK and −INSERT commands have a query (?) option which will list all the blocks in the current drawing. At the command line enter **−BLOCK <R>** and:
prompt	Enter block name (or ?) and enter: **? <R>**
prompt	Enter block(s) to list<*> and enter: *** <R>**
prompt	AutoCAD Text Window
with	Defined blocks
	"SEAT"
	"TABLE1"
	"TABLE2"

User Blocks	External Blocks	Dependent Blocks	Unnamed Blocks
3	0	0	0.

5 *Making a block.*
   Blocks can be created by command line entry or by a dialogue box and the following is interesting:
   a) command line: the original shape disappears from the screen
   b) dialogue box: options are available to retain the shape.

6   *Nested blocks.*

These are 'blocks within blocks' and to demonstrate this concept:

a) With the menu bar selection **Tools-Inquiry-List** then pick any 7 seat table and:

    prompt   `AutoCAD text Window`
    with   details about the selected object

b) This information tells the user that the selected object is a block with name TABLE2, created on layer OUT – Fig. 40.6(a).

c) With the EXPLODE icon, pick the previous selected 7 seat table then right-click

d) With the LIST command, pick any seat of the exploded block and the text window will display details about the object, i.e. it is a block with name SEAT, created on layer OUT – Fig. 40.6(b).

e) This is an example of a nested block, i.e. block SEAT is contained within block TABLE2.

f) If you exploded one of the seats of the exploded table, what would you expect to be displayed with the LIST command?

```
(a)          BLOCK REFERENCE   Layer   "OUT"
                    Space   Model space
             Handle = 3E4
             "TABLE2"
          at point, X=    110.00   Y=    150.00  Z=      0.00
             X scale factor       0.75
             Y scale factor       0.75
      rotation angle      0.0
             Z scale factor       1.00


(b)          BLOCK REFERENCE   Layer   "OUT"
                    Space   Model space
             Handle = 3F1
             "SEAT"
          at point, X=    132.50   Y=    150.00  Z=      0.00
             X scale factor       0.75
             Y scale factor       0.75
      rotation angle      0.0
             Z scale factor       1.00
```

Figure 40.6   The AutoCAD Text Window list for NESTED blocks.

Using blocks

1 Open your A3PAPER standard sheet with layer OUT current.

2 Refer to Fig. 40.7 and draw a 20 sided square with diagonals and then make a block with:
  a) name: BL1
  b) insertion point: diagonal intersection
  c) objects: window the shape – no dimensions.

3 Draw an inclined line, a circle, an arc and a polyline shape of line and arc segments – discretion for sizes, but use Fig. 40.7 as a guide for the layout.

4 The DIVIDE command.
  Menu bar with **Draw-Point-Divide** and:
  prompt   Select object to divide
  respond   **pick the line**
  prompt   Enter the number of segments or [Block] and enter: **B <R>**
  prompt   Enter name of block to insert and enter: **BL1 <R>**
  prompt   Align block with object and enter: **N <R>**
  prompt   Enter the number of segments and enter: **4 <R>** – fig. (a).

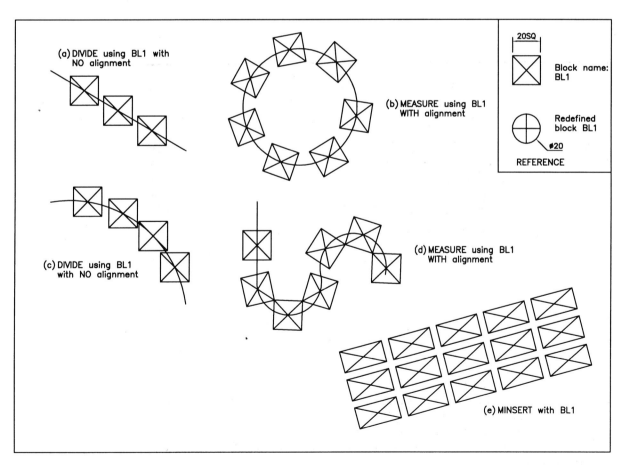

Figure 40.7 Using blocks exercise.

5 The MEASURE command.
    Menu bar with **Draw-Point-Measure** and:
    prompt   Select object to measure and: **pick the circle**
    prompt   Specify length of segment or [Block] and enter: **B <R>**
    prompt   Enter name of block to insert and enter: **BL1 <R>**
    prompt   Align block with object and enter: **Y <R>**
    prompt   Specify length of segment and enter: **35 <R>** – fig. (b).

6 Using the DIVIDE and MEASURE commands:
    a) divide the arc: with block BL1, no alignment and with five segments – fig. (c)
    b) measure the polyline: with block BL1, with alignment and with a segment length
        of 35 – fig. (d).

7 The MINSERT or multiple insert command.
    At the command line enter **MINSERT <R>** and:
    prompt   Enter block name and enter: **BL1 <R>**
    prompt   Specify insertion point and **pick a suitable point on screen**
    prompt   Enter X scale factor and enter: **1.5 <R>**
    prompt   Enter Y scale factor and enter: **0.75 <R>**
    prompt   Specify rotation angle and enter: **15 <R>**
    prompt   Enter number of rows and enter: **3 <R>**
    prompt   Enter number of columns and enter: **5 <R>**
    prompt   Enter distance between rows and enter: **20 <R>**
    prompt   Specify distance between columns and enter: **35 <R>**

8 The block BL1 is arrayed in a 3x5 rectangular matrix – fig. (e).

9 Using the EXPLODE icon, pick any of the minsert blocks and:
    prompt   1 was minserted
    i.e. you cannot explode minserted blocks.

10 *Redefining a block*
    a) Draw a circle of radius 10 and add the vertical and horizontal diagonals.
    b) Change the colour of the circle and lines to blue.
    c) At the command line enter **–BLOCK <R>** and:
        prompt   Enter block name and enter: **BL1 <R>**
        prompt   Block "BL1" already exist. Redefine it? [Yes/No]<N>
        enter   **Y <R>**
        prompt   Specify insertion base point
        respond **pick the centre of the circle**
        prompt   Select objects
        respond **window the circle and lines then right-click.**
    d) Interesting result? – all the red squares should be replaced by blue circles with the
        same alignment, scales, etc.

Layer 0 and blocks

Blocks have been created with the objects drawn on their 'correct' layers. Layer 0 is the AutoCAD default layer and can be used for block creation with interesting results. We will use the command line entry to demonstrate this example.

1 Erase all objects from the screen and draw:
   a) a 50 unit square on layer 0 – black
   b) a 20 radius circle inside the square on layer OUT – red
   c) two centre lines on layer CL – green.

2 With the command line entry –BLOCK, make a block of the complete shape with block name TRY, using the circle centre as the insertion point.

3 Make layer OUT current and with the command line –INSERT, insert block TRY at 55,210, full size with no rotation. The square is inserted with red continuous lines.

4 Make layer CL current and –INSERT block TRY at 135,210, full size and with no rotation. The square has green centre lines.

5 With layer HID current, –INSERT the block at 215,210, full size with no rotation and the square will be displayed with coloured hidden lines.

6 Make layer 0 current and –INSERT at 295,210 – square is black.

7 With layer 0 still current, freeze layer OUT and:
   a) no red circles
   b) no red square from first insertion – when layer OUT was current.

8 Thaw layer OUT and insert block TRY using the Insert dialogue box with:
   a) explode option active, i.e. tick in box
   b) insertion point: 175,115
   c) full size with 0 rotation
   d) square is black – it is on layer 0.

9 Explode the first three inserted blocks and the square should be black, i.e. it has been 'transferred' to layer 0 with the explode command.

10 Finally make layer OUT current and freeze layer 0 and:
    a) no black squares
    b) no objects from fourth insertion – why?

11 This completes this block exercises – no save necessary.

Task

1 Open the **LARGSC** drawing created in Chapter 38.

2 Refer to Fig. 40.8 and erase the arrayed square columns.

3 With layer OUT current, modify the factory outline, positioning the circular outline at the midpoint of the right vertical line.

4 Draw the BEAM shape using the sizes given at any suitable part of the screen then make a block of the shape with name BEAM, using the insertion point indicated

5 Insert block BEAM twice using the following:

	first	*second*
insertion point	4000,3750	16750,5000
scale	full size	full size
rotation	45	0.

6 Rectangular array the first inserted block:
a) for 4 rows and 4 columns
b) row offset 2500 and column offset 3250.

7 Polar array the second inserted block:
a) about the circular outline centre
b) 5 items
c) 180 degree angle to fill with rotation.

8 Erase the two columns wrongly positioned.

9 Now save the layout as **C:\BEGIN\LARGESCMOD.**

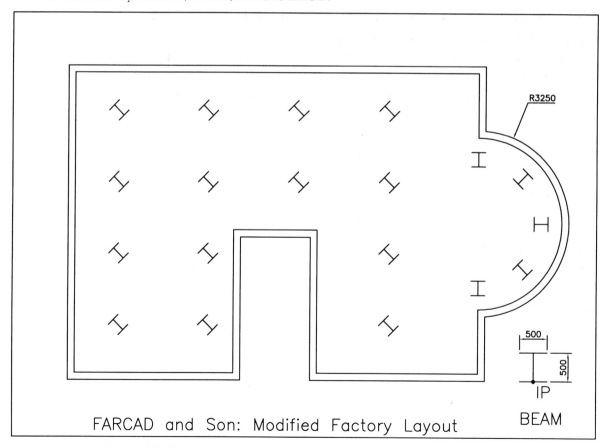

Figure 40.8 The modified large scale drawing with block insertions.

Summary

1 A block is a single object 'stored' for recall in the drawing in which it was created. We will return to this statement in a later chapter.

2 Blocks are used for the insertion of frequently used 'shapes'.

3 An inserted block is a single object.

4 Blocks can be inserted full size (X=Y) or with differing X and Y scale factors and varying rotation angles.

5 The explode command 'restores' an inserted block to its original objects.

6 The explode command can be used after insertion or during insertion.

7 Blocks are inserted with 'layers intact'.

8 Blocks can be used with the divide and measure commands.

9 Multiple block inserts are permissible with the MINSERT command. The insertion gives a rectangular array pattern.

10 Existing blocks can be redefined. The current drawing will be 'updated' to display the new block definition.

11 Unused blocks can be purged from a drawing.

Assignment

One activity requiring blocks to be created has been included for you to attempt.

Activity 36: In Line Cam and Roller Follower
1 Draw the two shapes using the reference sizes given, using your discretion for any omitted size.

2 Make a block of each shape with the block names CAM and FOL. Use the insertion point given.

3 Insert the CAM block using the information given.

4 Insert the FOL block:
   a) insertion point: 40,120
   b) X and Y scale: 0.75 with 0 rotation.

5 Rectangular array the inserted FOL block:
   a) for 1 row and 8 columns
   b) column offset: 50.

6 The followers have to be moved vertically downwards until they 'touch' the cams. This sounds easier than it may seem.

7 Complete the assignment task.

Chapter 41

WBLOCKS

Blocks are useful when shapes/objects are required for repetitive insertion in a drawing, but they are **drawing specific**, i.e. they can only be used within the drawing in which they were created. There are however, blocks which can be created and accessed by all AutoCAD users, i.e. they are **global**. These are called WBLOCKS (write blocks) and they are created in a similar manner to 'ordinary' blocks. Wblocks are stored and recalled from a 'named folder/directory' which we will assume to C:\BEGIN.

Creating Wblocks

1 Open your A3PAPER standard sheet with layer OUT current.

2 Refer to Fig. 41.1 and draw a rectangular polyline shape using the fig. (a) information:
 a) the start point is to be 5,5
 b) the polyline has to have a constant width of 5
 c) close the shape with the close option
 d) the rectangle sizes will become apparent during insertion.

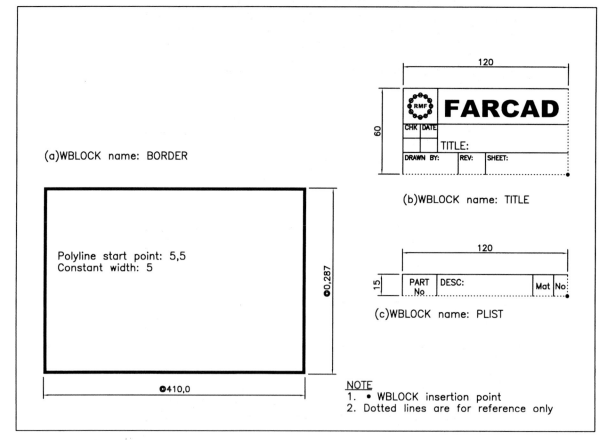

Figure 41.1 Layout and sizes for creating the WBLOCKS.

3   At the command line enter **WBLOCK <R>** and:

prompt   Write Block dialogue box – similar to Block dialogue box

respond   1. Source: ensure Objects active (dot)
            2. Base point: alter X: 2.5; Y:2.5; Z: 0
            3. Objects: pick Select objects

prompt   Select objects at the command line

respond   **pick any point on the polyline then right-click**

prompt   Write Block dialogue box

respond   1. Objects: Delete from drawing active
            2. Description:
               a) File name: BORDER
               b) Location: C:\BEGIN
               c) Insert units: millimeters – Fig. 41.2
            3. pick OK

prompt   as the file is created, the WBLOCK preview is displayed.

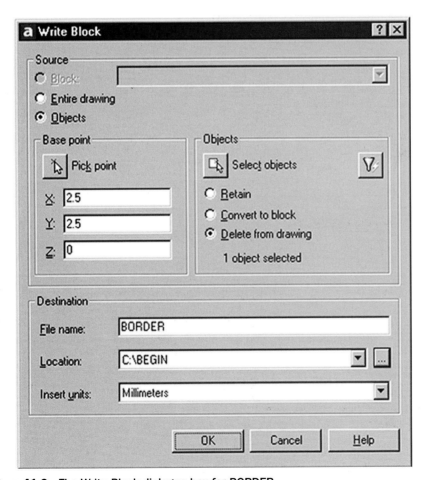

Figure 41.2   The Write Block dialogue box for BORDER.

4   Construct the title box from the information in fig. (b), the actual detail being to your own design, e.g. text style, company name and logo, etc. The only requirement is to use the given sizes – the donut and dotted lines are for 'guidance' only and should not be drawn. Use layers correctly for lines, text, etc.

5   Command line with **WBLOCK <R>** and:
      prompt   Write Block dialogue box
      respond   1. ensure Source: Objects
                  2. Base point: select Pick points
      prompt   Specify insertion base point at command line
      respond   pick point indicated by donut then right-click
      prompt   Write Block dialogue box
      respond   Objects: pick Select objects
      prompt   Select objects at command line
      respond   window the title box then right-click
      prompt   Write Block dialogue box
      respond   at Description alter:
                  1. File name: TITLE
                  2. Location: C:\BEGIN
                  3. Objects: Delete from drawing active
                  4. Insert units: millimeters
                  5. pick OK
      and   WBLOCK preview of saved file.

6   Create the parts list table headings using the basic layout in fig. (c). Again use your own design for text, etc.

7   With the WBLOCK command:
      a) Source: Objects
      b) Base point: Pick points and pick point indicated
      c) Objects: Select objects and window the shape
      d) File name: PLIST
      e) Location: C:\BEGIN
      f) Insert units: millimeters
      g) pick OK.

8   Now proceed to the next part of the exercise.

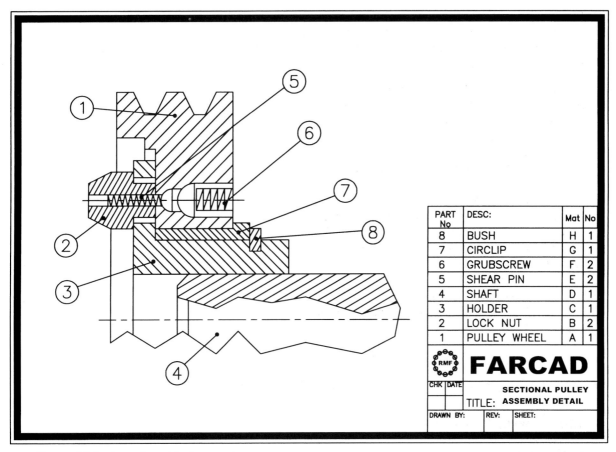

Figure 41.3 Inserting wblocks exercise.

Inserting Wblocks

1 Open A3PAPER – no to any save changes prompt – why?

2 Refer to Fig. 41.3 and draw the sectional pulley assembly. No sizes have been given, but you should be able to complete the drawing – use the snap to help. Add the hatching and the 'balloon' effect using donut-line-circle-middled text.

3 Select the INSERT icon from the Draw toolbar and:

prompt	Insert dialogue box
respond	**pick Browse**
prompt	Select Drawing File dialogue box
respond	1. scroll and pick C:\BEGIN
	2. scroll and pick BORDER
	3. pick Open
prompt	Insert dialogue box with BORDER listed
respond	1. ensure Insertion point, Scale and Rotation are Specific to On-screen, i.e. tick in three boxes
	2. pick OK
prompt	Specify insertion point and enter: **2.5,2.5 <R>**
prompt	Enter X scale and enter: **1 <R>**
prompt	Enter Y scale and enter: **1 <R>**
prompt	Specify rotation angle and enter: **0 <R>**

4   The polyline border should be inserted within the 'drawing frame' of the A3PAPER standard sheet.

5   Menu bar with **Insert-Block** and:

prompt	Insert dialogue box
respond	**pick Browse**
prompt	Select Drawing File dialogue box
respond	1. scroll and pick C:\BEGIN
	2. scroll and pick TITLE
	3. pick Open
prompt	Insert dialogue box
respond	1. cancel the three on-screen prompts – no tick
	2. Alter insertion point to X: 412.5; Y: 7.5; Z: 0
	3. ensure scale:- X: 1; Y: 1; Z: 1
	4. rotation:- angle: 0
	5. pick OK.

6   The wblock TITLE will be inserted into the lower right corner of the BORDER wblock.

7   At the command line enter **–INSERT <R>**

prompt	Enter block name and enter: **PLIST <R>**
prompt	Specify insertion point and enter: **412.5,67.5 <R>**
prompt	Enter X scale and enter: **1 <R>**
prompt	Enter Y scale and enter: **1 <R>**
prompt	Specify rotation angle and enter: **0 <R>**

8   The parts list wblock will be added 'on top' of the title box.

9   Now complete the drawing by:
     a) adding a parts list similar to that shown
     b) complete the title box.

10   When these additions have been completed, save the drawing.

11   *Notes*
     a) Three saved wblocks (BORDER, TITLE and PLIST) have been inserted into the A3PAPER drawing, these wblocks having been 'stored' in the C:\BEGIN folder. If the computer is networked, then all CAD users could access these three wblocks
     b) The border and title box could be permanently added to the A3PAPER standard sheet – you decide on this?
     c) The three wblocks could be used in every exercise and activity from this point onwards – again you decide on this.

About WBLOCKS

*** EVERY DRAWING IS A WBLOCK AND EVERY WBLOCK IS A DRAWING ***

The above statement is true – think about it!

1   Open your A3PAPER standard sheet and refer to Fig. 41.5 which displays five saved activity drawings, inserted at varying scales and rotation angles. The drawings are:

Drawing	IP	Xscale	Yscale	Rot
ACT8	5,5	0.5	0.5	0
ACT19	25,155	0.4	0.3	5
ACT25	220,5	0.25	0.5	0
ACT29	330,5	0.15	0.75	0
ACT31	200,190	0.2	0.3	–5

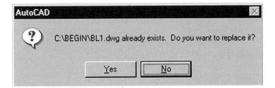

Figure 41.4 The AutoCAD message box for WBLOCKS having the same name as a drawing.

2   Note that I have frozen layers TEXT and DIMS for clarity.

3   Insert some other previously saved drawings.

4   When creating a wblock, if the path name entered (drive-folder-file name) has already been used, AutoCAD will display a message similar to Fig. 41.4.

5   *WBLOCK options.*
    When a wblock is being created, the user has three options for selecting the source. These are:
    a) Block: allows a previously created block to be 'converted' to a wblock.
    b) Entire drawing: the user can enter a new file name for the existing drawing.
    c) Objects: allows parts of a drawing to be saved as a wblock, i.e. as a drawing file. This is probably the most common source selection method.

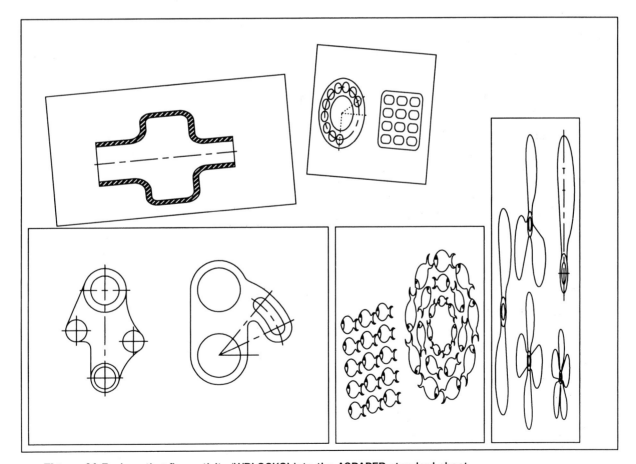

Figure 41.5 Inserting five activity 'WBLOCKS' into the A3PAPER standard sheet.

Exploding wblocks

1 Open A3PAPER and draw anywhere on the screen a circle of radius 25 and a square of side 50 – do not use the rectangle command.

2 Make WBLOCKS of the circle and square using:
   a) circle: Base point at the circle centre
                File name: CIR
   b) square: Base point at any corner of the square
                File name: SQ.

3 Now close the drawing (no save) and re-open A3PAPER

4 Menu bar with Insert-Block and:
   a) Browse and select CIR from C:\BEGIN
   b) Insert at a suitable point, full size with 0 rotation.

5 Menu bar with Insert-Block and:
   a) Browse and pick SQ from your named folder
   b) Pick Explode from the dialogue box and note the scale
   c) Insert at any suitable point – no rotation.

6 At the command line enter **–BLOCK <R>** and:
   prompt   Enter block name or [?]
   enter   **? <R>** – the query option
   prompt   Enter block(s) to list <*>
   enter   **<R>**
   prompt   AutoCAD Text Window
   with   Defined blocks: "CIR"

User Blocks	External References	Dependent Blocks	Unnamed Blocks
1	0	0	0

   respond cancel the text window.

7 With **LIST <R>** at the command line:
   a) pick the circle and:
      BLOCK REFERENCE listed for the object selected
   b) pick any line of square and:
      LINE listed for the selected object.

8 Menu bar with **Insert-Block** and only CIR will be listed

9 This exercise has demonstrated that:
   a) unexploded wblocks become blocks within the current drawing
   b) exploded wblocks do not become blocks.

10 This exercise is complete. Do not save.

Summary

1 Wblocks are global and could be accessed by all AutoCAD users.

2 Wblocks are usually saved to a named folder.

3 Wblocks are created with the command line entry WBLOCK <R> which results in the Write Block dialogue box.

4 Wblocks are inserted into a drawing in a manner similar to ordinary blocks, the user selecting:
   a) the named folder
   b) the drawing file name.

5 Wblocks can be exploded after/during insertion.

6 Unexploded wblocks become blocks in the current drawing.

7 All saved drawings are WBLOCKS.

Assignment

A single activity for you to attempt.

Activity 37: Coupling arrangement.
1 Complete the drawing using your discretion for sizes not given. I have deliberately not given all sizes in this drawing.

2 Insert the wblock BORDER at 5,5 – full size with no rotation.

3 Insert the wblock TITLE at 412.5,7.5 and customise to suit.

4 Add all text and dimensions.

5 Save?

Chapter 42

Attributes

An attribute is an item of text attached to a block or a wblock and allows the user to add repetitive type text to frequently used blocks when they are inserted into a drawing. The text could be:

a) weld symbols containing appropriate information
b) electrical components with values
c) parts lists containing coded, number off, material, etc.

Attributes used as text items are useful, but their main advantage is that attribute data can be **extracted** from a drawing and stored in an attribute extraction file. This data could then be used as input to other computer packages, e.g. databases, spreadsheets, etc. for creating a Bill-of-Material, an Inventory for example.

This chapter is only a 'taster' as the topic will not be investigated fully. The editing and extraction features of attributes are beyond the scope of this book. The purpose of this chapter is to introduce the user to:

a) attaching attributes to a block
b) inserting an attribute block into a drawing.

Getting started

The attribute example for demonstration is a fisherman's trophy and will use a previously created and saved drawing. The fish 'symbol' will represent the block for adding the attributes. The added attributes will give information about the type of fish, the year it was caught and the river.

1 Open the C:\BEGIN\FISH drawing created during Chapter 34 and refer to Fig. 42.1.

2 Erase any centre lines, text and dimensions.

3 *a*) Move the complete shape from 'the nose' to the point 50,50.
   b) Scale the shape about the point (50,50) by 0.5 – fig. (a).
   c) Note that the 50,50 point is essential for positioning the text items.

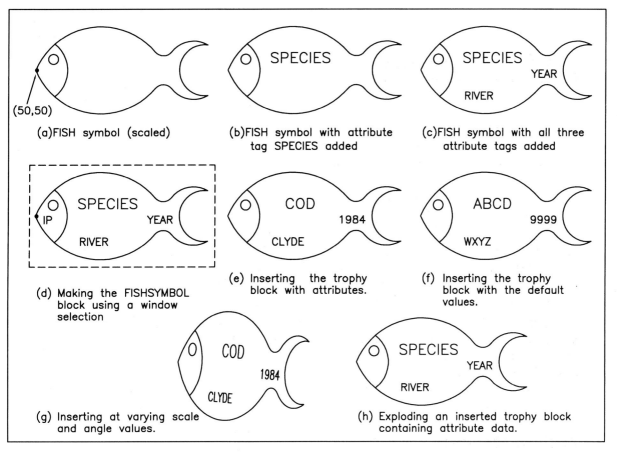

Figure 42.1 Making and using attributes with the block TROPHY.

Defining the attributes

Before a block containing attributes can be inserted into a drawing, the attributes must be defined, so:

1 Make layer TEXT current.

2 Menu bar with **Draw-Block-Define Attributes** and:

prompt   Attribute Definition dialogue box

with   selections for Mode, Attribute, Insertion point and Text

respond   with the following:

     1. *At Mode:*

        leave all four options un-selected – no tick

     2. *At Attribute enter:*

        a) Tag: SPECIES

        b) Prompt: What type of fish displayed?

        c) Value: ABCD

     3. *At Insertion point enter:*

        a) X: 95; Y: 60; Z: 0

     4. *At Text options alter:*

        a) Justification: scroll and pick Center

        b) Text Style: ST1

        c) Height: 7

        d) Rotation: 0 – dialogue box as Fig. 42.2

then   pick OK.

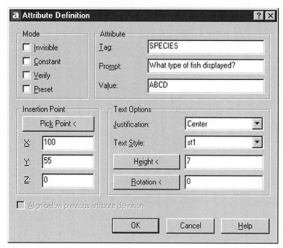

Figure 42.2　Attribute Definition dialogue box.

3　The attribute tag SPECIES will be displayed in the fish symbol as Fig. 42.1(b)

4　Activate the Attribute Definition command two more times and enter the following attribute information in the Attribute Definition dialogue box in the same way as step 2

	First entry	*Second entry*
Attribute modes	blank	blank
Attribute tag	RIVER	YEAR
Attribute prompt	Where caught?	What was the year?
Default value	WXYZ	9999
Insertion Pt X	80	145
Insertion Pt Y	30	45
Insertion Pt Z	0	0
Justification	Left	Right
Text style	ST1	ST1
Height	5	5
Rotation	0	0.

5　When all the attribute information has been entered, the fish symbol will display the three tags – fig. (c).

6　*Note:*

a) When attributes are used for the first time, the word Tag, Prompt and Value can cause confusion. The following description may help to overcome this confusion:

　1. tag:　　is the actual attribute 'label' which is attached to the drawing at the specified text start point. This tag item can have any text style, height and rotation.

　2. prompt:　is an aid to the user when the attribute data is being entered with the inserted block

　3. value:　is an artificial name/number for the attribute being entered. It can have any alpha-numeric value.

b) The Insertion point in the Attribute Definition dialogue box refers to the attribute text tag and not to a block.

7　In our first attribute definition sequence, we created the SPECIES tag with the following attribute information:

a) Tag: SPECIES.

b) Prompt: What type of fish displayed?

c) Default value: ABCD.

d) Text insertion point for SPECIES, centred on 100,55 with height 7 and 0 rotation angle.

Creating the attribute block

1   Menu bar with **Draw-Block-Make** and:
    prompt   Block Definition dialogue box
    respond   enter/activate the following:
        a) Name: TROPHY
        b) Base point: X: 50 Y: 50 Z: 0
        c) Objects: Select and window symbol and attributes as fig. (d) then right-click
        d) Objects: Delete active
        e) Preview: Create icon from Block geometry active
        f) Insert units: Millimeters
        g) Description: TROPHY block with three attributes
    and   dialogue box as Fig. 42.3
    then   pick OK.

2   The symbol and attributes have been made into a block and should disappear from the screen as we activated this option from the dialogue box.

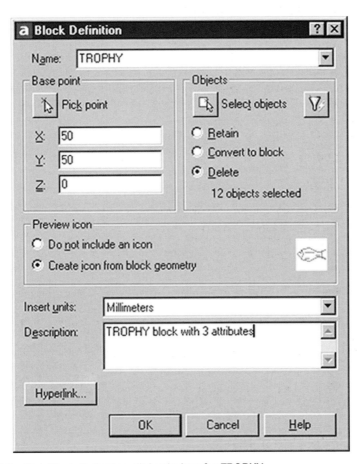

Figure 42.3   The Block Definition dialogue box for TROPHY.

Testing the created block with attributes

Now that the block with attributes has been created, we want to 'test' the attribute information it contains. This requires the block to be inserted into the drawing, and this will be achieved with both command line and dialogue box entries.

1 Make layer OUT current.

2 At the command line enter ATTDIA <R> and:

prompt	Enter new value for ATTDIA <?>
enter	**0 <R>**

3 ATTDIA is a system variable, and when set to 0 will only allow attribute values to be entered from the keyboard.

4 At the command line enter **–INSERT <R>** and:

prompt	Enter block name and enter: **TROPHY <R>**
prompt	Specify insertion point and: **pick any point to suit**
prompt	Enter X scale and enter: **1 <R>**
prompt	Enter Y scale and enter: **1 <R>**
prompt	Specify rotation angle and enter: **0 <R>**
prompt	What type of fish displayed?<ABCD> and enter: **COD<R>**
prompt	Where caught?<WXYZ> and enter: **CLYDE <R>**
prompt	What was the year?<9999> and enter: **1984 <R>**

4 The fish trophy symbol will be displayed with the attribute information as fig. (e).

5 *Note*

a) the prompt and defaults values are displayed as entered

b) the order of the last three prompt lines (i.e. type, caught and year) may not be in the same order as mine. Don't worry if they are not the same.

6 Now insert the trophy block twice more with –INSERT from the command line using:

a) at any suitable point, full size with 0 rotation and accept the default values, i.e. right-click or <R> at the prompt line – fig. (f)

b) at another point on the screen with the X scale factor as 0.75, the Y scale factor as 1.25, the rotation angle –5. Use the same attribute entries as step 4, i.e. COD, CLYDE and 1984. The result should be as fig. (g).

7 Explode any inserted block which contains attribute information and the tags will be displayed – fig. (h).

8 We are now ready to insert the 'real' attribute data.

Attribute information

The fisherman's trophy cabinet contains five prime examples of what he has caught over the past few years, and each catch is represented in the trophy cabinet by the block symbol containing the appropriate attribute information. The attribute data to be displayed is:

Species	*River*	*Year*
SALMON	SPEY	1995
TROUT	TAY	1996
PIKE	DART	1997
CARP	NENE	1998
EELS	DERWENT	1999.

Attribute information can be added to an inserted block:
a) from the keyboard – as previous example
b) via a dialogue box which will now be discussed.

1   Erase all objects from the screen and make layer OUT current

2   At the command line enter **ATTDIA <R>** and:
   prompt   Enter new value for ATTDIA<0>
   enter   **1 <R>**

3   Menu bar with **Insert-Block** and:
   prompt   Insert dialogue box
   with   Block name: TROPHY – from previous insertion
   respond   1. ensure all on-screen prompts not active, i.e. no tick
              2. insertion point:- X: 40; Y: 230; Z: 0
              3. scale: X: 1.2; Y: 1.2; Z: 1
              4. rotation: angle 0
              5. pick OK
   prompt   Edit Attributes dialogue box
   with   Entered prompts and default values as Fig. 42.4(a)
   respond   1. alter What type to: SALMON
              2. alter Where caught to: SPEY
              3. alter What year to: 1995
              4. dialogue box as Fig. 42.4(b)
              5. pick OK.

4   The trophy block will be inserted with the attribute information displayed.

5   Using step 3 as a guide with the attribute data listed above, refer to Fig. 42.5 and insert the TROPHY block to complete the cabinet – use your imagination with the scales

6   Complete the cabinet and save?

This completes our brief 'taster' into attributes.

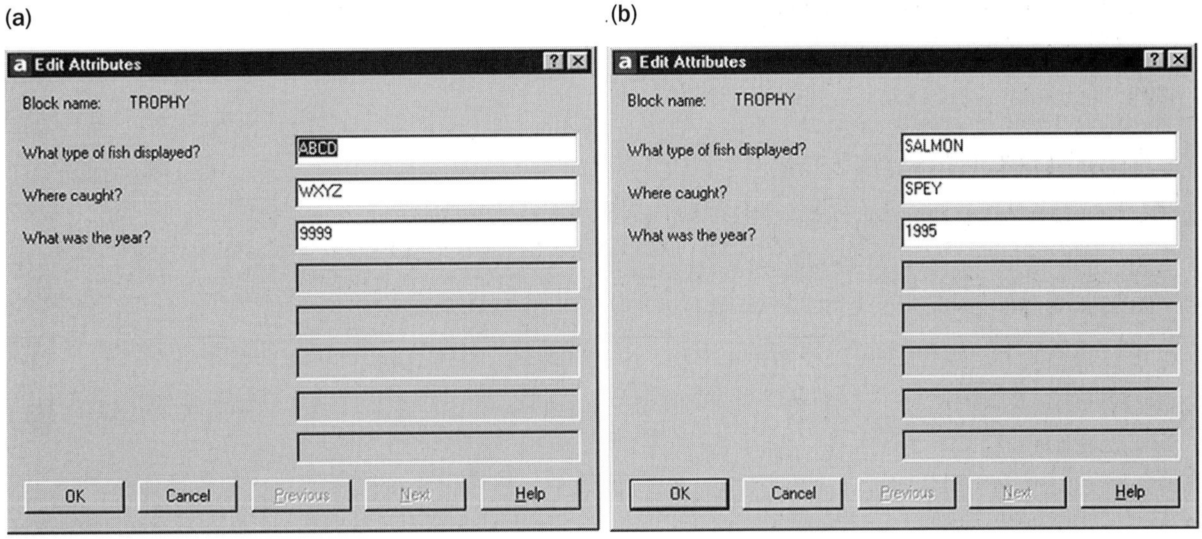

(a)   .(b)

Figure 42.4   The Edit Attributes dialogue box.

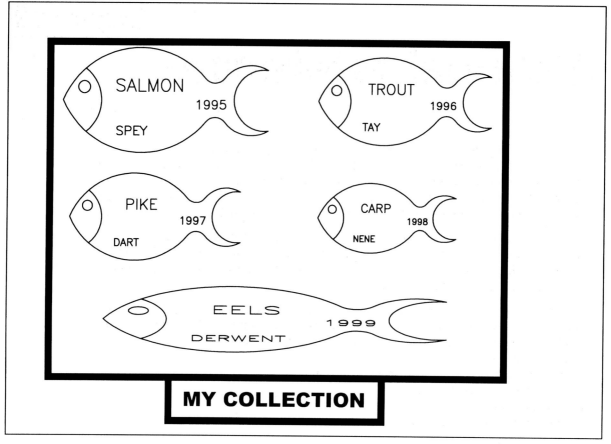

Figure 42.5 The fisherman's cabinet using block TROPHY.

Point of interest?

In the previous chapter we created a title box as a wblock. This title box had text items attached to it, e.g. drawing name, date, revision, etc. These text items could have been made as attributes. When the title box was inserted into a drawing, the various text items (attributes) could have been entered to the drawing requirements. Think about this application of attributes!

Summary

1 Attributes are text items added to BLOCKS or WBLOCKS.

2 Attribute must be defined by the user.

3 Attribute data is added to a block when it is inserted into a drawing.

4 Attributes can be edited and extracted from a drawing, but these topics are beyond the scope of this book.

Chapter 43

External references

Wblocks contain information about objects, colour, layers, linetypes, dimension styles, etc. and all this information is inserted into the drawing with the wblock. All this information may not be required by the user, and it also takes time and uses memory space. Wblocks have another disadvantage, this being that drawings which contain several wblocks are not automatically updated if one of the original wblocks is altered.

External references (or **xrefs**) are similar to wblocks in that they are created by the user and can be inserted into a drawing, but they have one major advantage over the wblock. Drawings which contain external references are automatically updated if the original external reference 'wblock' is modified.

A worked example will be used to demonstrate external references. The procedure may seem rather involved as it requires the user to save and open several drawings, but the final result is well worth the effort. For the demonstration we will:
a) create a wblock
b) use the wblock as an xref to create two drawing layouts
c) modify the original wblock
d) view the two drawing layouts.
e) use the existing C:\BEGIN folder

Getting started

1   Open your A3PAPER standard drawing sheet and refer to Fig. 43.1.

2   Make a new current layer with:
name: XREF; colour: red; linetype: continuous.

3   Draw:
a) a circle of radius 18
b) a item of text, middled on the circle centre with height 5 and rotation angle 0. The item of text is to be AutoCAD and is to be colour blue – fig. (a).

Creating the xref (a wblock)

1   At the command line enter **WBLOCK <R>** and:
prompt   Write Block dialogue box
respond   1. Source: Objects
2. Base point: Pick point and pick circle centre point
3. Objects: Select objects, pick circle and text then right-click
4. Objects: Delete from drawing active
5. File name: XREFEX
6. Location: C:\BEGIN
7. Insert units: Millimeters
8. pick OK.

2   A preview of the wblock will be displayed and a blank screen returned, due to the delete from drawing option being active.

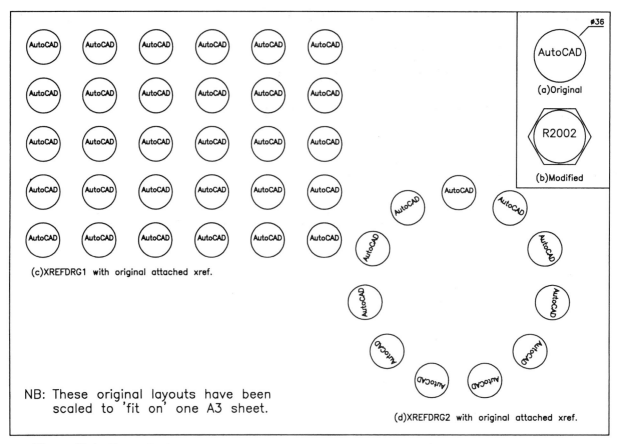

(c)XREFDRG1 with original attached xref.

NB: These original layouts have been scaled to 'fit on' one A3 sheet.

(d)XREFDRG2 with original attached xref.

Figure 43.1 External reference example.

Inserting the xref (drawing layout 1)

1 Menu bar with **File-Close** (no to save changes) then menu bar with **File-Open** and select your A3PAPER standard sheet again with layer OUT current.

2 Menu bar with **Insert-External Reference** and:

prompt   Select Reference File dialogue box (looks familiar?)

respond   1. scroll and pick C:\BEGIN
           2. scroll and pick XREFEX
           3. pick Open

prompt   External Reference dialogue box

with   Name: XREFEX and Path: C:\BEGIN\XREFEX.dwg

respond   1. ensure Reference Type: Attachment (black dot)
           2. Retain path active
           3. all on-screen options active, i.e. tick
           4. dialogue box similar to Fig. 43.2
           5. pick OK

prompt   Attach Xref "XREFEX": C:\BEGIN\XREFEX.dwg

and   "XREFEX" loaded

then   Specify insertion point and enter: **50,50 <R>**

prompt   Enter X scale and enter: **1 <R>**

prompt   Enter Y scale and enter: **1 <R>**

prompt   Specify rotation angle and enter: **0 <R>**

3 The named external reference (XREFEX) will be displayed at the insertion point entered. The complete process seems similar to inserting a wblock?

4 Now rectangular array the inserted attached xref for:
   a) 5 rows with row offset: 50
   b) 6 columns with column offset: 60.

5 Save the layout as **C:\BEGIN\XREFLAY1** – Fig. 43.1(c).

Inserting the xref (drawing layout 2)

1 Close the existing drawing then re-open A3PAPER

2 At the command line enter **XREF <R>** and:
   prompt   Xref Manager dialogue box
   respond **pick Attach**
   prompt   Select Reference File dialogue box
   respond 1. scroll and pick XREFEX from your C:\BEGIN folder
              2. pick OK
   prompt   External Reference dialogue box
   respond 1. Reference Type: Attachment
              2. De-activate the three on-screen prompts (no tick)
              3. Insertion point:- X: 200; Y: 250; Z: 0
              4. X,Y,Z scales: 1
              5. Rotation angle: 0
              6. pick OK

3 Now polar array the inserted attached xref with:
   a) centre point: 200,150
   b) number of items: 11
   c) angle to fill: 360
   d) rotate items as copied active

4 Save the layout as **C:\BEGIN\XREFLAY2** – Fig. 43.1(d).

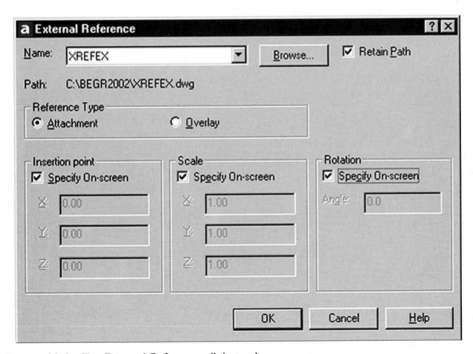

Figure 43.2 The External Reference dialogue box.

Modifying the original xref

1 Close the current drawing.

2 Open the original XREFEX drawing from C:\BEGIN. This is the original wblock.

3 *a*) Change the text item to R2002 and colour green
   b) Draw a hexagon, centred on the circle and circumscribed in a circle of radius 19. Change the colour of this hexagon to blue. These modifications are shown in Fig. 43.1(b).

4 Menu bar with **File-Save** to automatically update C:\BEGIN\XREFEX.

Viewing the original layouts

1 Close the existing drawing.

2 Menu bar with **File-Open** and:
   a) pick XREFLAY1
   b) note the Preview then pick Open
   c) interesting result?

3 Menu bar with **File-Open** and:
   a) pick XREFLAY2
   b) note the Preview then pick Open
   c) again an interesting result?

4 The layout drawings should display the modified XREFEX without any 'help' from us. This is the power of external references. Surely this is a very useful (and dangerous concept)?

5 Menu bar with **Format-Layer** and note the layer: **XREFEX|XREF**. This indicates that an external reference (xrefex) has been attached to layer xref. The (|) is a pipe symbol indicating an attached external reference.

6 This completes our simple investigation into xrefs.

Summary

1 External references are wblocks which can be attached to drawings

2 When the original xref wblock is altered, all drawing files which have the external reference attached are automatically updated to include the modifications to the original wblock.

Chapter 44

Isometric drawings

An isometric is a 2D representation of a 3D drawing and is useful as it can convey additional information about a component which is not always apparent with the traditional orthographic views. Although an isometric appears as a 3D drawing, the user should never forget that it is a 'flat 2D' drawing without any 'depth'.

Isometric drawings are created by the user with polar coordinates and AutoCAD has the facility to display an isometric grid as a drawing aid.

Setting the isometric grid

There are two methods for setting the isometric grid these being by using a dialogue box and via the keyboard.

Dialogue box method

1 From the menu bar select **Tools-Drafting Settings** and:
   prompt   Drafting Settings dialogue box
   respond pick the Snap and Grid tab and:
       1. Snap On(F9) active – black dot
       2. Grid On(F7) active
       3. Snap type & Style with:
         a) Grid snap active
         b) Isometric snap active
       4. set Grid Y Spacing: 10
       5. set Snap Y Spacing: 5
       6. dialogue box as Fig. 44.1
       7. pick OK.

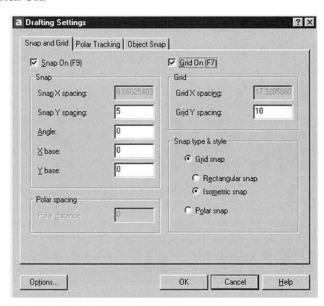

Figure 44.1 The Drafting Settings dialogue box.

2 The screen will display an isometric grid of 10 spacing, with the on-screen cursor 'aligned' to this grid with a snap of 5.

3 Use the Drafting Settings dialogue box to 'turn the isometric grid off', i.e. pick the Rectangular snap

4 The screen will display the standard grid pattern.

Keyboard entry method

1 At the command line enter **SNAP <R>** and:
   prompt Specify snap spacing or [ON/OFF/Aspect/Rotate/Style/Type]
   enter   **S <R>** – the style option
   prompt Enter snap grid style [Standard/Isometric]
   enter   **I <R>** – the isometric option
   prompt Specify vertical spacing<5.00>
   enter   **10 <R>**

2 The screen will again display the isometric grid pattern with the cursor 'snapped to the grid points'

3 Leave this isometric grid on the screen.

Isoplanes

1 AutoCAD uses three 'planes' called **isoplanes** when creating an isometric drawing, these being named top, right and left. The three planes are designated by two of the X, Y and Z axes as shown in Fig. 44.2(a) and are:
   a) isoplane top: XY axes
   b) isoplane right: XZ axes
   c) isoplane left: YZ axes

2 When an isoplane is 'set' or 'current', the on-screen cursor is aligned to that isoplane axis – Fig. 44.2(b).

3 The isoplane can be set by one of three methods:
   a) at the command line enter **ISOPLANE <R>** and:
      prompt Enter isometric plane setting[Left/Top/Right]
      enter   **R <R>** – right plane
   b) Using a 'toggle' effect by:
      1. holding down the **Ctrl** key (control)
      2. pressing the **E** key
      3. toggles to isoplane left
      4. Ctrl E again – toggles to isoplane top
   c) Using the F5 function key:
      1. press F5 – toggles isoplane right
      2. press F5 – toggles isoplane left, etc.

4 *Note:*
   It is the users preference as to what method is used to set the isometric grid and isoplane, but my recommendation is:
   a) set the isometric grid ON from the Drafting Settings dialogue box with a grid spacing of 10 and a snap spacing of 5
   b) toggle to the required isoplane with Ctrl E or F5
   c) isoplanes are necessary when creating 'circles' in an isometric 'view'.

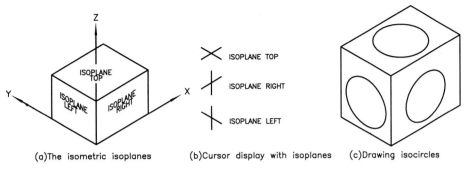

(a)The isometric isoplanes   (b)Cursor display with isoplanes   (c)Drawing isocircles

Figure 44.2 Three isometric concepts.

Isometric circles

Circles in isometric are often called isocircles and are created using the ellipse command and the correct isoplane **MUST** be set. Try the following exercise:

1   Set the isometric grid on with spacing of 10 and toggle to isoplane top.

2   Using the isometric grid as a guide and with the snap on, draw a cuboid shape as Fig. 44.2(c). The size of the shape is not important at this stage – only the basic shape.

3   Select the ELLIPSE icon from the Draw toolbar and:
    prompt   Specify axis endpoint or [Arc/Center/Isocircle]
    enter   **I <R>** – the isocircle option
    prompt   Specify center of isocircle
    respond **pick any point on top 'surface'**
    prompt   Specify radius of isocircle
    respond **drag and pick as required.**

4   Toggle to isoplane right with Ctrl E.

5   At the command line enter **ELLIPSE <R>** and:
    prompt   Specify axis endpoint or [Arc/Center/Isocircle]
    enter   **I <R>**
    prompt   Specify center of isocircle and **pick a point on 'right side'**
    prompt   Specify radius of isocircle and **drag/pick to suit**

6   Toggle to isoplane left, and draw an isocircle on the left side of the cuboid.

7   The cuboid now has an isometric circle on the three 'sides'.

8   Now continue with the example which follows.

Isometric example

1   Open your A3PAPER standard sheet and refer to Fig. 44.3.

2   Set the isometric grid on, with a grid spacing of 10 and a snap spacing of 5.

3   With the LINE icon draw:
    First point: pick towards lower centre of the screen
    Next point and enter: @80<30 <R>
    Next point and enter: @100<150 <R>
    Next point and enter: @80<–150 <R>
    Next point and enter: @100<–30 <R>
    Next point and enter: @50<90 <R>
    Next point and enter: @80<30 <R>
    Next point and enter: @50<–90 <R><R> – fig. (a).

4 With the COPY icon:
   a) objects: pick lines D1, D2 and D3 then right-click
   b) base point: pick intersection of pt 1
   c) displacement: pick intersection of pt 2.

5 Draw the two lines (endpoint-endpoint) to complete the sides and top then draw a top diagonal line – fig. (b).

6 Toggle to isoplane top.

7 With the ELLIPSE-Isocircle command:
   a) pick midpoint of diagonal as the centre
   b) enter a radius of 30.

8 Copy the isocircle:
   a) from the diagonal midpoint
   b) by: @50<90 – fig. (c).

9 Erase the diagonal.

10 Draw in the two 'cylinder' sides using the quadrant snap and picking the top and bottom isometric circles.

11 Trim objects to these lines and erase unwanted objects to give the complete isometric

12 *Task:*
   Draw two additional 30 radius isometric circles and modify to give the completed isometric as fig. (d).

13 Save if required.

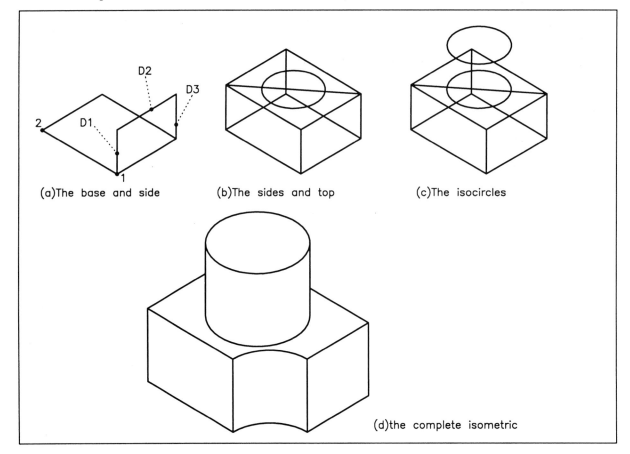

(a)The base and side   (b)The sides and top   (c)The isocircles

(d)the complete isometric

Figure 44.3 Isometric exercise.

Summary

1   An isometric is a 'flat' 2D drawing having 3D visualisation.

2   An isometric grid is available as a drafting aid.

3   Isometrics are generally constructed with polar coordinates.

4   Circles are drawn with the ellipse-isocircle option and the correct isoplane must be set.

5   The objects snaps (endpoint, midpoint, etc.) are available with isometric drawings.

6   The modify commands (e.g. copy, trim, etc.) are available with isometrics.

7   The OFFSET command does not give the effect that the user would expect, and should not be used.

8   The isocircle option of the ellipse command is only available when the isometric grid is 'set on'. Try this for yourself with a normal rectangular grid.

Assignment

Activity 38: 6 into 1
This activity requires that isometric circles are drawn with the three isoplanes. The recommended procedure is:

1   Draw an isometric cube of side 50.

2   Draw six isometric circles, one at the centre of each face. The radius is 25.

3   Copy each isometric circle by 50 at the appropriate angle, i.e. 30, 90, 150, –150, –90 and –30.

4   Draw in the 'cylinder sides' using the quadrant snap.

5   Trim as required with care.

6   Erase unwanted objects.

7   The sectional isometric is left for you to complete.

Chapter 45

Model space and paper space

AutoCAD has two drawing environments:
a) model space: used to draw the component
b) paper space: used to layout the drawing paper for plotting.

The two drawing environments are independent of each other and while the concept is particularly applicable to 3D modelling we will demonstrate its use with a previously saved 2D drawing.

Until now all work has been completed in model space.

Notes

1   The user must realise that this chapter is an introduction to the model/paper space concept. Paper space is particularly targeted for plotting and as I have no idea what type of plotter the reader has access to, I have assumed that no plotter is available. This does not affect the exercise.

2   When the paper space environment is entered, a new icon will be displayed. The model space and paper space icons are shown in Fig. 45.1.

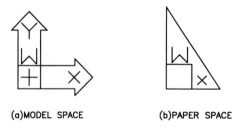

(a)MODEL SPACE   (b)PAPER SPACE

Figure 45.1   The model space and paper space icons.

3   Before paper space can be used, the paper space environment must be 'entered'. This can be achieved by:
a) picking MODEL from the status bar
b) picking one of the layout tabs from the drawing screen
c) entering TILEMODE then 0 at the command line.

Getting started

1   Open your A3PAPER standard drawing sheet and:
   a) Erase the rectangular border
   b) Make two new layers, VP and SHEET both with your own colour selection and with
       continuous linetype
   c) Make layer VP current.

2   AutoCAD generally defaults with a Model tab, a Layout1 tab and a Layout2 tab. While
     the layout tabs could be used for this exercise, we will create a new tab.

3   Menu bar with **Tools-Wizards-Create Layout** and:

prompt	Create Layout - Begin dialogue box
respond	**alter Layout name to FACTORY then pick Next** – Fig. 45.2
prompt	Create Layout - Printer dialogue box
respond	**pick None then Next**
prompt	Create Layout – Paper Size dialogue box
respond	1. Drawing units: Millimeters
	2. scroll at paper size and pick **ISOA3 (420.00x297.00)**
	3. pick Next
prompt	Create Layout - Orientation dialogue box
respond	**pick Landscape then Next**
prompt	Create Layout - Title Block dialogue box
respond	**pick None then Next**
prompt	Create Layout - Define Viewports dialogue box
respond	1. Viewport setup: Single
	2. View scale: Scaled to Fit
	3. pick Next
prompt	Create Layout - Pick Location dialogue box
respond	**pick Select location**
prompt	Drawing screen returned
with	Specify first corner at the command prompt
enter	**10,10 <R>**
prompt	Specify opposite corner
enter	**210,145 <R>**
prompt	Create Layout - Finish dialogue box
respond	**pick Finish.**

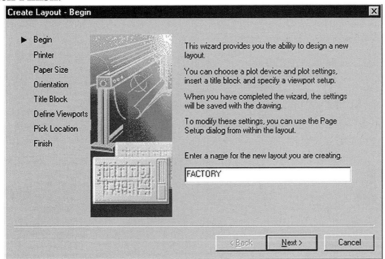

Figure 45.2   The Create Layout (Begin) dialogue box.

4   The drawing screen will be returned in Paper Space (note icon) with:
   a) a white area – the A3 drawing paper
   b) a dotted line area – the plottable area
   c) a coloured rectangle – the created viewport
   d) a new tab – FACTORY
   e) the paper space icon displayed
   f) PAPER is displayed in the Status bar.

5   As the paper space icon is displayed, the user is 'in paper space'.

The sheet layout

Our A3 drawing paper has a paper space viewport and we now want to create another three viewports. Refer to Fig. 45.3 and:

1   With layer VP current, menu bar with **View-Viewports-1 Viewport** and:
   prompt   Specify corner of viewport and enter: **230,75 <R>**
   prompt   Specify opposite corner and enter: **380,170 <R>**

2   Repeat the menu bar View-Viewports-1 Viewport selection and create another viewport from **190,180** to **280,250**

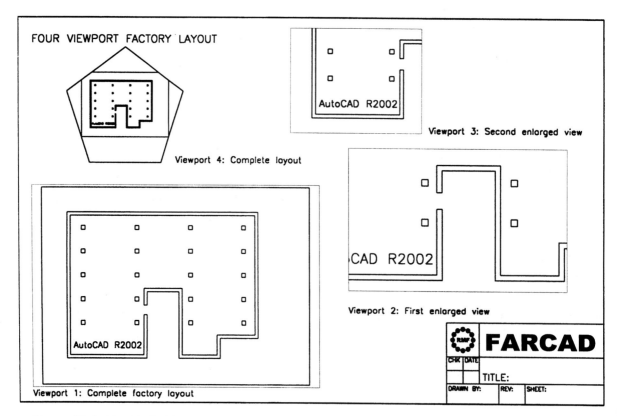

Figure 45.3   The LARGESC model/paper space exercise.

3   Menu bar with **View-Viewports-Polygonal Viewport** and:
   prompt   Specify start point and enter: **50,160 <R>**
   prompt   Specify next point and enter: **@50<0 <R>**
   prompt   Specify next point and enter: **@50<72 <R>**
   prompt   Specify next point and enter: **@50<144 <R>**
   prompt   Specify next point and enter: **@50<–144 <R>**
   prompt   Specify next point and enter: **C <R>**

4   We have now created three rectangular paper space viewports and one pentagonal paper space viewport..

The model

Rather than draw a new component we will insert an already completed and saved (I hope) drawing into our created viewports. This drawing is the large scale factory layout (LARGESC) from Chapter 38.

1   Enter model space with a left-click on PAPER in the Status bar.

2   The model space environment will be entered and the traditional model space icon will be displayed in the rectangular viewports.

3   Make the first created viewport (lower left) active by:
   a) moving the pointer into the rectangular area
   b) left-click
   c) cursor cross-hairs displayed and the viewport border appears highlighted.

4   Menu bar with **Insert-Block** and:
   prompt   Insert dialogue box
   respond 1. pick Browse
            2. scroll and pick C:\BEGIN folder
            3. scroll and pick LARGESC (or your entered name)
            4. pick Open
   prompt   Insert dialogue box with Name: LARGESC
   respond 1. ensure Specify On-screen active, i.e. ticks
            2. ensure Explode active, i.e. tick
            3. pick OK
   prompt   Specify insertion point for block and enter: **0,0 <R>**
   prompt   Specify scale factor for XYZ axes and enter: **1 <R>**
   prompt   Specify rotation angle and enter: **0 <R>**

5   Now select from the menu bar **View-Zoom-All.**

6   The active viewport will display the complete factory layout while the other three viewports may display some lines.

7   Make each viewport active in turn by moving the pointing arrow into the viewport and left-click, then View-Zoom-All from the menu bar and the layout will be displayed in each viewport.

8   With the large rectangular viewport active:
   a) erase all text
   b) freeze layer DIMS.

9   Any text and dimensions 'disappear' from all the viewports.

Using the viewports

1 With the lower right viewport active select **View-Zoom-Window** from the menu bar and:
   prompt Specify first corner and enter: **5000,1500 <R>**
   prompt Specify opposite corner and enter: **13000,8000 <R>**

2 With the top right viewport active, zoom a window from 1000,1000 to 9000,8000

3 With the pentagonal viewport active, menu bar with **View-Zoom-Scale** and:
   prompt Enter a scale factor
   enter **0.015 <R>**

4 Make layer TEXT current and the lower left viewport active and select from menu bar **Draw-Text-Single Line Text** and:
   prompt Specify start point of text and enter: **2500,2750 <R>**
   prompt Specify height and enter: **450 <R>**
   prompt Specify rotation angle and enter: **0 <R>**
   prompt Enter text and enter: **AutoCAD R2002 <R>**

5 The entered item of text will be displayed in all viewports.

6 Make layer OUT current with the lower right viewport active.

7 Draw two lines:
   a) first point: 7250,5250   next point: 8750,5250
   b) first point: 7250,6000   next point: @1500,0.

8 Use the TRIM command to trim these lines to give an opening into the factory – displayed in all viewports. You may have to zoom in on the wall area to complete the trim, but remember to zoom previous.

Using the paper space environment

1 Enter paper space with a left-click on MODEL in the status bar and try and erase any object from a viewport – you cannot as they were created in model space.

2 Make layer TEXT current and add the following text items to the layout:

	Start	Ht	Rot	Text item
a)	10,4	4	0	Viewport 1: Complete factory layout
b)	230,60	4	0	Viewport 2: First enlarged view
c)	285,180	4	0	Viewport 3: Second enlarged view
d)	110,160	4	0	Viewport 4: Complete layout
e)	10,240	5	0	FOUR VIEWPORT FACTORY LAYOUT.

3 Now enter model space with a left-click on PAPER in the status bar and try to erase any of the added text items – you cannot as they were created in paper space.

Completing the layout

1 Enter paper pace and make layer SHEET current.

2 With the LINE command draw:
first point: 0,0 next point: @405,0 next point: @0,257
next point: @–405,0 next point: close.

3 Menu bar with **Insert-Block** and:
a) select Browse
b) pick your C:\BEGIN folder
c) scroll and pick TITLE then Open
d) activate Explode
e) insertion point: X 405; Y 0; Z 0
f) scale: uniform at 0.9
g) rotation: 0
h) pick OK.

4 The factory layout – created in paper space is now complete.

5 *Task:*
If you have access to a printer/plotter:
a) plot from model space with any viewport active
b) plot from paper space
c) in model space, investigate the coordinates of the top right corner of the left viewport
   – about 22000,15000?
d) in paper space investigate the coordinates of:
   1. top right corner of our sheet: 405,257?
   2. top right corner of white area: 410,275?
e) *Question:* how can a drawing area on 22000,15000 be displayed on an A3 sheet of
   paper? This is the 'power' of model/paper space.

Model/Paper space example 2

The previous exercise created the viewports prior to 'inserting' a drawing into the layout.
In this exercise, we will open a previously created drawing, and then adapt the paper
space layout.

1 Open WORKDRG (which has not been used for some time) and erase any text and
dimensions to leave the original red outline, two red circles and four green centre lines.

2 Make a new layer, named VP with continuous linetype, colour to suit and current.

3 Left-click on the Layout1 tab and:
prompt Page Setup - Layout1 dialogue box
with   details about the Plotter to be used
respond **pick Cancel** – as I will assume no plotter.

4 The screen will be returned and display:
a) a paper space layout with the paper space icon
b) a white area – the drawing paper
c) a dotted area – the plottable area
d) a coloured outline – an active viewport
e) the WORKDRG and the black border within the coloured viewport.

5 Erase the coloured viewport and WORKDRG 'disappears'.

6 Layer VP still current.

7   Menu bar with **View-Viewports-New Viewports** and:

    prompt   `Viewports dialogue box` with two tabs:
            a) New Viewports
            b) Named Viewports

    respond   1. New Viewports tab active
            2. pick Three: Right
            3. Viewport Spacing: 0.1
            4. Setup: 2D – dialogue box as Fig. 45.4
            5. pick OK

    prompt   `Specify first corner` and enter: **5,5 <R>**
    prompt   `Specify opposite corner` and enter: **260,195 <R>**

8   The drawing screen will be returned (in paper space) with a three viewport configuration, each viewport displaying the WORKDRG and the black border.

9   Enter model space with one of the following methods:
    a) pick PAPER from the Status bar
    b) enter **MS <R>** at the command prompt
    c) refer to Fig. 45.5.

10   Note the model space icon in all three viewports.

11   Make the lower left viewport active, and menu bar with **View-Zoom-Extents.**

12   *a*) Make the large right viewport active by moving the cursor into it and left-click
    b) Erase the black border – it disappears from all viewports.

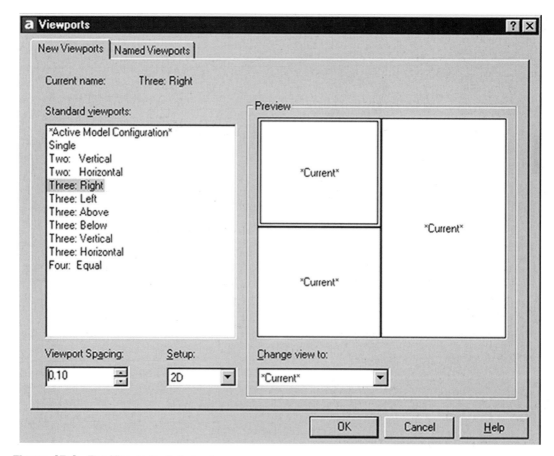

Figure 45.4   The Viewports dialogue box.

13 *a*) Make the top left viewport active
   b) Zoom-window from: 60,90 to 140,160
   c) The circle and centre lines 'fill the viewport'.

14 Still with the top left viewport active:
   a) Make layer OUT current
   b) Draw a circle with centre: 100,140 and radius: 3
   c) Polar array this circle, about the large circle centre, for 13 items with 360 angle to fill
      and rotated as copied.

15 Make the lower left viewport active and layer TEXT current.

16 Menu bar with **Draw-Text-Single Line Text** and:
   a) start point: enter C <R> – the centre text option
   b) centre point of text: 100,175
   c) height: 5 and rotation angle: 0
   d) text: AutoCAD <R>, then Release <R> then 2002 <R><R>

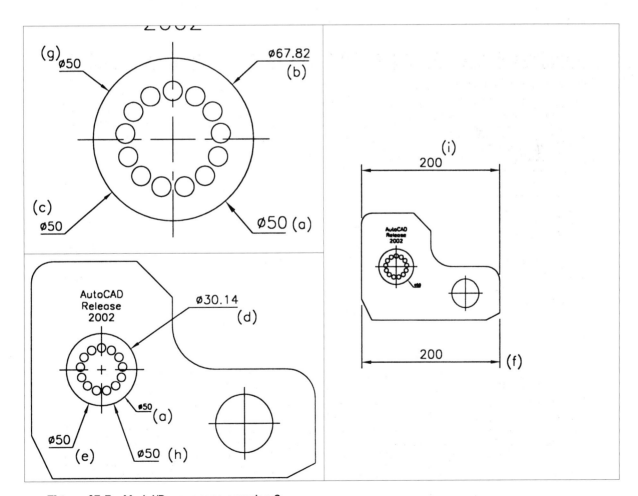

Figure 45.5   Model/Paper space exercise 2.

Dimensioning in model and paper space

When a multi-viewport layout has been created in paper space, many users are unsure whether dimensions should be added in model space or paper space. We will investigate this concept with our screen layout.

1 In model space with the Layout1 tab still active.

2 Make the top left viewport active and layers DIMS current.

3 Diameter dimension the circle, illustrated by dimension (a).

4 This diameter dimension is displayed in all three viewports, and is one of the main 'drawbacks' of model space dimensioning with a multi-viewport layout.

5 Enter paper space with PS <R> or select MODEL from the Status bar.

6 Select the diameter dimension command and pick the circle in top left viewport. The value is illustrated by dimension (b) and is 67.8. The value is obviously wrong as the circle has a diameter of 50. This is one of the main 'drawbacks' of paper space dimensioning.

7 We seem to have a slight problem. Dimensioning in model space will display the dimensions in all viewports, while in paper space the wrong dimension is obtained.

8 AutoCAD overcomes these two apparent problems with two entirely different concepts:
   a) With model space dimensions, viewport specific layers are used. This concept is considered beyond the scope of this book
   b) With paper space dimensions, a system variable **DIMLFAC** is modified to suit the model layout in a particular viewport.

9 At the command line enter **DIMLFAC <R>** and:
   prompt   Enter new value for DIMLFAC<1.0000>
   enter   **0.737246 <R>**

10 You should still be in paper space, so diameter dimension the same circle as before and it should be 50. This is illustrated by dimension (c) in Fig. 45.5.

11 Why enter 0.737246 for DIMLFAC? The reason is that the true dimension is 50, the paper space value was 67.8, hence it has to be scaled down by 50/67.8 which is the DIMLFAC value.

12 *Task*
   a) Return DIMLFAC to 1
   b) Diameter dimension the same circle, but pick the one which is displayed in the lower left viewport – (d)
   c) Calculate the DIMLFAC for this dimension to be correct
   d) Dimension with this DIMLFAC value – (e)
   e) Can you add the correct 200 linear dimension (f) as shown?
   f) Hint: Unit precision may need to be altered.

13 Now right-click the Layout1 tab name and:
   prompt   Shortcut menu
   respond   pick Rename
   prompt   Rename Layout dialogue box
   respond   1. Name: alter to MYTRY – Fig. 45.6
              2. pick OK.

14 The layout tab will now display the entered name.

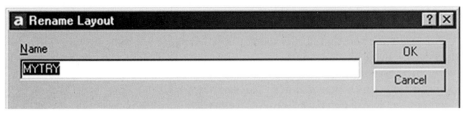

Figure 45.6 The Rename Layout dialogue box.

DIMLFAC

DIMLFAC (Dimension Linear scale Factor) is a system variable which when set to the correct value will scale linear and radial measurements and allow paper space dimensioning of model space objects to have the 'correct' value. In the previous exercise we set the DIMLFAC variable by a 'rather crude' method, but the result was correct.

We will now investigate another method for setting DIMLFAC.

1 Still with the three viewport configuration displayed with several diameter dimensions.

2 In Model space, set DIMLFAC to 1 with command line entry.

3 Enter paper space.

4 At the command line enter **DIM <R>** and:
prompt   Dim:
enter   **DIMLFAC <R>**
prompt   Enter new value for dimension variable, or Viewport
enter   **V <R>** – the viewport option
prompt   Select viewport to set scale
respond   **pick the border of the top left viewport**
prompt   DIMLFAC set to -0.73723
then   Dim:
respond   **ESC** to end the command line dimension sequence.

5 *Task*
Using the procedure described in step 4, refer to Fig. 45.5 and use the command line DIM command to select the other viewports and dimension the objects as follows:

	First	*Second*
a) viewport	lower left	right
b) DIMLFAC value	–1.65904	–3.36154
c) object to dimension	circle	linear A to B
d) reference	(h)	(i).

6 Decide which of the methods to 'set' the DIMLFAC system variable is easier for adding the correct dimensions when in paper space. The second?

Our second model/paper space exercise is now complete, but before leaving the exercise, left-click on the Model tab name and:
a) traditional 2D drawing screen returned with original WORKDRG layout
b) There is no black border (erased earlier) but the arrayed circles, text items and one diameter dimension are displayed
c) Are these what you would have expected?
d) If you offset the right circle by 10 outwards, what would you expect if you enter paper space?

You can now save the exercise but we will not use this layout again.

True Associative Dimensioning

This chapter has so far demonstrated that model space and paper space are two drawing environments which can be used to:

a) layout the drawing sheet to user requirements
b) set viewports in paper space to display different parts of a drawing
c) dimension the component in model space
d) add dimensions to a paper space layout with the use of the DIMLFAC system variable.

In a previous chapter we also demonstrated that dimensions added in model space are **associative**, i.e. they will 'alter' when the component geometry is altered. This was demonstrated using the STRETCH command.

The DIMLFAC method of adding 'true' dimensions in paper space has been superseded in AutoCAD 2002, and we will now demonstrate this concept.

1 Close any existing drawing and open C:\BEGIN\WORKDRG.

2 Erase any dimensions, text and the border.

3 Make a new layer named VP, colour to suit with continuous linetype. Make this layer current and refer to Fig. 45.7.

4 Left-click on the Layout1 tab and:
   prompt Page Setup – Layout1 dialogue box
   respond 1. Plot Device tab and set:
       a) Plotter Configuration: None
       b) Plot Style: monochrome
     2. Layout Settings tab and set:
       a) Paper size: ISO A3 (420x297)
       b) Drawing Orientation: Landscape
       c) Plot area: Layout
       d) Plot scale: 1:1
     3. pick OK
   and paper space returned with WORKDRG in a viewport.

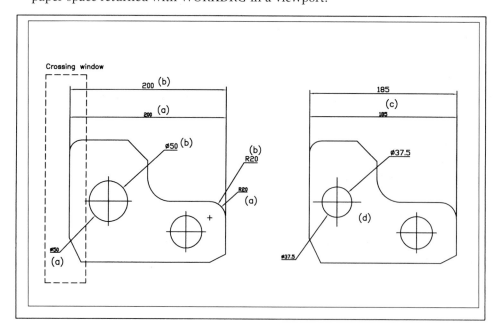

Figure 45.7 Model/paper space dimension demonstration.

5 Erase this viewport by selecting any part of it to leave a blank paper space 'sheet' with:
   a) white area: the drawing paper
   b) dotted area: the plotable area.

6 Menu bar with **View-Viewports-1 Viewport** and:
   prompt   Specify corner of viewport and enter: **10,10 <R>**
   prompt   Specify opposite corner and enter: **300,220 <R>**
   and   WORKDRG displayed in this new viewport.

7 Pick the Model tab and WORKDRG is still displayed as originally 'opened'. Note that my Fig. 45.7 has two WORKDRG's displayed. This is solely for demonstration purpose with this exercise.

8 At the command line enter **DIMASSOC <R>** and:
   prompt   Enter new value for DIMASSOC
   enter   **2 <R>**

9 With the Model tab active, make layer DIMS current and:
   a) linear dimension any 'part' of the model
   b) diameter dimension a circle
   c) radius dimension one of the filleted 'corners'.

10 Pick the Layout1 tab and the added dimensions should be displayed as indicated by (a) in Fig. 45.7.

11 With the Layout1 tab active and in paper space, dimension the same three objects as before, indicated by (b). These should have values the same as the model space dimensions.

12 Still with the Layout1 tab active and in model space, use the STRETCH command with:
   a) a crossing option as indicated
   b) base point: 0,0
   c) second point: @15,0.

13 Both the 200 value model and paper space linear dimension will be replaced by a value of 185 to reflect the stretch command. This is denoted by dimensions (c).

14 With Layout1 active and in model space, use the SCALE command and:
   a) select the dimensioned circle as the object
   b) pick the circle centre as the base point
   c) enter a scale factor of 0.75.

15 The model and paper space dimensions will reflect this scale effect – dimensions (d).

16 Select the model tab and freeze layer DIMS.

17 With the MOVE command:
   a) window the complete WORKDRG
   b) base point: select any line endpoint
   c) second point: enter @0,50.

18 Pick the Layout1 tab and thaw layer DIMS.

19 The dimensions should have moved with the WORKDRG shape. Note that this move is not displayed in Fig. 45.7.

20 This exercise should have demonstrated that:
   a) dimensions can be added in model and paper space
   b) both 'types' of dimensions are **REAL**, i.e. give the true value of the object to be dimensioned
   c) both model and paper space dimensions are truly associative, i.e. will reflect any change in the object geometry
   d) the dimensions 'move' with selected objects, even if the dimension layer is frozen.

21 *DIMASSOC*
   DIMASSOC is a system variable which controls dimension associativity, i.e. its value determines whether any added dimension will change when the object it is associated with is changed. DIMASSOC can have one of three values as follows:
   0 : dimensions are displayed exploded, i.e. any part of the dimension can be selected
   1 : the complete dimension is a single object and model space associativity applies
   2 : the complete dimension is a single object and paper space associativity applies.

   This exercise is now complete and can be saved if required.

Summary

1 Model and paper space are two drawing 'environments' which 'exist together'.

2 Model space is used for traditional draughting purposes.

3 Paper space is used to lay out the paper to user specifications.

4 While the model/paper space concept is especially suitable for 3D and Solid modelling, the techniques can also be used in 2D.

5 Viewports are created in paper space and all drawing work completed in active model space viewports.

6 Viewports can be created to user requirements.

7 Layout tabs can be used to 'set' viewport configurations and renamed.

8 Dimensions are generally added in model space using viewport specific layers – not considered in this book.

9 The DIMLFAC system variables allows paper space dimensioning of model space objects to 'give the correct dimension value'.

10 AutoCAD 2002 has true model and paper space associativity which supersedes the need to use DIMLFAC.

11 DIMASSOC is a system variable which determines the 'type' of associativity available and:
   1 : model space associativity
   2 : paper space associativity.

Final thought

Should all drawing work be completed using a layout tab, and dimensions added in paper space with DIMASSOC set to 2.

I will leave this thought for the user to think over!

Chapter 46

Templates and standards

This topic has been left to the end, when it should probably have been included nearer the beginning. The reasons for this were to allow the user to:
a) become proficient at draughting with AutoCAD
b) understand attributes
c) understand the concepts of paper space layouts.

What is a template?

1   A template is a prototype drawing, i.e. it is similar to our A3PAPER standard sheet which has been used when every new exercise/activity has been started. The terms prototype/standard drawing are used whenever a drawing requires to be used with various default settings, e.g. layers, text styles, dimension styles, etc.

2   With AutoCAD, all drawings are saved with the file extension **.dwg** while template files have the extension **.dwt.** Any drawing can be saved with the .DWG or .DWT extensions.

3   Template drawings (files) are used to 'safeguard' the prototype drawing being mistakenly overwritten – have you ever saved work on your A3PAPER standard sheet by mistake? Templates help overcome this problem.

4   AutoCAD has templates which conform to several drawing conventions, including:

Standard	Paper sizes
a) ANSI	A, B, C, D, E, F
b) DIN	A0, A1, A2, A3, A4
c) Gb	A0, A1, A2, A3, A4
d) ISO	A0, A1, A2, A3, A4
e) JIS	A0, A1, A2, A3, A4.

5   Other templates are also available.

6   Template files can be opened:
a) from the Startup dialogue box
b) from the TODAY window.

In this chapter we investigate:
a) both methods of opening a template file
b) completing and saving a drawing using the opened template file
c) creating our own template file.

Using an AutoCAD template file from Startup

1 Close any existing drawing

2 Menu bar with **File-New** and:

    prompt   Create New Drawing dialogue box
    respond **pick Use a Template**
    prompt   Use a Template dialogue box
    respond 1. scroll at Select a Template
            2. pick **Iso a3-named plot styles.dwt**
            3. note the preview and description – Fig. 46.1
            4. pick OK.

Figure 46.1 The Create New Drawing (Use a Template) dialogue box.

3 The screen will display:
    a) a layout with a title box
    b) the paper space icon
    c) a new tab – ISO A3 Title Block.

4 *Investigate*:
    a) Format-Layers and note the layer names (especially FRAMES), colours and linetypes
        (all continuous)
    b) Dimension-Style and note the style names and the current style – ISO-25?

5 *a*) Left-click on PAPER from the Status bar to enter model space and note the 'thick black
        outline'. This is the drawing area
    b) Draw a line from: 0,0 to: 420,290, i.e. A3 paper available
    c) Erase the line.

6 Rather than start a new drawing we will insert a previously created drawing so from the
   menu bar **Insert-Block** and:

    prompt   Insert dialogue box
    respond **pick Browse**
    prompt   Select Drawing File dialogue box
    respond 1. scroll and pick the C:\BEGIN folder
            2. scroll and pick a saved activity, e.g. act19
            3. pick Open
    prompt   Insert dialogue box
    with   Name: act19
    respond 1. activate Explode
            2. de-activate the On-screen prompts
            3. enter the Insertion point as X: 0; Y: 0; Z: 0
            4. enter the Scale as 1 and the Rotation angle as 0
            5. pick OK.

7 The selected drawing (file or wblock?) will be inserted into the ISO A3 template file/drawing.

8 *a*) Erase unwanted objects (e.g. the border) which have been inserted and are not required
   b) move the drawing to a suitable area of the screen
   c) optimise your drawing layout
   d) make a new text style, name: **sta** with text font: **Arial Black.**

9 *a*) Enter paper space with **PS <R>** at the command line
   b) Zoom a window around the title block.

10 From the menu bar select **Modify-Object-Attribute-Single** and:

prompt	`Select a block`
respond	**pick any XXX text item in the 'title block'**
prompt	`Enhanced Attribute Editor dialogue box`
with	*a*) Tab options: Attribute, Text Options, Properties
	b) Attribute Tag, Prompt and Value details
	c) dialogue box similar to Fig. 46.2(a)
respond	1. resize the dialogue box by dragging the lower edge downwards until eleven attributes (FILENAME-SHEET) are displayed
	2. pick FILENAME
	3. alter value to: **R2002/AB/01**
	4. pick DRAWING_NUMBER
	5. alter value to: **456-BDF**
	6. continue to pick the tag lines and alter the values as follows:

Tag	*Value*
OWNER	FARCAD
CHECKED_BY	RMF
DESIGNED_BY	H.T.CAMPBELL
APPROVED_BY_DATE	CEO: HM on 04/03/01
DATE	03/02/01
TITLE	STEAM EXPANSION BOX
SCALE	1:1
EDITION	REV 3
SHEET	1 of 3

and	dialogue box at this stage as Fig. 46.2(b)
now	1. pick the OWNER tag – highlights
	2. pick the Text Options tab
prompt	`Text Options dialogue box`
respond	1. Text Style: scroll and pick sta
	2. Height: alter to 10
	3. Oblique Angle: alter to 10 – Fig. 46.3
	4. pick the Properties tab
prompt	`Properties dialogue box`
respond	1. scroll and color and pick Red
	2. pick Apply then OK.

(a)   (b)

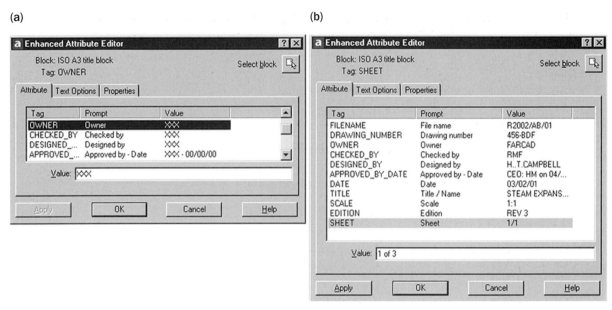

Figure 46.2 The Enhanced Attribute Editor dialogue box: (a) original and (b) with altered attribute values.

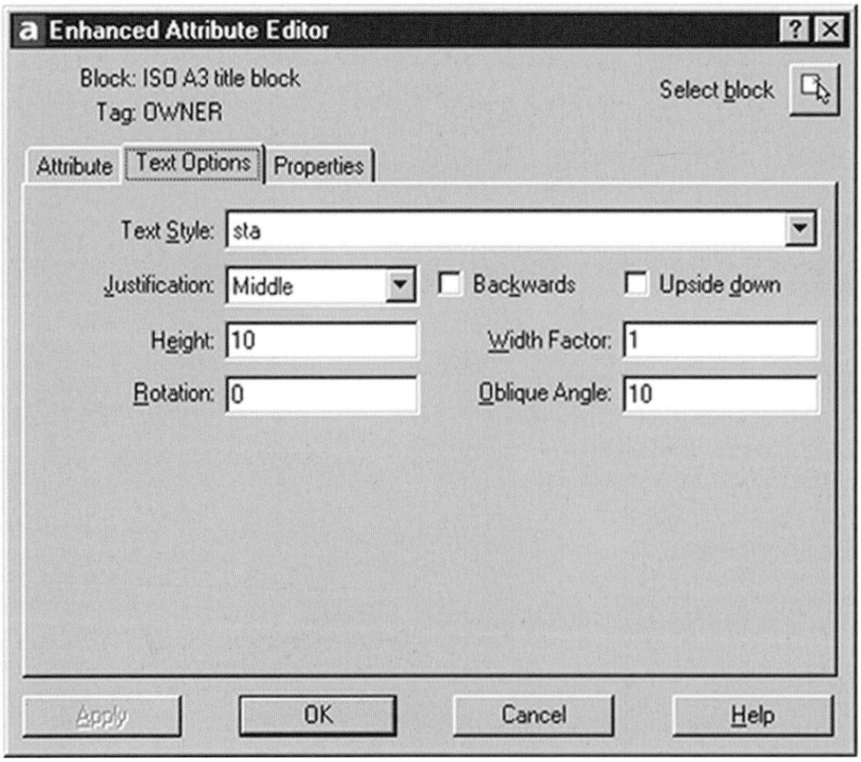

Figure 46.3 The Enhanced Attribute Editor dialogue box with the Text Options tab active.

11 The drawing screen will be returned and the attribute values entered in step 20 will be displayed. The original XXX OWNER attribute will have been replaced by the value FARCAD in red at a height of 10 and with a 10 obliquing angle.

12 Using the sequence **Modify-Object-Attribute-Single**, and your imagination, modify some of the other entered attributes for colour, height, text style, etc.

13 When you have completed the attribute 'editing', zoom-previous and return to model space. Your drawing should now resemble Fig. 46.4.

14 Finally, right-click on ISO A3 Title Block from the Layout tabs and:
    prompt   Shortcut menu
    respond **pick Rename**
    prompt   Rename Layout dialogue box
    respond 1. alter name to **MYLAYOUT**
               2. pick OK.

15 Menu bar with **File-Save As** and:
    a) Save Drawing As dialogue box
    b) File type extension is *.dwg
    c) scroll to your C:\BEGIN folder and enter any suitable drawing name.

16 *Note*
    a) We began the exercise with a template file – extension .dwt
    b) We saved the drawing as a drawing file – extension .dwg
    c) The original template file is thus unchanged
    d) This is surely useful to the user?

17 This first template exercise is now complete.

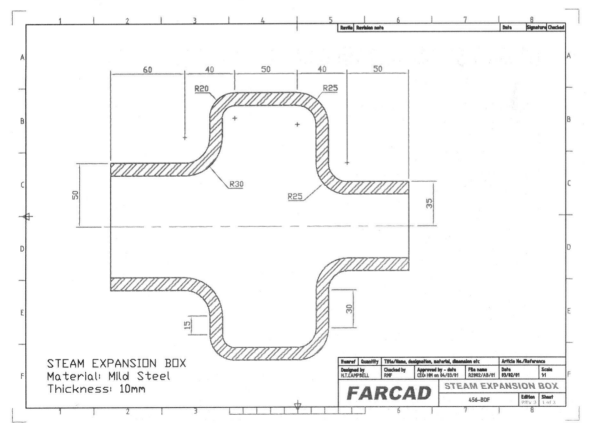

Figure 46.4   Template exercise 1 using Activity 19.

Using an AutoCAD template file from the TODAY window

1   Close the existing drawing then menu bar with **File-New** and pick Start from Scratch-Metric-OK to 'get us started'.

2   Pick the TODAY icon from the Standard Toolbar and:
    prompt   AutoCAD 2002 Today window/dialogue box
    respond   1. My Drawings: pick the Create Drawings tab
                2. Select how to begin: scroll and pick Template
                3. pick ▶I line to display a list of templates
                4. scroll until the ISO A0 names – Fig. 46.5
                5. pick ISO A0-Named Plot Styles.dwt.

3   The screen will return a paper space A0 drawing.

4   Move the cursor to lower left corner of white area and the coordinates will be approx 0,0. At the upper right corner of the white area the coordinates are approx 1180,2400.

5   Enter model space with **MS <R>** or pick **PAPER** from Status bar.

6   As with the previous exercise we will not create a new drawing but will insert the modified large scale drawing of the factory layout with I beams – completed and saved in Chapter 40.

7   Menu bar with **Insert-Block** and using the Insert dialogue box:
    a) scroll and pick the C:\BEGIN folder
    b) scroll and pick the modified large scale drawing from Chapter 40 (if not, pick LARGESC from Chapter 38)
    c) activate the on-screen prompts (all tick)
    d) activate Explode
    e) insertion point: 0,0
    f) XYZ scale: 0.02
    g) rotation angle: 0.

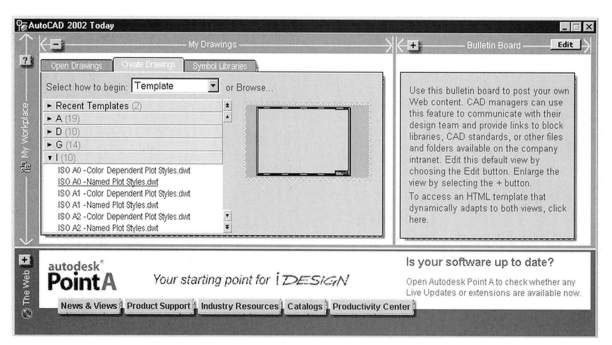

Figure 46.5   Opening an AutoCAD template file from the TODAY window.

8　Hopefully the modified factory layout will be displayed within the A0 sheet layout.

9　Erase the black border and position the factory layout for maximum effect.

10　*Task 1*
   a) In paper space, zoom in on the title block
   b) Modify the attributes using the same procedure as the first exercise, entering your own values
   c) When completed, return to model space.

11　*Task 2*
   If your layout is the modified one with the circular wall and I beams, then redefine the block BEAM to your own spec, but remember that the drawing was inserted at a scale of 0.02. This from the original factory layout being drawn at a scale of 50. Think about this before attempting the task. Figure 46.6 is my layout.

12　The exercise is now complete and can be saved as a .dwg file with a suitable name.

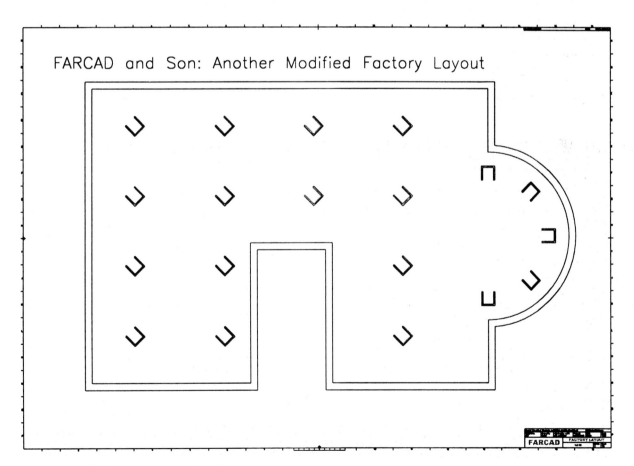

Figure 46.6　Template exercise 2.

Creating our own A3 template file

We will now create our own A3 sized template file and will use our existing A3PAPER drawing file as it has layers, linetypes, text styles and dimension styles already created.

1 Close all existing drawings then open your A3PAPER standard sheet from your BEGIN folder.

2 *a*) Erase the black border
   b) Make a new layer named VP, continuous linetype, any colour
   c) Make this new layer current.

3 Menu bar with **Tools-Wizards-Create Layout** and:
   prompt   Create Layout – Begin dialogue box
   respond alter the various dialogue boxes as follows:
       1. Begin Layout name: MY A3 LAYOUT
       2. Printer: None
       3. Paper size: Millimeters ISO A3 (420x297)
       4. Orientation: Landscape
       5. Title Block: none for now
       6. Define viewports: none for now
       7. Finish: pick Finish.

4 Still with layer VP current, select **View-Viewports-Polygonal Viewport** from the menu bar and:
   prompt   Specify start point and enter: **10,10 <R>**
   prompt   Specify next point and enter: **275,10 <R>**
   prompt   Specify next point and enter: **275,70 <R>**
   prompt   Specify next point and enter: **395,70 <R>**
   prompt   Specify next point and enter: **395,250 <R>**
   prompt   Specify next point and enter: **10,250 <R>**
   prompt   Specify next point and enter: **C <R>** – to close the viewport.

5 Menu bar with **Insert-Block** and:
   a) pick Browse
   b) scroll and pick BORDER from your C:\BEGIN folder then Open
   c) activate the three on-screen prompts then OK
   d) insertion point: 0,0
   e) X scale: 0.976; Y scale: 0.895; rotation: 0.

6 Menu bar with Insert-Block and open TITLE from your C:\BEGIN folder and insert with:
   a) insertion point: 400,5
   b) X scale: 1; Y scale: 1; rotation: 0.

7 *a*) Enter model space with command line **MS <R>**
   b) Make layer OUT current.

8 Menu bar with **File-Save** and:
   prompt   Save Drawing As dialogue box
   respond 1. scroll at Files of type
       2. pick **AutoCAD Drawing Template File (*.dwt)**
   prompt   Save Drawing As dialogue box
   with   Template file active (i.e. Save in)
   respond 1. File name: enter **A3LAYOUT**
       2. pick Save
   prompt   Template Description dialogue box
   enter   The following lines of text, but **DO NOT PRESS RETURN**
       This is my A3PAPER standard sheet layout created in paper space. My wblocks BORDER and TITLE have been inserted. The layout sheet has layers, text style, units and dimension styles customised to my own requirements. Saved in TEMPLATE
   and   dialogue box as Fig. 46.7
   respond pick OK

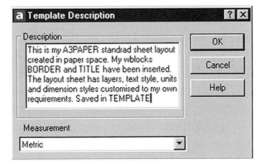

Figure 46.7   The Template Description dialogue box for the A3LAYOUT standard sheet.

9   Your A3PAPER standard sheet will be saved as a template file and:
   a) added to the list of existing AutoCAD templates
   b) be able to be 'opened' as a template
   c) will not be 'over-written' when a drawing is saved

10   Repeat step 8, but save your A3LAYOUT template file in your C:\BEGIN folder. You should have the same template description dialogue box as before.

Task

1   Close all existing files.

2   Menu bar with **File-New** and:
   prompt   Create New Drawing dialogue box
   respond **pick Use a Template**
   prompt   Select a Template list
   respond 1. scroll and pick **A3layout.dwt**
           2. pick Open.

3   The screen will display your A3 layout template file.

4   *a*) insert any drawing (e.g. ACT25) full size with 0 rotation
   b) modify and reposition the inserted drawing
   c) layout similar to Fig. 46.8.

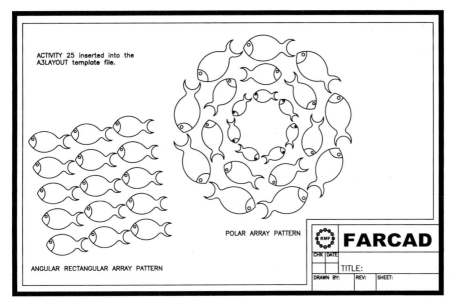

Figure 46.8   Using the A3LAYOUT template file.

5   Menu bar with **File-Save As** and:
    a) File type: *.dwg?
    b) Save as any suitable name.

6   Close any existing file then re-select your A3LAYOUT template file which should be displayed 'as new'.

7   Thus a template file can be opened, used and a drawing file saved, leaving the original template file 'untouched'.

8   Try opening your A3LAYOUT template file from the C:\BEGIN folder. Does the same concept apply when saving?

CAD standards

CAD standards allow the user to create a file that defines certain properties, these being layers, dimension styles, text styles and linetypes. The file is saved as a 'template' file with the extension **.dws** (drawing standard).

As all of our exercises have been completed using the A3PAPER drawing, all the 'standards properties' are the same in every one of our saved drawings. We therefore have to modify our existing standard sheet to demonstrate how standards work.

1   Close any existing drawings then menu bar with **File-New** and from the Create New Drawing dialogue box:
    a) Select Use a Template
    b) pick Browse, scroll and select the C:\BEGIN folder
    c) pick A3LAYOUT then Open.

2   Immediately the file is displayed on the screen, menu bar with **File-Save As** and with the Save Drawing As dialogue box:
    a) file type: scroll-pick **AutoCAD 2000 Drawing Standard (*.dws)**
    b) file name: enter **A3LAYOUT**
    c) save in: scroll and pick C:\BEGIN folder
    d) pick Save.

3   Menu bar with **Format-Lauer** and with the Layer Properties Manager dialogue box alter the following layer properties:

Layer name	property to alter
CONS	linetype: HIDDEN
TEXT	colour: green
DIMS	lineweight: 0.30.

4   Menu bar with **File-Save As** and:
    a) file type: AutoCAD 2000 Drawing Standard (*.dws)
    b) file name: TEST-1
    c) save in: C:\BEGIN
    d) pick Save.

5   *a*) We have thus saved our original A3LAYOUT template file as an AutoCAD standards file with the name A3LAYOUT.dws.
    b) The modified template file has been saved as an AutoCAD standards file with the name TEST-1
    c) both standard files have been saved in the C:\BEGIN folder.

6   Close all existing files.

7   Menu bar with **File-Open** and:
  prompt   Select File dialogue box
  respond   1. file type: Standards (*.dws)
      2. Look in: C:\BEGIN
      3. pick TEST-1 then Open.

8   The modified template/standards file with green text items will be displayed.

9   Menu bar with **Tools-CAD Standards-Configure** and:
  prompt   Configure Standards dialogue box
  with   Two tabs – Standards and Plug-ins
  respond   1. ensure Standards tab active
      2. **pick +**
  prompt   Select Standards File dialogue box
  respond   1. ensure C:\BEGIN folder current
      2. file type: Standard (*.dws)
      3. pick A3LAYOUT then Open
  prompt   Configure Standards dialogue box
  with   1. Standards tab active
      2. **A3LAYOUT** listed
      3. Description of File, Last Modified, Format
      4. dialogue box as Fig. 46.9, but Note:
        My Fig. 46.9 lists the Standards files with C:\BEGR2002. This is due to the
        fact that my computer system already had a BEGIN folder.
  respond   **pick Check Standards**
  prompt   Check Standards dialogue box
  with   1. Problem: Layer 'CONS'
      2. Replace with details
      3. Preview of changes information – Fig. 46.10(a)
  respond   **pick the Fix tick**, i.e. we are replacing the HIDDEN linetype on layer CONS
      with the Continuous Standard value from the A3LAYOUT standards file
  then   Check Standards dialogue box
  with   1. Problem: Layer 'DIMS'
      2. Replace with details
      3. Preview of changes information
  respond   **pick the Fix tick**, i.e. we are replacing the 0.30mm lineweight value on layer
      DIMS with the default standard value from the A3LAYOUT standards file
  then   Check Standards dialogue box
  with   1. Problem: Later 'TEXT'
      2. Replace with details
      3. Preview of changes information
  respond   **pick the Fix tick**, i.e. we are replacing the green colour on layer TEXT with
      the Blue standard value from A3LAYOUT
  then   Check Standards dialogue box with details about the checking that has been
      carried out – Fig. 46.10(b)
  respond   1. read the dialogue box information
      2. pick Close.

This exercise is now complete. Do not save any changes.

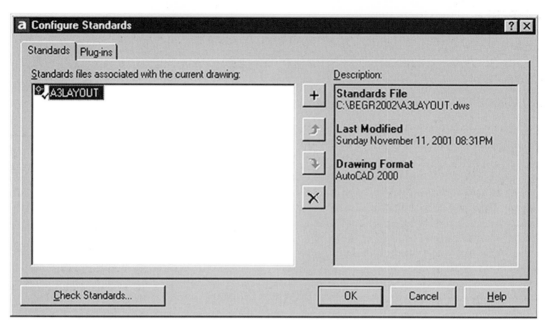

Figure 46.9 The Configure Standards dialogue box for A3LAYOUT.

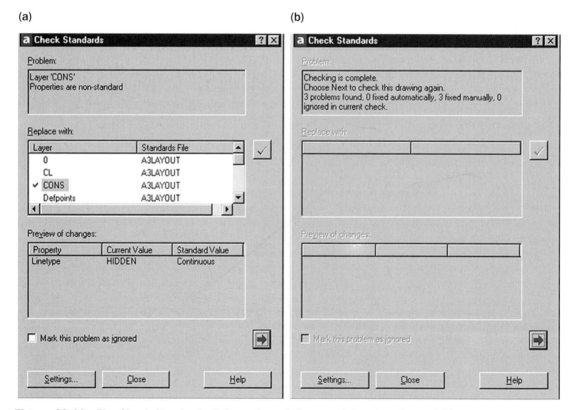

Figure 46.10 The Check Standards dialogue box at the start (a) and at the end (b).

Summary

1 A template file is a prototype drawing with various defaults set to user requirements.

2 AutoCAD has several template files conforming to different drawing standards, e.g. ANSI, ISO, etc.

3 Template files can be created and saved by the user.

4 Template files 'safeguard' the prototype drawing being 'overwritten'.

5 CAD standards allow the user to 'check' layers, linetypes, dimension styles and text styles between several templates.

6 Template files have the extension .dwt and can be saved in the AutoCAD Template folder or in user-defined folders.

7 Standards files have the extension .dws and can be saved in an AutoCAD folder or a user-defined folder.

Chapter 47

The AutoCAD design centre

The AutoCAD Design Centre is basically a drawing management system with several powerful advantages to the user including:
a) the ability to browse different drawing sources
b) viewing object definitions prior to opening
c) shortcuts to commonly used drawings and folders
d) searching for specific drawing content
e) opening drawings by drag mode
f) viewing and attaching raster image files into the drawing area
g) palette control

In this chapter we will investigate several of these topics.

Accessing the Design Centre

1 Open your A3PAPER standard sheet and cancel any floating toolbars

2 Menu bar with **Tools-AutoCAD Design Centre** and:
   prompt DesignCenter dialogue box.

3 The first time that the design center is activated, it is usually docked at the left of the drawing screen area. Move and resize by:
   a) left-click in the design center title bar and hold down the button
   b) drag into the drawing area
   c) resize to suit.

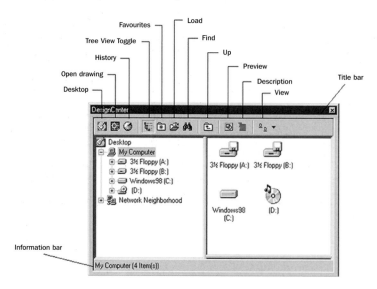

Figure 47.1   The basic Design Center dialogue box with toolbar descriptions.

4 The design center dialogue box (Fig. 47.1) consists of the following:
   a) The Design Center title bar
   b) A toolbar with several icon selections
   c) The tree view and hierarchy area on the left
   d) The palette with icons on the right
   e) An information bar below the tree view and palette.

Using the tree view and hierarchy

The tree view side of the design center dialogue box consists of a list of names with a (+) or a (−) beside them. If a name in the tree view side is selected (left-click) then the palette side of the dialogue box will reveal the contents of the selected item. If the (+) beside an item is selected then that item is **'expanded'** and the tree view side of the dialogue box will reveal the 'contents' of the selected item. The selected item's (+) will be replaced with a (−) indicated that it has been expanded. If the (−) is then selected, the expanded effect is removed and the (+) is returned. The selection of the (−) is termed **'collapsing'**.

To demonstrate using the tree view, refer to Fig. 47.2 and:

1 Note the Design Center layout as opened – fig. (a). Your display may differ from mine, but this is not important.

2 Expand My Computer (if available) by picking the (+) to give a tree expansion similar to fig. (b).

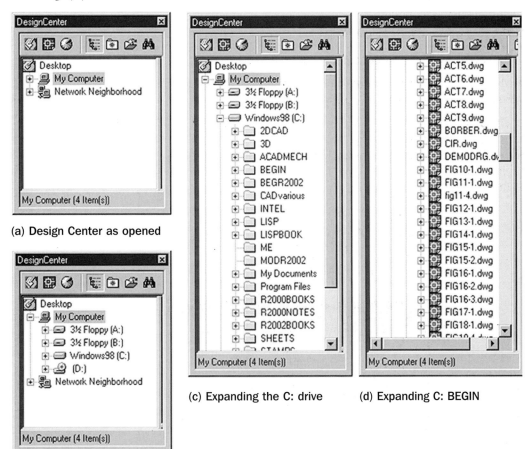

(a) Design Center as opened

(b) Expanding 'My Computer'

(c) Expanding the C: drive

(d) Expanding C: BEGIN

Figure 47.2 Tree view expansions.

3 Expand the C: drive by selecting the (+) to give the folders on the C drive of your system – this should be entirely different from mine?

4 Expand C:\BEGIN by selecting the (+) to give a listing of the files you have saved, similar to fig. (d).

5 Note that in Fig. 47.2, I have only displayed the tree view side of the Design Centre. This was to ensure all the displays 'fitted' into a single layout.

6 Move the pointer arrow onto BEGIN and right-click
prompt   Shortcut menu
respond **pick Explore**
and   palette side of the Design Center dialogue box will display icons of the contents of the named folder.

7 Figure 47.3 displays part of my named folder selection.

8 *Task*
   a) Collapse your named folder by picking the (−) at its name
   b) Collapse the C: drive
   c) Collapse My computer and the Design Center should be displayed 'as opened' – Fig. 47.4.

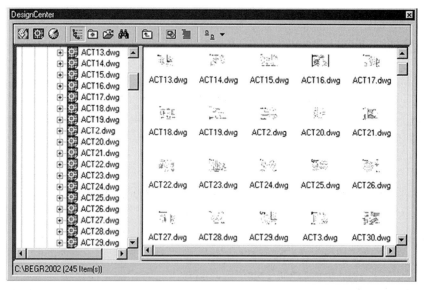

Figure 47.3 The 'explored' C:\BEGIN folder with tree view and palette displays.

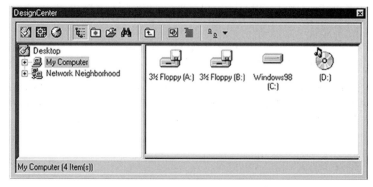

Figure 47.4 The Design Center and the collapsing the BEGIN, C: and My Computer trees.

The Design Center menu bar

The menu bar of the Design Center allows the user access to the several icons. These are displayed and listed in Fig. 47.1 and are:

Icon	*Description*
Desktop	list local/network drives similar to Fig. 47.4
Open Drawing	lists all drawings currently opened in AutoCAD and displays a drawing icon in both the tree view and palette
History	displays a list of files previously opened with Design Center with their paths
Tree	
View Toggle	toggles between the tree/palette screen and the palette screen
Favourites	lists the contents of AutoCAD's favourites folder
Load	allows access to the Load Design Center palette and will display details about a selected item in the palette area
Find	allows searches for named files using user entered names
Up	will toggle the Design Center 'up a level'
Preview	displays a preview of a selected item in the palette area
Description	details information about a selected item if applicable in the palette area
Views	allows the user to access large/small icons as well as lists and details.

Exercise 1

1   Still have AutoCAD opened with A3PAPER displayed.

2   Activate the Design Center and position the dialogue box in centre of screen

3   Expand the following:
   a) My Computer
   b) the C: drive
   c) the BEGIN folder.

4   Right-click on BEGIN and pick Explore and the palette area will display information about the selected folder. This may not be in the 'form' we want.

6   From the Design Center menu bar, scroll at Views and pick Large icons and the palette area will display icons for the contents of the BEGIN folder

7   The various selections from View in the menu bar are displayed in Fig. 47.5 and are:
   a) Large icons   *b*) Small icons   *c*) Listing   *d*) Details.

8   From the Design Center menu bar select Preview and Description and the palette side will display two additional areas. These can be resized by dragging the lower 'border' up or down.

9   Scroll at right of palette and pick (left-click) any drawing icon and a preview and description (if applicable) will be displayed similar to Fig. 47.6.

10   Scroll and pick another item (icon) If possible try and pick another type of file format. Figure 47.7 displays a BMP file of Fig. 40.4 which was the Block Definition dialogue box. This display is a screen dump pasted and saved in the Paint application package.

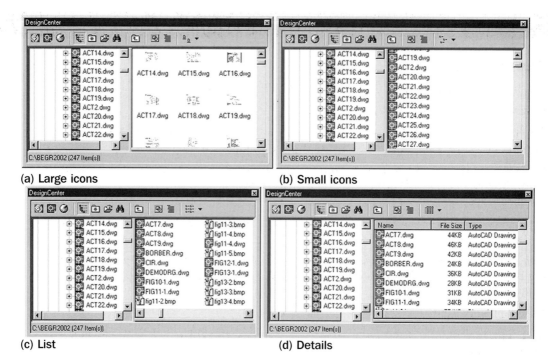

(a) Large icons (b) Small icons

(c) List (d) Details

Figure 47.5 Menu bar views.

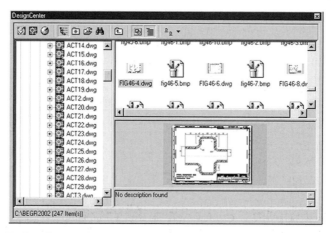

Figure 47.6 The Design Center tree view and palette for a selected drawing icon with the drawing Preview and Description displayed.

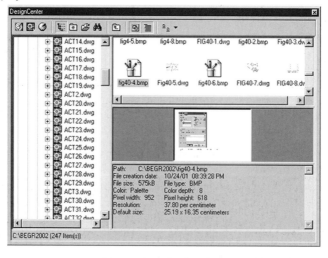

Figure 47.7 The Design Center tree view and palette with Preview and Description active for an 'opened' BMP file.

Exercise 2

1   Still with the Design Center displayed and A3PAPER opened?

2   Ensure that Preview and Description are 'active'.

3   Scroll in the tree area until the saved drawing from Chapter 41 is displayed. If you are unsure about this, it is Fig. 41.3 which had three wblocks inserted.

4   Expand this named drawing (your equivalence to Fig. 41.3) by picking the (+) at its name. The expansion will display the following names:
    Blocks, Dimstyles, Layers, Layouts, Linetypes, Textstyles and Xrefs

5   Right-click on the drawing name and pick Explore to display icons in the palette area of the seven named items in the tree view.

6   Pick Blocks from the palette area, right-click and Explore to display information about the blocks in the named drawing. As stated earlier, there should be three – BORDER, PLIST and TITLE.

7   Pick the block TITLE and note the display – Fig. 47.8.

8   Investigate selecting (left-click) the following from the expanded tree view side of the Design center:

Selection	*Result*
Dimstyles:	A3DIM listed and perhaps others?
Layers:	0 CL CONS Defpoints DIMS HID OUT SECT TEXT
Layouts:	not used but Layout1 and Layout2 displayed
Linetypes:	ByBlock ByLayer CENTER Continuous HIDDEN
Textstyles:	Standard ST1 and perhaps ST2, ST3 and ST4?
Xrefs:	none listed (obviously?).

9   Design Center menu bar with:
    a) Up: displays contents from the Blocks selection
    b) Up: displays the expanded BEGIN folder icons
    c) Up: displays the C: drive expansion
    d) Up: displays the My Computer icons.

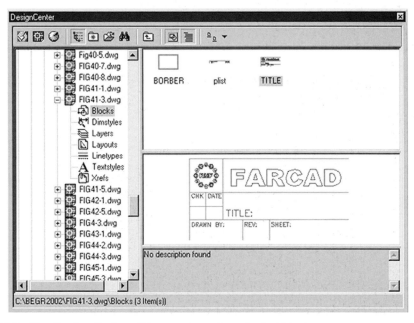

Figure 47.8   The Design Center with a preview of a selected block.

Exercise 3

1   Still have the Design Center displayed from exercise 2?

2   From the Design Center menu bar pick the **Find** icon and:
prompt   Find dialogue box
respond   1. scroll at **Look for**
          2. pick **Xrefs**
          3. scroll at In
          4. pick Local hard drives (C:)
          5. pick Browse
prompt   Browse for Folders dialogue box
respond   1. scroll and pick C:BEGIN
          2. pick OK
prompt   Find dialogue box
with   named folder name displayed
respond   1. at Search for the name, enter *
          2. pick **Find Now**
and   the search will begin
then   Find dialogue box
with   a list of the 'found' xrefs – Fig. 47.9.

3   Note that my Fig. 47.9 had the folder name C:\BEGR2002 and not BEGIN

4   Now cancel the Find dialogue box and cancel the Design Center.

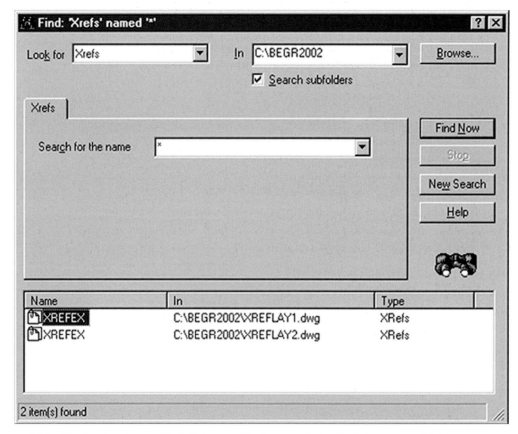

Figure 47.9   The Design Center Find dialogue with the search for Xrefs in a named folder.

Using the Design Center example 1

1 Close any existing drawings then menu bar with **File-New** and select Start from Scratch, Metric then OK.

2 Menu bar with Format-Layers, Format-Text Styles, Format-Dimension Styles and note the layers, text styles and dimension styles in this current drawing.

3 Activate the Design Center and position to suit, i.e. docked or floating.

4 Expand the following by picking the (+) at the appropriate folder/file:
   a) the C: drive
   b) the BEGIN folder
   c) the A3PAPER drawing file.

5 From the expanded A3PAPER tree, explore Layers to display the layers in the A3PAPER drawing in the palette side of the design centre.

6 From the palette side:
   a) left-click on the OUT layer
   b) hold down the left button
   c) drag into the drawing area
   d) repeat 'pick-and-drag' for the other layers.

7 Explore Dimstyles to display the dimension styles current in the A3PAPER drawing – Fig. 47.10.

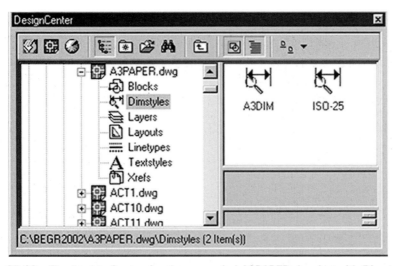

Figure 47.10 The Design Center for the expanded A3PAPER drawing with Dimstyles explored.

8 Left-click on A3DIM and drag into the drawing area.

9 Explore Textsystles and drag created text styles into the drawing area.

10 Now check the layers, Text styles and Dimension Styles in the current opened drawing. All those from A3PAPER are now available to the user in the opened drawing.

11 Collapse the expanded A3PAPER.

12 Menu bar with Insert-Block and there are no block names listed in the current drawing – obviously?

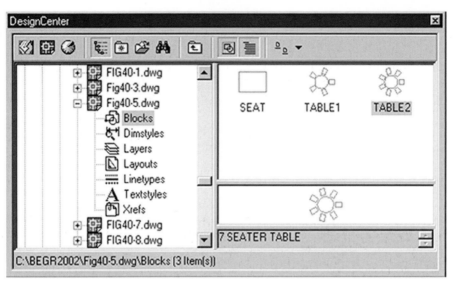

Figure 47.11 The Design Center for the expanded Fig. 40.5 with Blocks explored.

13 *a*) Expand the conference room drawing from Chapter 40 – this was Fig. 40.5 which
       (hopefully) you saved
    b) explore Blocks – Fig. 47.11
    c) left-click on each of the three blocks (SEAT, TABLE1 and TABLE2) and drag into the
       drawing area, positioning to suit
    d) menu bar with Insert-Blocks and the three 'pick and drag' blocks are listed.

14 This means that we have used the Design Center to 'insert' blocks from a previously saved
    drawing into the current opened drawing. This is ***contrary*** to the statement made in
    Chapter 40, that blocks are can only be used in the drawing in which they were saved.
    Thus the Design Center allows **any** blocks, wblocks, layers, linetypes, text styles,
    dimension styles, layouts and xrefs to be 'inserted' from any drawing into the current
    drawing. This is a very powerful aid to the CAD user.

15 This exercise is complete and can be saved.

16 Collapse all the expanded tree hierarchies and close the Design Center.

Using the Design Center example 2

In this example we will use the Design Center to investigate the AutoCAD blocks available to the user in various drawings which come with the package.

1   Close all existing drawing then open your A3PAPER standard sheet with layer OUT current

2   Activate the Design Center and position to suit.

3   Expand the following trees:
    a) C: drive
    b) Program Files – read note **
    c) AutoCAD 2002
    d) Sample
    e) Design Center.

4   The Design Center tree hierarchy should display a list of AutoCAD drawing names as Fig. 47.12.

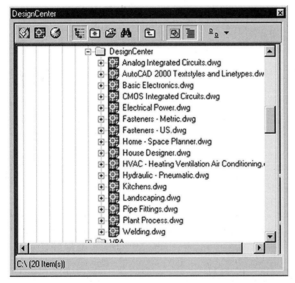

Figure 47.12   The Design Center tree hierarchy for the expanded C:\Program Files\AutoCAD 2002\Sample.

5   Note: Program Files is the folder into which my AutoCAD 2002 has been installed. Your system may use a different folder name.

6   Refer to Fig. 47.13 and 'insert' into the opened A3PAPER some of AutoCAD 2002's blocks. Remember that as they are blocks, you may have to scale after insertion. The following table is a list of the blocks I have inserted:

Ref	File name	Block	Scale
(a)	Analog Integrated Circuits	10104 Analog IC	1.5
(b)	Basic Electronics	Full Wave Bridge	1.5
(c)	CMOS Integrated Circuits	14160 CMOS IC	1
(d)	Electrical Power	Alarm	1.5
(e)	Fasteners – Metric	Cross Oval Screw	1
(f)	Fasteners – US	Hex Bolt	1
(g)	Home – Space Planner	Computer Terminal	0.1
(h)	Home Designer	Staircase – Spiral	0.05
(i)	HVAC	Fan and Motor	0.1
(j)	Hydraulic–Pneumatic	Valve–Servo	2
(k)	Kitchens	Range Oven	0.05
(l)	Landscaping	Tree–deciduous	0.025
(m)	Pipe Fittings	Elbow–Double Branch	1
(n)	Plant Process	Compressor–Recip	2
(o)	Welding	Seam Weld	3

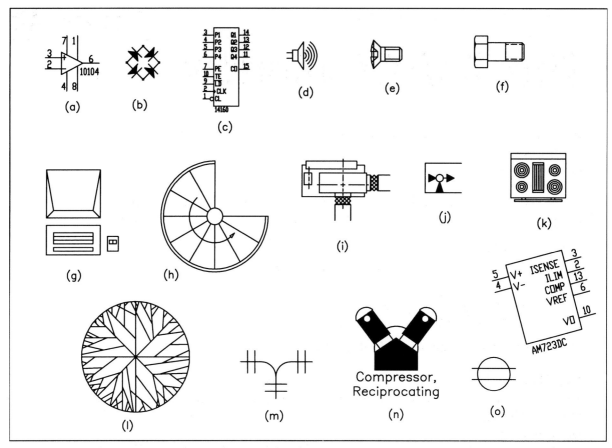

Figure 47.13 Insert blocks from various AutoCAD drawings using the Design Center.

7 If you are really feeling adventurous, you could create a drawing using any/or all of the AutoCAD drawing blocks.

8 The method of inserting blocks from the Design Center has so far only considered the Pick-and-drag method. There is another method which is more accurate than that used. To demonstrate this:

a) expand the AutoCAD Analog Integrated Circuits file

b) explore Blocks

c) from the palette right-click on the AM723DC block and:

  prompt   Shortcut menu
  respond pick Insert Block
  prompt   Insert dialogue box
  respond 1. Insertion at X: 340 Y: 110 (or enter own values)
       2. Scale with X: 40 Y:40
       3. Rotation angle: 15

d) The selected block will be inserted at the desired point.

9 *Question:* Why the large scale value using the dialogue box, when the pick-drag method did not need it?

The exercise is now complete and can be saved if required.

This chapter has introduced the user to the AutoCAD 2002 Design Center. Hopefully having tried the exercises, you will be able to navigate your way through it. It is fairly easy, but takes some patience and practice.

Chapter 48

The TODAY window

In Chapter 4, the TODAY window was mentioned with its three distinct sections of:
a) My Drawings
b) Bulletin Board
c) autoDesk Point A.

The TODAY window was to be only displayed on request and the AutoCAD traditional dialogue box was 'set' as the default. In this chapter, we will now investigate the TODAY topics

Getting started

1   Close any existing drawings then open your A3PAPER standard sheet
2   Either:
     a) enter **TODAY <R>** at the command line
     b) select the Today icon from the Standard toolbar

3   The complete TODAY window will be displayed

My Drawings

The My Drawings section allows the user to open a drawing by accessing one of three tabs, these being:

1   Open Drawings
     Allows access to previously opened drawings. The following Options (Fig. 48.1) are available:
     a) Most recently used
     b) History (by date)
     c) History (by filename)
     d) History (by location)
     e) Browse.

2   Create Drawings
     This tab allows the user to create a new drawing from one of three Options:
     a) Template: opens AutoCAD's template file with access to the A, D, G, I, J and M templates (Fig. 48.2) or the user can Browse for templates in other folders.
     b) Start from Scratch: allows access to the English or Metric default AutoCAD drawings – Fig. 48.3.
     c) Wizard: lets the user select the Quick or Advanced setup for starting drawings – Fig. 48.4. The Quick setup allows to user to set the units and drawing area, while with the advanced setup, the user sets units, angle settings, the drawing area, title block and actual drawing layout.
     d) Note: at this stage, I would expect the reader to understand the various tab options in this section.

3   Symbol Libraries
     The symbol libraries tab allows the user to start a new drawing and access the library of symbols via the Design Centre. Figure 48.5 displays a typical example of using this option.

Note: I would suggest that at this stage, the reader is familiar with the various tab options available from the My Drawings section of the TODAY window. We have discussed in previous chapters how to open drawings and templates as well as how to use the Design Center. The TODAY window is another method of activating these options.

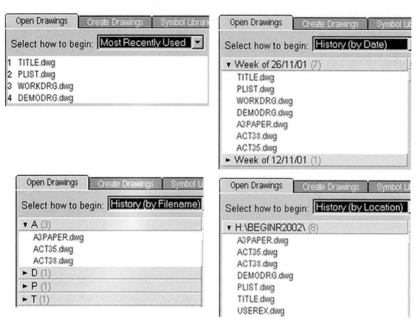

Figure 48.1 The My Drawings – Open Drawings tab selections.

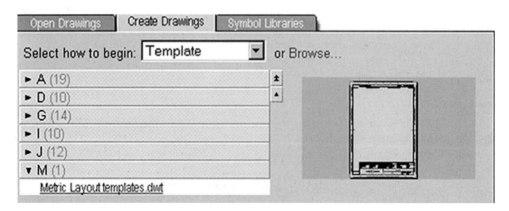

Figure 48.2 My Drawings (Create-Template tab) options.

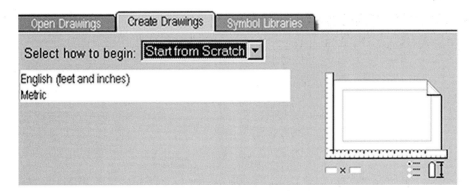

Figure 48.3 My Drawings (Create-Start from Scratch tab) options.

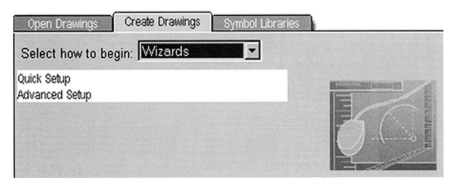

Figure 48.4 My Drawings (Wizards tab) options.

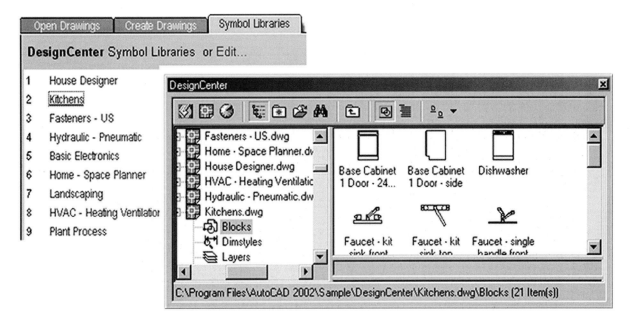

Figure 48.5 My Drawings (Symbol Libraries tab) with Kitchens selected and explored.

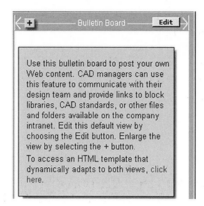

Figure 48.6   The Bulletin Board 'as opened'.

Bulletin Board

The Bulletin Board is a CAD management tool and allows information to be passed to various 'departments' in an organisation. This information could be a memo, information about a new contract, alterations to an existing drawing, etc. The 'message' contained in the Bulletin Board can be 'written' using any word processor. The Bulletin Board 'as opened' in the TODAY window is displayed in Fig. 48.6.

To demonstrate how to create a Bulletin Board message:

1   Start AutoCAD or begin a new drawing from scratch.

2   From the Windows taskbar select **Start-Programs-Accessories-Notepad** and:
    prompt   Notepad screen which is blank
    respond enter the following lines of text, remembering <R> at end of each line:
        From: Bob McFarlane, Curriculum Manager
        To: Helen Lawson, Senior Lecturer CAD and Media
              Todd Dunlop, Lecturer CAD
              Barry Skea, Lecturer CAD
              Michael McGuire, Lecturer CAD
        Message: Please attend a meeting in R1443 at 08.30 on Friday 7th
                  December.
        Agenda: 1. Course Reviews
                  2. Block 2 results
                  3. Student progress
                  4. Block 3 timetables
                  5. AOCB.

3   Menu bar with **File-Save As** and:
    prompt   Save As dialogue box
    respond 1. at Save in, scroll and pick a suitable folder, e.g. Windows-Temp
           2. enter filename: MEETING3
           3. save as type: Text document
           4. pick Save
           5. minimise Notepad to return to AutoCAD.

4   From the TODAY window pick the **Edit** button and:
   prompt   Program Files dialogue box
   respond 1. pick **Browse**
              2. scroll (navigate) to folder of saved text file
              3. pick MEETING3 then Open
   prompt   Program Files dialogue box, similar to Fig. 48.7
   respond **pick Save Path**
   and   Message displayed in Bulletin Board – Fig. 48.8.

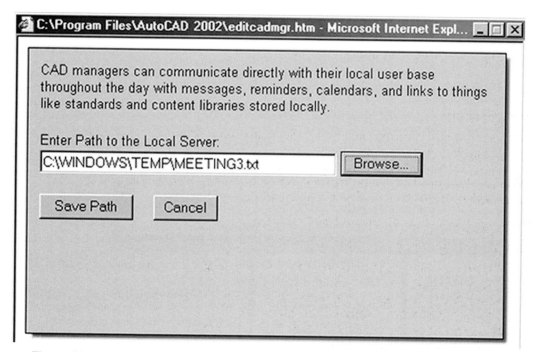

Figure 48.7   Program Files dialogue box with path displayed.

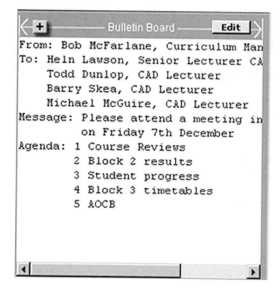

Figure 48.8   The Notepad text file 'posted' to the Bulletin Board.

autoDESK Point A

The autodesk point A is a facility which allows users to access information specific to AutoCAD, e.g. industry specific news, resources, catalogues, links, etc. To use the facility, the user must have Internet connections. Figure 48.9 displays part of the main point A screen.

If you have Internet access on your system:

1 Left-click on the (+) to open autodesk point A

2 Scroll at right until New Books is displayed

3 Pick Search and:
    prompt   `Internet Search screen`
    respond   add the following data:
           1. Author: Bob McFarlane
           2. Title: AutoCAD*
           3. Publisher: Arnold Butterworth Heinemann (Fig. 48.10)
    then   **pick Search Now**
    and   screen displays search results similar to Fig. 48.11.

4 Now cancel the various opened internet sites.

This demonstration of the TODAY window is now complete.

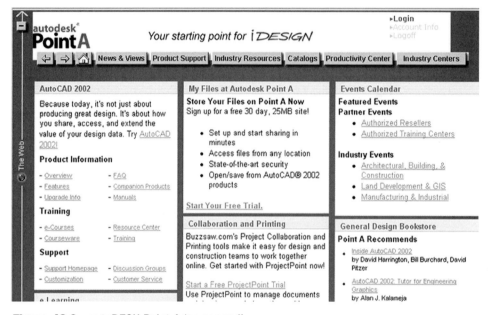

Figure 48.9 autoDESK Point A 'as opened'.

Figure 48.10 Search data information.

3. **Beginning AutoCAD 2000**
   by McFarlane, Bob McFarlane (Paperback)
   Usually ships in 2 to 3 days
   Other Editions: Paperback | All Editions

   List Price: $44.99
   Our Price: $44.99

   Add to cart

   Or buy used: $32.00

4. **Modelling With Autocad 2000**
   by McFarlane, Bob McFarlane (Paperback - October 2000)
   Usually ships in 24 hours

   List Price: $39.99
   Our Price: $39.99

   Add to cart

5. **Modelling With Autocad Release 14 for Windows Nt 8**
   by Bob McFarlane (Paperback - July 1998)
   Special Order
   Other Editions: Paperback | All Editions

   List Price: $18.99
   Our Price: $18.99

   Add to cart

   Or buy used: $12.00

Figure 48.11   Part of the screen display with search data.

Chapter 49

'Electronic' AutoCAD

AutoCAD 2002 allows the user to generate electronic drawing files, these files being in Drawing Web Format (DWF).These DWF files can be opened, viewed or plotted by third party persons having access to Volo View or Volo View Express.

In this final chapter, we will investigate several of these electronic generating devices.

Creating a DWF file

DWF files are 'at the heart' of AutoCAD' electronic generation, and to demonstrate how these are created:

1   Close any existing drawing files.

2   Open any of your saved AutoCAD activities, e.g. ACT30.

3   Menu bar with **File-Plot** and:
    prompt   Plot dialogue box
    respond   activate the named tab and set as detailed:
    A Plot Device tab active
        1. Plotter configuration name:
           DWF eView (optimized for viewing) pc3
        2. Plot to file:
           File name: **TRIAL**
           Location: **named folder** – dialogue box similar to Fig. 49.1
    B Plot Setting tab active
        1. Paper size: ISOA3(420x297)
        2. Plot area: Extents
        3. Drawing orientation: Landscape
        4. Plot scale: Scaled to fit
        5. Full preview and:
           prompt   Preview of activity drawing
           respond   right-click and plot.

4   The command line will display 'normal plot' information

5   Menu bar with **File-Save As** and:
    a) alter file name to ACT30(ET) – for future use
    b) ensure saving as a *.dwg file
    c) pick Save.

6   Now exit AutoCAD.

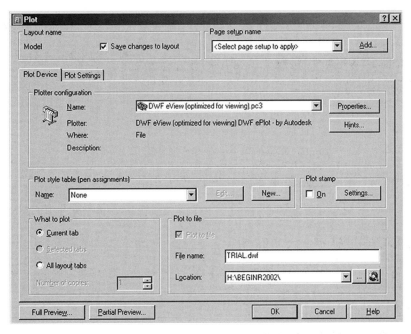

Figure 49.1 The Plot dialogue box (Plot Device tab) for creating a DWF file from a DWG file.

Volo View Express

If you have access to Volo View (normally loaded when AutoCAD 2002 is installed) then:

1 Open Volo View with the taskbar sequence **Start-Programs-Volo View Express-Volo View Express** (or similar sequence)

2 Menu bar with **File-Open** and:

prompt   Open dialogue box

respond   1. scroll and select your named folder

        2. scroll and select **Drawing Web Format (*.dwf)** file type

        3. dialogue box similar to Fig. 49.2

        4. pick TRIAL then Open

and   screen should display the saved activity 'drawing'

3 Menu bar with **View-Zoom-All**

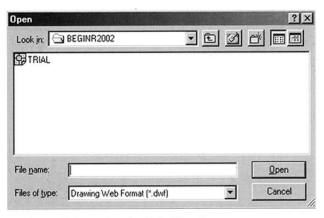

Figure 49.2 The Open dialogue box for Volo View Express.

4 Why use Volo View?
   a) Volo View is 'tool' which allows third parties to view AutoCAD drawings without the need for the full AutoCAD draughting package to be loaded on their system. This is useful to managers and also saves money as Volo View is cheaper than AutoCAD
   b) With Volo View, it is possible to add text and sketch. This could be notes relating to a wrong dimension or suggestions to improve the design. Again this is a useful tool for managers, designers, marketing, etc.
   c) The Markup menu bar item in Volo View allows the user access to the text and sketch facilities
   d) A modified Markup screen can be saved in the following file formats:
       1. RedlineXML (*.rml)
       2. Drawing Web Format (*.dwf)
       3. AutoCAD DXF (*.dxf)
   e) Figure 49.3 displays a 'screen dump' of the ACT30 drawing opened as TRIAL in Volo View with annotations (text and sketch) added.
   f) Markup 'drawings' can be inserted into existing AutoCAD drawings with the menu bar sequence **Insert-Markup.**
   g) Note:
       1. I found it impossible to insert a Volo View markup drawing into an existing AutoCAD drawing – perhaps it was 'my system'.
       2. I found that if I opened a Volo View saved DXF file in AutoCAD it was then possible to insert an AutoCAD drawing. Figure 49.4 is an opened Volo View DXF file with ACT30 inserted, with the suggested moifications.

5 If you have tried this exercise, close Volo View and return to AutoCAD with the opened activity (ACT30) displayed.

This exercise is now complete.

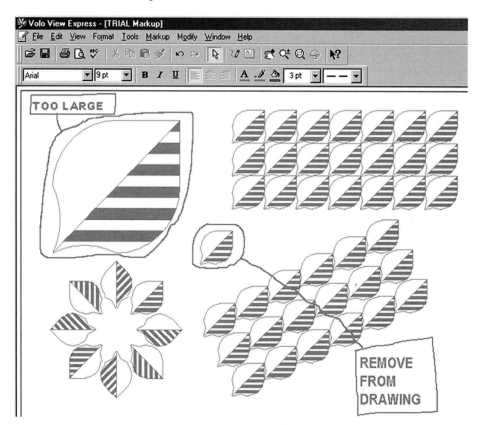

Figure 49.3 Volo View screen with text and sketching added.

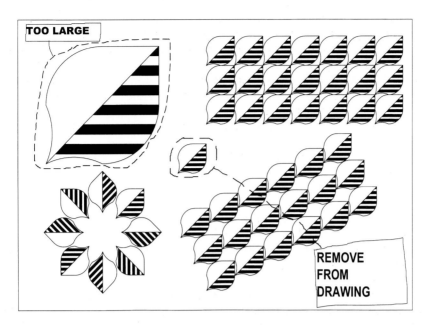

Figure 49.4 ACT30 with Volo View details inserted.

e-Transmit

eTransit allows the user to create a set of AutoCAD drawings (only DWG or DWT formats) which can be posted to the Internet or sent to others as an e-mail attachment. The process generates a report file which allows the user to add notes and a password if required. The files to be transmitted can be stored by the user:
1. in a named folder
2. in a created self-executable or zip file.

We will demonstrate the concept by example (which you should only attempt if you have e-mail access), so:

1 Close any existing drawings then open ACT30(ET).

2 Menu bar with **File-eTransmit** and:
   prompt   Create Transmittal dialogue box
   respond 1. Notes: add as required
              2. Type: Self-extracting executable (*.exe)
              3. Browse: pick Browse and:
   prompt   Specify self-extracting executable dialogue box
   respond 1. scroll and pick named folder
              2. file name should be ACT(30)ET).exe
              3. pick Save
   and   dialogue box similar to Fig. 49.5
   respond pick OK.

3 The command line will return the message:
   Transmittal created: folder\ACT(30)ET.exe.

4 Exit AutoCAD, open Windows Explorer and:
   a) Navigate to your named folder
   b) Arrange the icons by name
   c) Right-click on the new ACT(30)ET icon and pick Properties from the displayed Shortcut menu
   d) Note the dialogue box display similar to Fig. 49.6 then Cancel

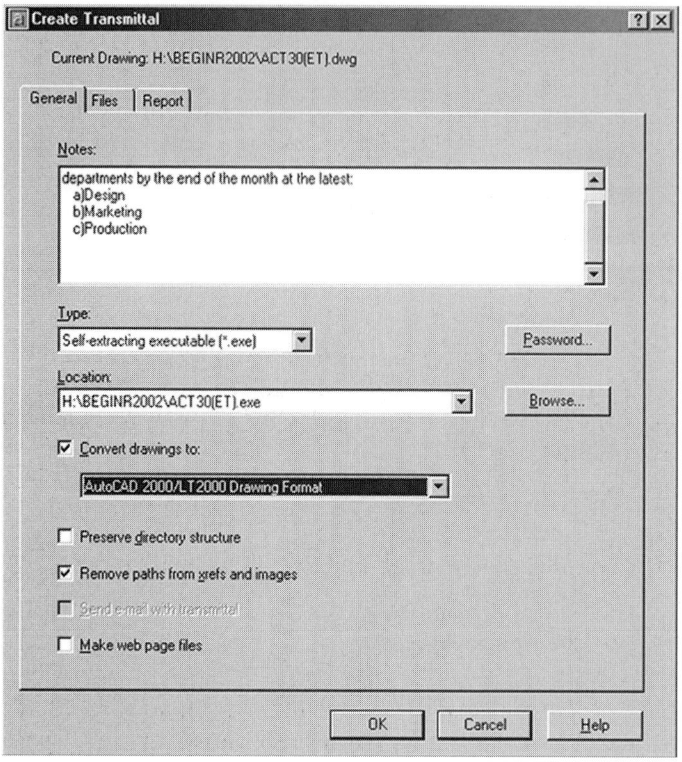

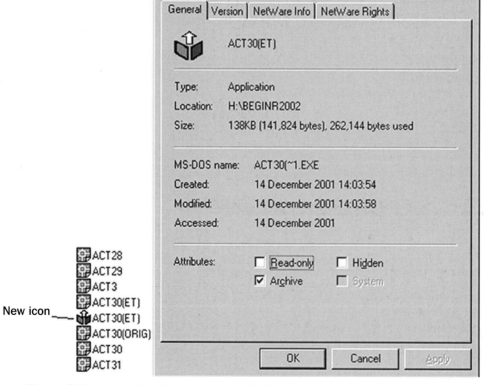

Figure 49.5 Create Transmittal dialogue box.

Figure 49.6 Properties dialogue box for ACT(30)ET.

5   To demonstrate how an eTransmit file is 'unpacked', double left-click on the ACT30(ET) executable file and:

prompt   eTransmit dialogue box
respond   1. pick Browse
          2. scroll and pick a new folder for extracting the file
          3. pick OK
prompt   eTransmit dialogue box similar to Fig. 49.7
respond   pick OK
prompt   % display of extraction status
then   eTransmit dialogue box similar to Fig. 49.8
respond   pick OK and Explorer screen returned.

6   *a*) From Explorer, scroll and pick the named folder to which the eTransmit was extracted (in my case the folder was named CMG). The display should be similar to Fig. 49.9.
    b) Double left-click on the ACT30(ET) text file and Notepad will display the Transmittal report similar to Fig. 49.10. Cancel the text file.
    c) Double left-click on the ACT30(ET) drawing file and AutoCAD will be opened and display the transmitted file.

7   AutoCAD files can be attached and send as a e-mail, and Fig. 49.11 is a typical e-mail screen with two attachments:
    a) the ACT30(ET).dwg file
    b) the ACT30(ET).exe file.

8   While an AutoCAD drawing file can be sent as an attachment, the file may be very large and this can cause 'problems' when it is being 'opened' by the receiver. An exe (or zipped) file may be easier to send electronically.

    This exercise is now complete.

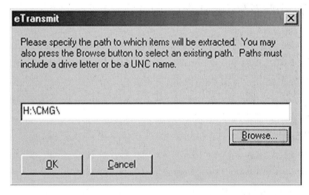

Figure 49.7   eTransmittal dialogue box with folder name for extraction.

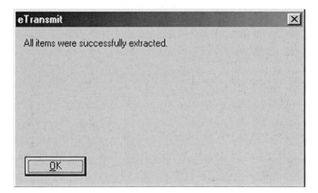

Figure 49.8   eTransmittal dialogue box at end of extraction.

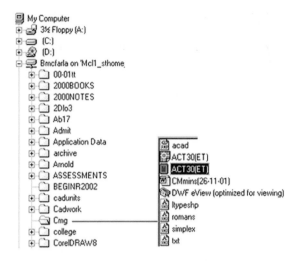

Figure 49.9   Expansion of CMG folder.

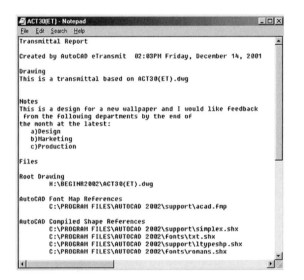

Figure 49.10   Part of the ACT(30)ET text file.

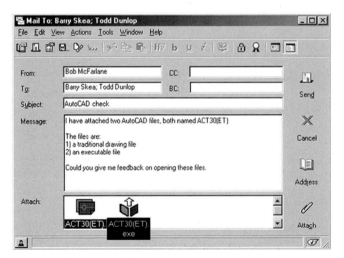

Figure 49.11   Sending files by e-mail.

Publishing to the web

AutoCAD 2002 allows the user to create web pages of existing drawings. To demonstrate the concept:

1   Start a new metric drawing from scratch to display the typical AutoCAD 2002 blank screen

2   Menu bar with **File-Publish to Web** and:
   prompt   Publish to Web (Begin) dialogue box (as Layout dialogue box)
   respond   1. pick Create New Web Page
           2. pick New
   prompt   AutoCAD message similar to Fig. 49.12
   respond   1. read the message
           2. pick Cancel

3   Now open any drawing, e.g. ACT28, then menu bar with File-Publish to Web and:
   prompt   Publish to Web (Begin) dialogue box
   respond   1. pick Create New Web Page
           2. pick New
   prompt   Publish to Web (Create Web Page) dialogue box
   respond   1. Web page name: enter MYPAGE
           2. Note parent directory
           3. Add any suitable description – Fig. 49.13
           4. pick Next
   prompt   Publish to Web (Select Image Type) dialogue box
   respond   1. select type from list : DWF
           2. pick Next
   prompt   Publish to Web (Select Template) dialogue box
   respond   1. select a template type, e.g. Array plus Summary
           2. pick Next
   prompt   Publish to Web (Apply Theme) dialogue box
   respond   1. scroll and select an element, e.g. Ocean Waves – Fig. 49.14
           2. pick Next
   prompt   Publish to Web (Enable i-drop) dialogue box
   respond   activate i-drop then pick Next
   prompt   Publish to Web (Select Drawings) dialogue box
   with   your opened ACT28 drawing listed
   respond   1. Layout: select Model (probably active)
           2. Label: alter to GEAR
           3. Description: enter to sent
           4. Pick Add →
   and   GEAR added to list
   respond   Pick the (. . .) at Drawing
   prompt   Publish to Web dialogue box
   with   *.dug file type
   respond   1. Scroll to named folder
           2. Pick another drawing or activity
           3. Layout: Model
           4. Label: alter to suit
           5. Description: alter to suit
           6. Pick Add →
   then   Pick (. . .) and select another 2 or 3 drawings and repeat the above six responses
   and   Pick Next

prompt   Publish to Web (Generate Images) dialogue box
respond   1. Regenerate images for drawing (etc.) active
           2. pick Next
prompt   Plot progress information displayed
then   Publish to Web (Preview and Post) dialogue box
respond   pick Preview
and   Internet Explorer with Images of drawings – Fig. 49.16
respond   1. view your images
           2. close the Internet to return to AutoCAD
and   select Finish

4   You have now create a web page which can be:
    a) edited to your requirements
    b) posted to the Internet

5   Note:
    a) I hope that in this chapter the user has realised that AutoCAD 2002 has uses other
        than drawing. The web page creation is very useful and relatively simple to create
    b) Figure 49.17 is a detailed listing the the web page data MYPAGE and it is surprising
        the information required for a simple web page creation?

    This exercise is now complete and it is the end of the book.

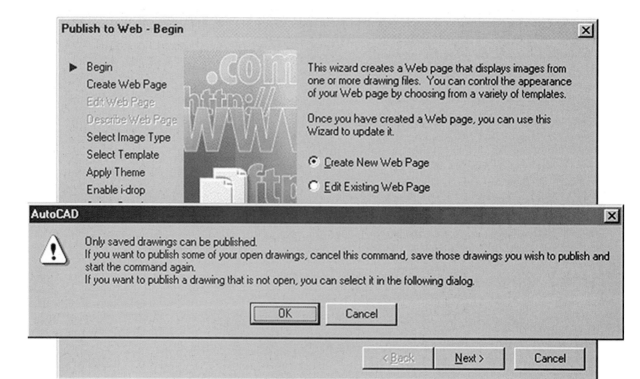

Figure 49.12 The AutoCAD message screen.

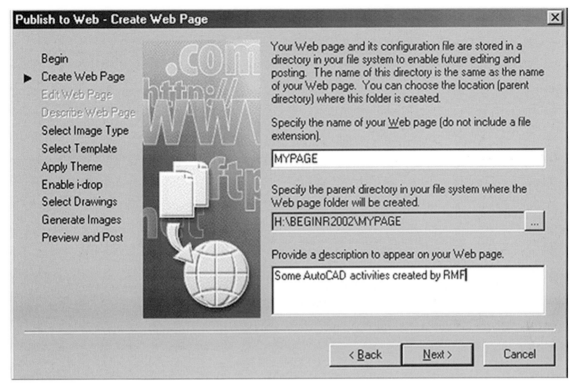

Figure 49.13   The Publish to Web (Create Web Page) dialogue box.

Figure 49.14   Publish to Web (Apply Theme) dialogue box.

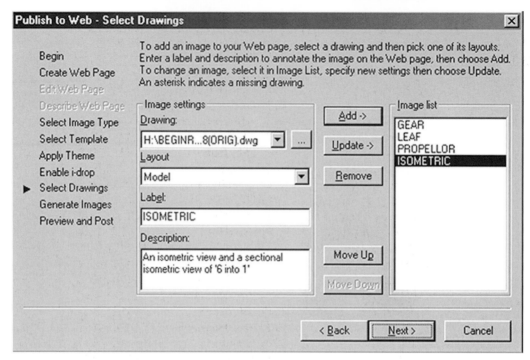

Figure 49.15 Publish to Web (Select Drawings) dialogue box.

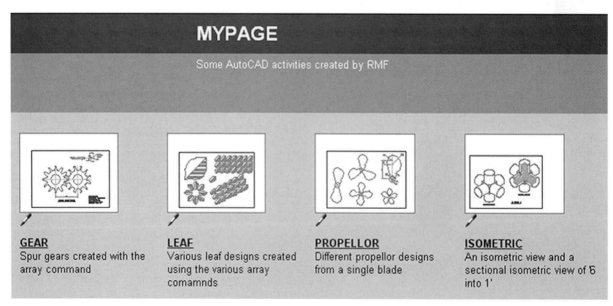

Figure 49.16 Preview of web page design with drawings.

Name	Size	Type	Modified
ACT28(ORIG)	42KB	AutoCAD Drawing	19/12/01 11:47
ACT29(ORIG)	37KB	AutoCAD Drawing	19/12/01 11:46
ACT30(ORIG)	96KB	AutoCAD Drawing	13/12/01 13:21
ACT38(ORIG)	32KB	AutoCAD Drawing	19/12/01 11:47
acwebpublish	2KB	Cascading Style She...	01/12/00 16:40
acwebpublish	5KB	HTML Document	18/01/01 16:51
acwebpublish	1KB	Internet Shortcut	19/12/01 15:46
acwebpublish	2KB	XML Document	19/12/01 16:06
acwebpublish_fra...	5KB	HTML Document	18/01/01 16:46
acwebpublish_fra...	3KB	HTML Document	18/01/01 17:26
adsk_ptw_array_o...	8KB	JScript File	12/01/01 14:05
adsk_ptw_content...	2KB	JScript File	04/12/00 15:47
adsk_ptw_page_d...	2KB	JScript File	05/12/00 11:08
adsk_ptw_page_title	2KB	JScript File	05/12/00 11:15
adsk_ptw_validate...	2KB	JScript File	15/12/00 15:22
iDrop_1	1KB	XML Document	19/12/01 16:06
iDrop_2	1KB	XML Document	19/12/01 16:06
iDrop_3	1KB	XML Document	19/12/01 16:06
iDrop_4	1KB	XML Document	19/12/01 16:06
iDropButton	1KB	GIF Image	17/11/00 15:15
IM1	7KB	DrawingWebFormat	19/12/01 15:23
IM1a	4KB	JPEG Image	19/12/01 15:23
IM2	4KB	DrawingWebFormat	19/12/01 15:23
IM2a	5KB	JPEG Image	19/12/01 15:23
IM3	3KB	DrawingWebFormat	19/12/01 15:23
IM3a	5KB	JPEG Image	19/12/01 15:23
IM4	8KB	DrawingWebFormat	19/12/01 15:23
IM4a	4KB	JPEG Image	19/12/01 15:23
MYPAGE	1KB	AutoCAD Publish to ...	19/12/01 16:06
template_preview	64KB	Bitmap Image	30/11/00 16:07
xmsg_adsk_ptw_all	2KB	JScript File	09/02/01 11:16

BEGINR2002
Mypage

Figure 49.17 Listing of the saved MYPAGE web data.

ACTIVITY 1

Draw the simple shapes using the LINE and CIRCLE commands.
Set the grid and snap spacing to suit.

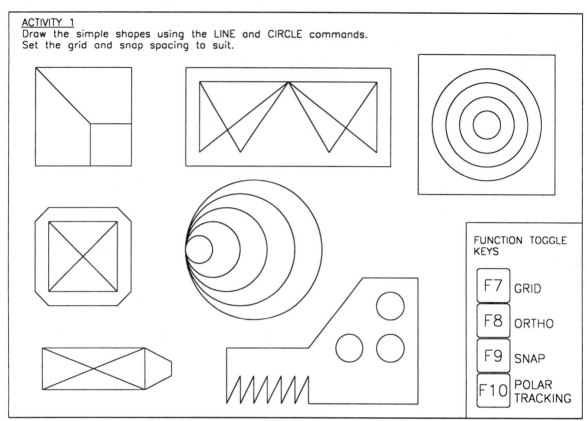

FUNCTION TOGGLE KEYS

F7	GRID
F8	ORTHO
F9	SNAP
F10	POLAR TRACKING

ACTIVITY 2

Draw the three templates using the sizes
given. Suggested start points are:
A(80,110), B(260,200) and C(380,40).

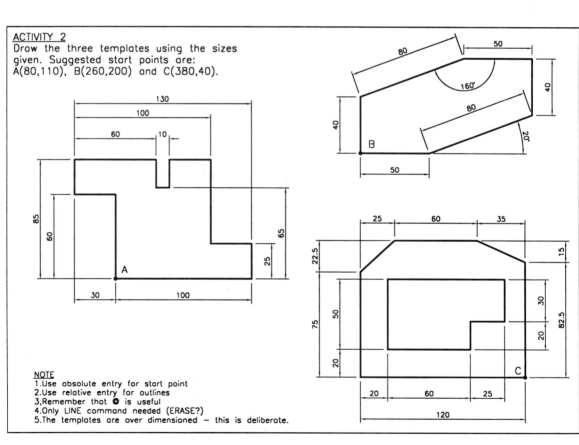

NOTE
1. Use absolute entry for start point
2. Use relative entry for outlines
3. Remember that ⊕ is useful
4. Only LINE command needed (ERASE?)
5. The templates are over dimensioned – this is deliberate.

ACTIVITY 3
Draw the three shapes using the information given. The recommended start points are:
A: (60,60)
B: (180,260)
C: (270,70)

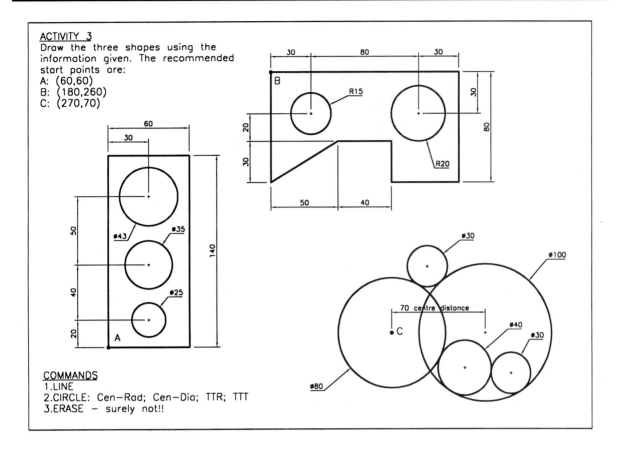

COMMANDS
1. LINE
2. CIRCLE: Cen-Rad; Cen-Dia; TTR; TTT
3. ERASE – surely not!!

ACTIVITY 4
Draw the three shapes using the object snap modes with the hints given.
The start points are:
A: (210,195)
B: (30,35)
C: (300,60)

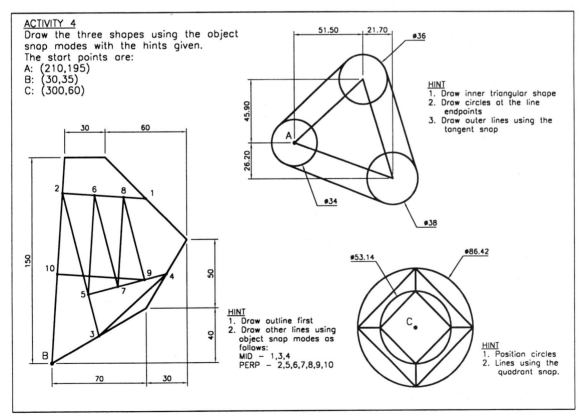

HINT
1. Draw inner triangular shape
2. Draw circles at the line endpoints
3. Draw outer lines using the tangent snap

HINT
1. Draw outline first
2. Draw other lines using object snap modes as follows:
MID – 1,3,4
PERP – 2,5,6,7,8,9,10

HINT
1. Position circles
2. Lines using the quadrant snap.

ACTIVITY 5
Draw full size the two templates.
Suggested start points are:
A: (80,160)
B: (370,90)

USE DISCRETION FOR SIZES NOT GIVEN

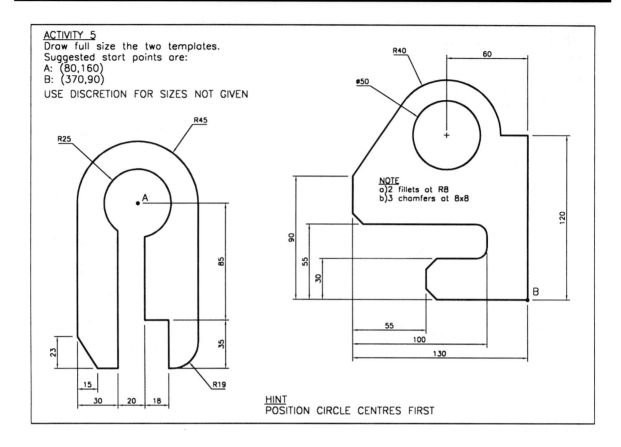

NOTE
a) 2 fillets at R8
b) 3 chamfers at 8x8

HINT
POSITION CIRCLE CENTRES FIRST

ACTIVITY 6
Draw the four shapes to the sizes given.
Make use of OFFSET and TRIM where possible.
Add all text, but DO NOT dimension
Discretion for start points and dimensions not given.

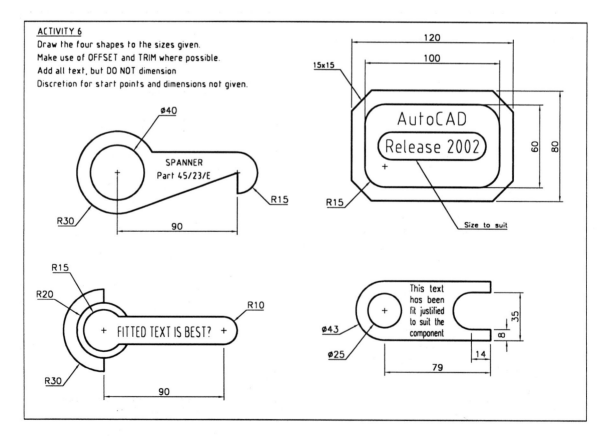

SPANNER
Part 45/23/E

AutoCAD
Release 2002

Size to suit

FITTED TEXT IS BEST?

This text
has been
fit justified
to suit the
component

ACTIVITY 7
1. Draw the two components full size
2. Add all text
3. Add all dimensions
4. Use layers correctly
5. Suggested start points are A: (50,70) and B: (380,220)

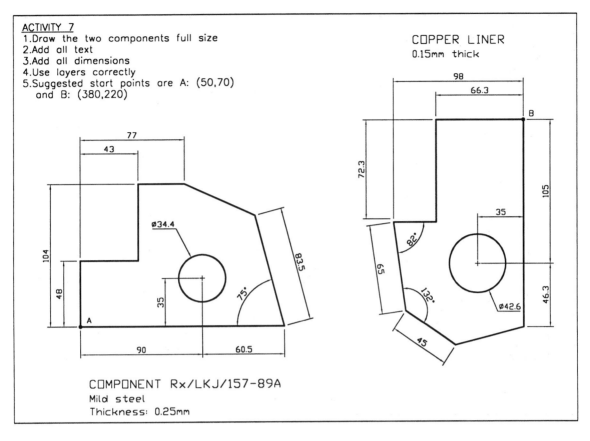

COMPONENT Rx/LKJ/157-89A
Mild steel
Thickness: 0.25mm

COPPER LINER
0.15mm thick

ACTIVITY 8
Draw, dimension and add text. Use layers correctly.

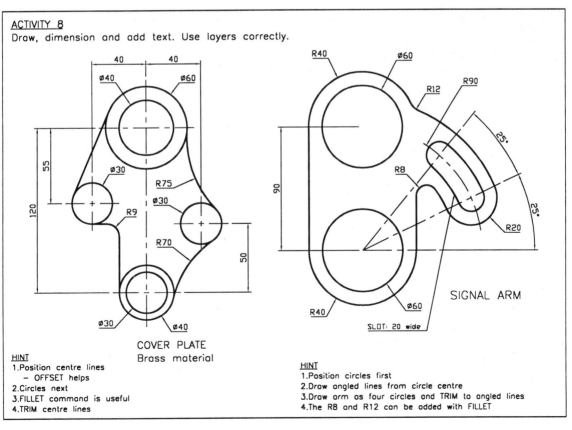

COVER PLATE
Brass material

SIGNAL ARM

HINT
1. Position centre lines
   – OFFSET helps
2. Circles next
3. FILLET command is useful
4. TRIM centre lines

HINT
1. Position circles first
2. Draw angled lines from circle centre
3. Draw arm as four circles and TRIM to angled lines
4. The R8 and R12 can be added with FILLET

ACTIVITY 9
Draw, fully dimension and add text
Easier than you think!!!

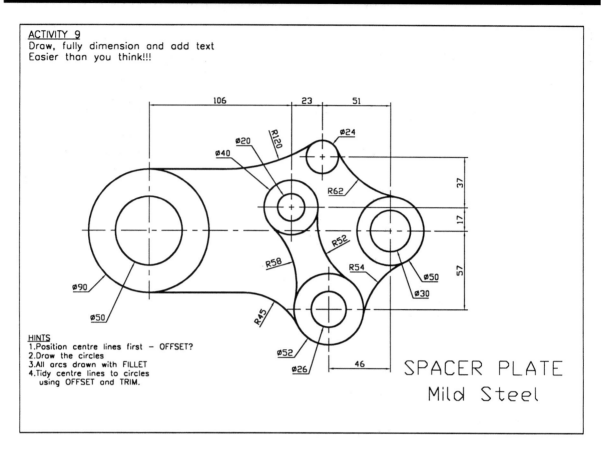

HINTS
1. Position centre lines first — OFFSET?
2. Draw the circles
3. All arcs drawn with FILLET
4. Tidy centre lines to circles
   using OFFSET and TRIM.

SPACER PLATE
Mild Steel

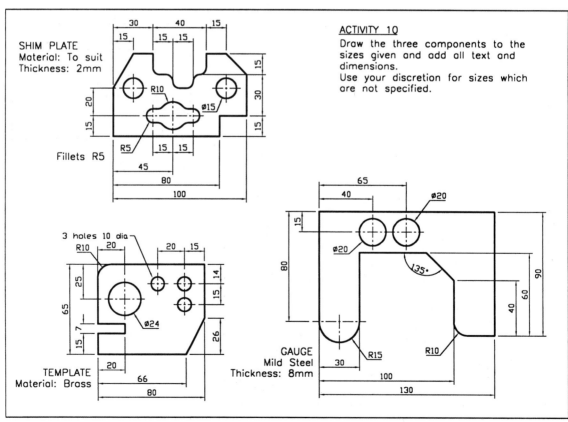

SHIM PLATE
Material: To suit
Thickness: 2mm

Fillets R5

ACTIVITY 10
Draw the three components to the
sizes given and add all text and
dimensions.
Use your discretion for sizes which
are not specified.

3 holes 10 dia

TEMPLATE
Material: Brass

GAUGE
Mild Steel
Thickness: 8mm

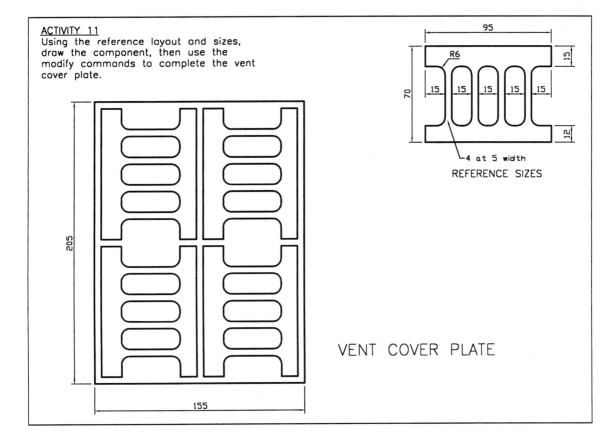

ACTIVITY 11
Using the reference layout and sizes,
draw the component, then use the
modify commands to complete the vent
cover plate.

REFERENCE SIZES

4 at 5 width

VENT COVER PLATE

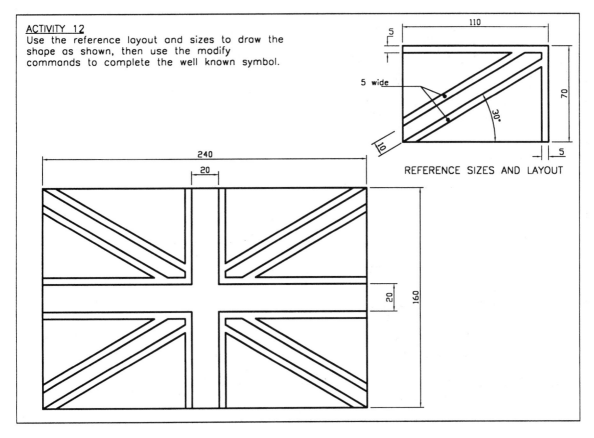

ACTIVITY 12
Use the reference layout and sizes to draw the
shape as shown, then use the modify
commands to complete the well known symbol.

REFERENCE SIZES AND LAYOUT

5 wide

ACTIVITY 13
Draw the template full size using the reference sizes given.

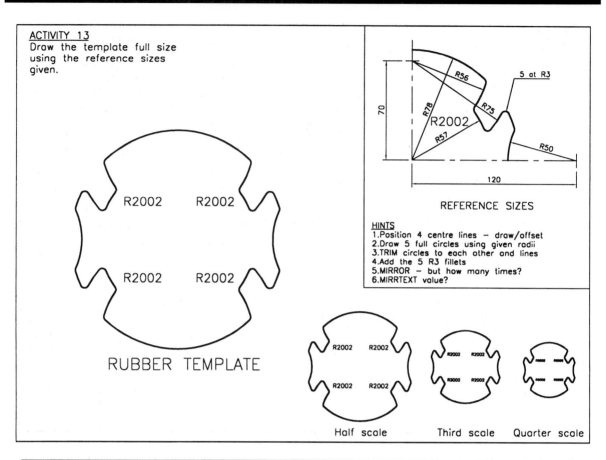

REFERENCE SIZES

HINTS
1. Position 4 centre lines – draw/offset
2. Draw 5 full circles using given radii
3. TRIM circles to each other and lines
4. Add the 5 R3 fillets
5. MIRROR – but how many times?
6. MIRRTEXT value?

RUBBER TEMPLATE

Half scale   Third scale   Quarter scale

ACTIVITY 14
Draw the two components using the sizes given. Use your layers correctly and add the dimensions for additional practice.

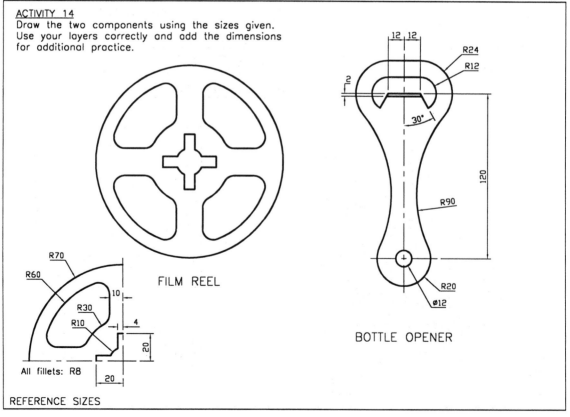

FILM REEL

BOTTLE OPENER

All fillets: R8

REFERENCE SIZES

ACTIVITY 15
Robotic arm re-alignment using only grips.

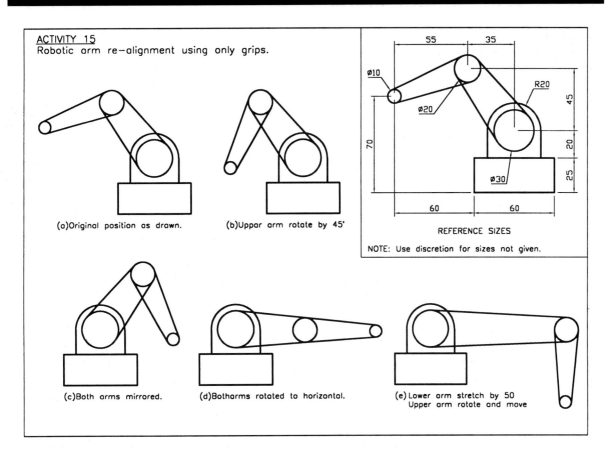

(a)Original position as drawn.

(b)Upper arm rotate by 45°

REFERENCE SIZES

NOTE: Use discretion for sizes not given.

Ø10 Ø20 R20 45 70 20 25 Ø30 55 35 60 60

(c)Both arms mirrored.

(d)Both arms rotated to horizontal.

(e) Lower arm stretch by 50
Upper arm rotate and move

ACTIVITY 16
Draw the three components using the sizes given, and add the hatching. Layers have to be used correctly, and use your discretion for any sizes not given.

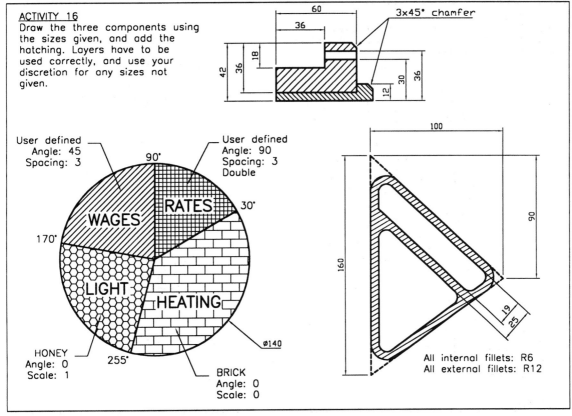

60 36 18 36 42 3x45° chamfer 30 36 12

User defined
Angle: 45
Spacing: 3

90°

User defined
Angle: 90
Spacing: 3
Double

RATES

30°

WAGES

170°

LIGHT

HEATING

HONEY
Angle: 0
Scale: 1

255°

Ø140

BRICK
Angle: 0
Scale: 0

100 90 160 19 25

All internal fillets: R6
All external fillets: R12

ACTIVITY 17
Draw the three views of the component
and add hatching, text and dimensions.
Use lyaers correctly.

COVER PLATE

Material: Mild Steel

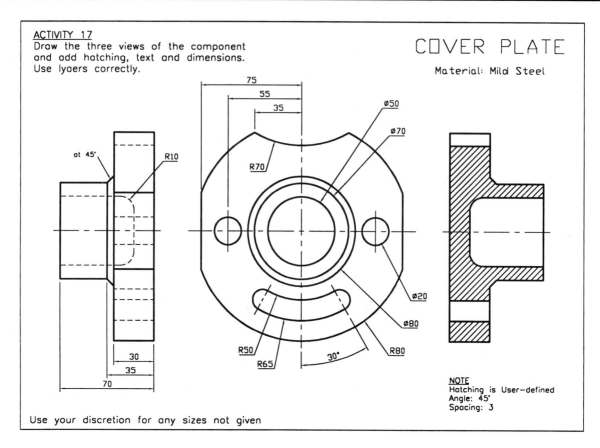

NOTE
Hatching is User-defined
Angle: 45°
Spacing: 3

Use your discretion for any sizes not given

ACTIVITY 18
Draw the four views and add the hatching, text and diimensions.

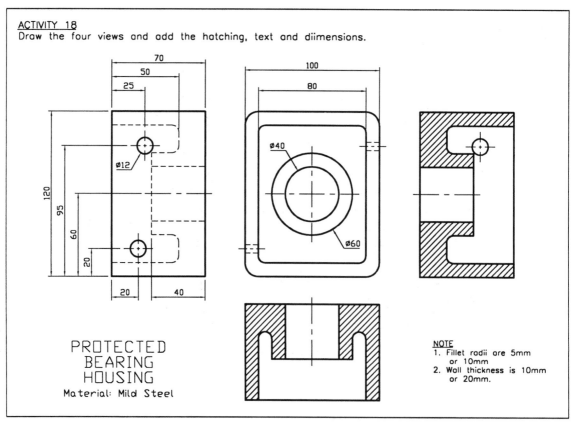

PROTECTED
BEARING
HOUSING
Material: Mild Steel

NOTE
1. Fillet radii are 5mm
   or 10mm
2. Wall thickness is 10mm
   or 20mm.

ACTIVITY 19
Draw the component and add text, hatching and dimensions.

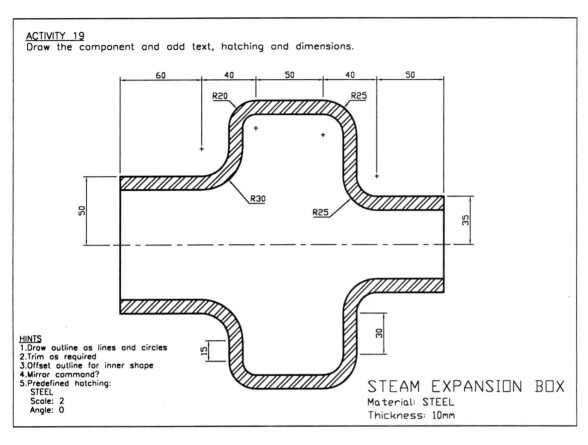

HINTS
1. Draw outline as lines and circles
2. Trim as required
3. Offset outline for inner shape
4. Mirror command?
5. Predefined hatching:
   STEEL
   Scale: 2
   Angle: 0

STEAM EXPANSION BOX
Material: STEEL
Thickness: 10mm

ACTIVITY 20
Draw full size, adding the hatching.

GASKET COVER

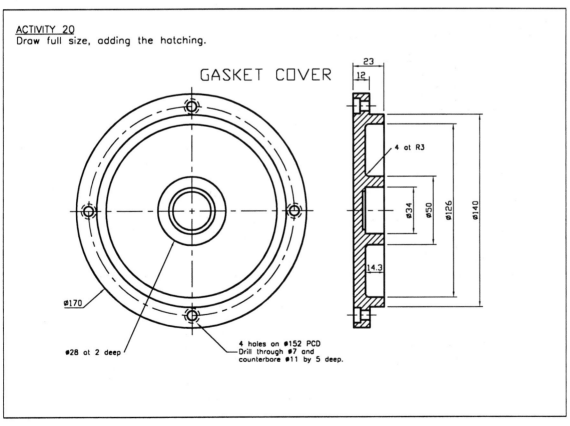

4 at R3

4 holes on ⌀152 PCD
Drill through ⌀7 and
counterbore ⌀11 by 5 deep.

⌀28 at 2 deep

⌀170

ACTIVITY 21
Draw the games board
using the information
given.

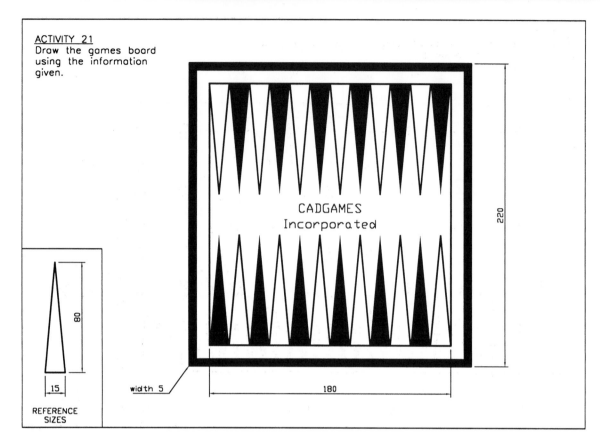

REFERENCE
SIZES

ACTIVITY 22
Draw the polyline shapes using the
sizes given.

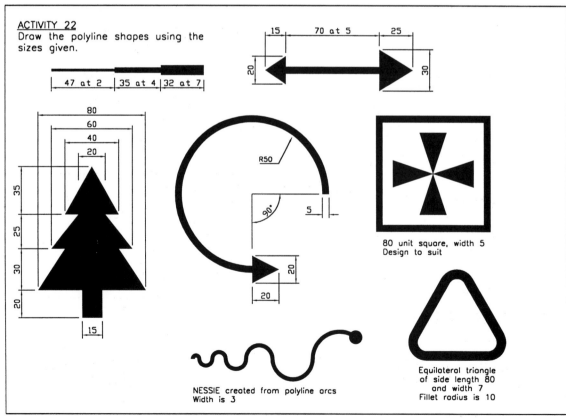

80 unit square, width 5
Design to suit

NESSIE created from polyline arcs
Width is 3

Equilateral triangle
of side length 80
and width 7
Fillet radius is 10

ACTIVITY 23
Draw the printed circuit board using the information given.
Add the ordinate dimensions.

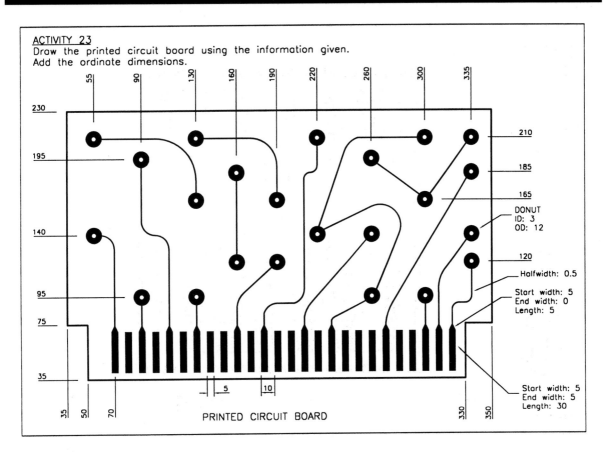

DONUT
ID: 3
OD: 12

Halfwidth: 0.5

Start width: 5
End width: 0
Length: 5

Start width: 5
End width: 5
Length: 30

PRINTED CIRCUIT BOARD

ACTIVITY 24
Draw the two components and add
text and dimensions.

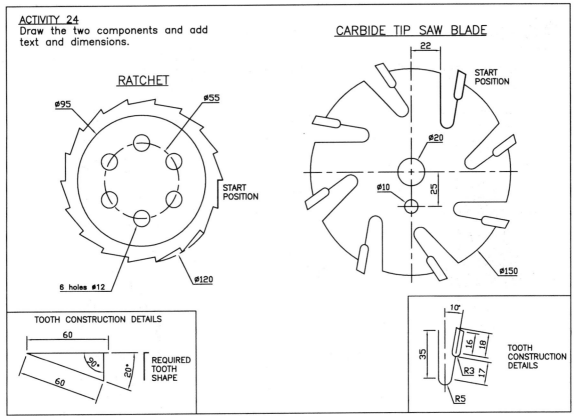

CARBIDE TIP SAW BLADE

RATCHET

Ø95   Ø55

START
POSITION

6 holes #12   Ø120

START
POSITION

Ø20

Ø10

25

Ø150

TOOTH CONSTRUCTION DETAILS

60

90°

20°

60

REQUIRED
TOOTH
SHAPE

10°

35   16   18

R3   17

R5

TOOTH
CONSTRUCTION
DETAILS

ACTIVITY 25

Using the FISH drawing, create the rectangular and polar array patterns using the following information:

a) A 5x3 angular rectangular array at 10° from 0.2 scale

b) polar arrays with:
   i) scale: 0.25; radius: 90; items: 9
   ii) scale: 0.2; radius: 60; items: 7
   iii) scale: 0.15; radius: 35; items: 7

POLAR ARRAY PATTERN

ANGULAR RECTANGULAR ARRAY PATTERN

ACTIVITY 26

Draw the basic bulb using the dimensions given then produce the array design.

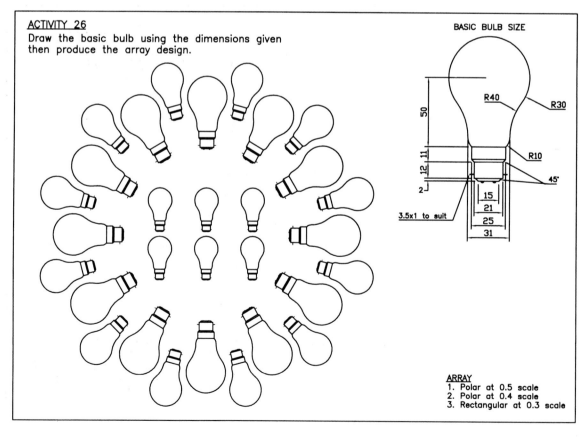

BASIC BULB SIZE

R40   R30

50

R10

11

12

45°

2

15

21

3.5x1 to suit

25

31

ARRAY
1. Polar at 0.5 scale
2. Polar at 0.4 scale
3. Rectangular at 0.3 scale

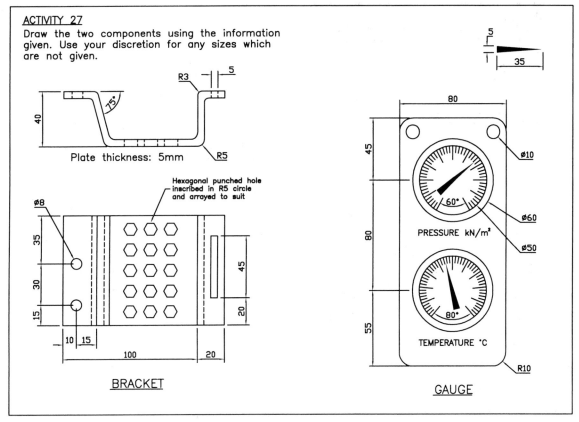

ACTIVITY 27

Draw the two components using the information given. Use your discretion for any sizes which are not given.

Plate thickness: 5mm

Hexagonal punched hole inscribed in R5 circle and arrayed to suit

BRACKET

PRESSURE kN/m²

TEMPERATURE °C

GAUGE

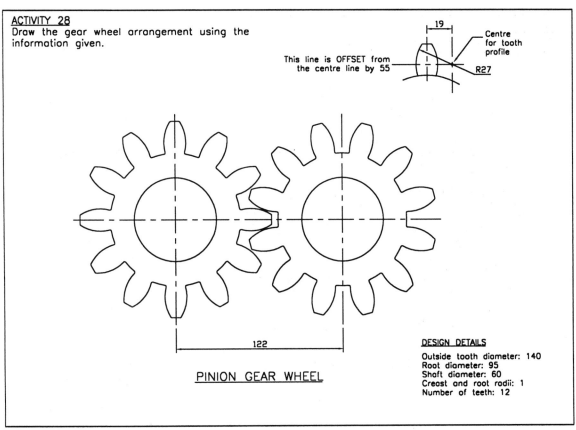

ACTIVITY 28

Draw the gear wheel arrangement using the information given.

This line is OFFSET from the centre line by 55

Centre for tooth profile

R27

PINION GEAR WHEEL

DESIGN DETAILS

Outside tooth diameter: 140
Root diameter: 95
Shaft diameter: 60
Creast and root radii: 1
Number of teeth: 12

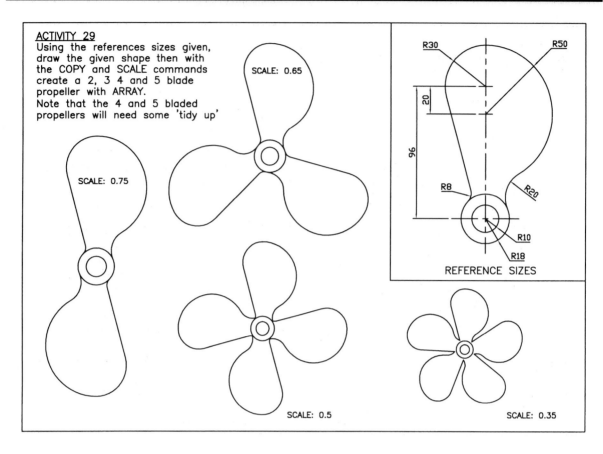

ACTIVITY 29
Using the references sizes given, draw the given shape then with the COPY and SCALE commands create a 2, 3 4 and 5 blade propeller with ARRAY.
Note that the 4 and 5 bladed propellers will need some 'tidy up'

SCALE: 0.65

SCALE: 0.75

SCALE: 0.5

SCALE: 0.35

R30

R50

R8

R20

R10

R18

20

96

REFERENCE SIZES

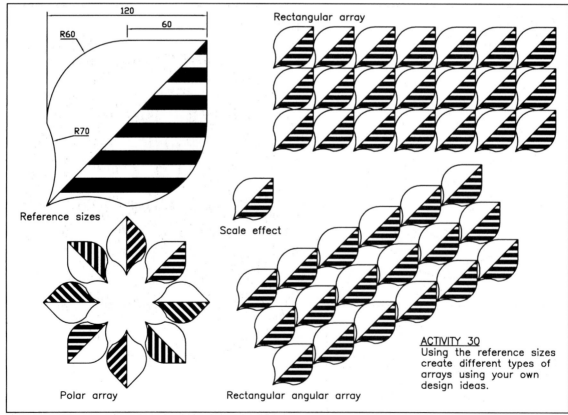

120

60

R60

R70

Reference sizes

Rectangular array

Scale effect

Polar array

Rectangular angular array

ACTIVITY 30
Using the reference sizes create different types of arrays using your own design ideas.

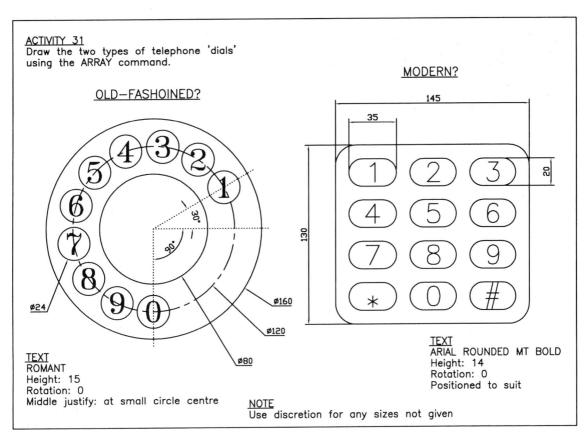

ACTIVITY 31
Draw the two types of telephone 'dials' using the ARRAY command.

MODERN?

OLD—FASHOINED?

Ø24

Ø160

Ø120

Ø80

30°

90°

145

35

130

20

1 2 3
4 5 6
7 8 9
* 0 #

TEXT
ROMANT
Height: 15
Rotation: 0
Middle justify: at small circle centre

TEXT
ARIAL ROUNDED MT BOLD
Height: 14
Rotation: 0
Positioned to suit

NOTE
Use discretion for any sizes not given

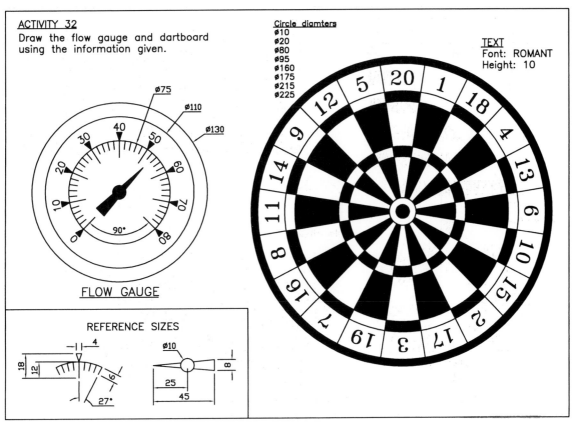

ACTIVITY 32
Draw the flow gauge and dartboard using the information given.

Circle diamters
Ø10
Ø20
Ø80
Ø95
Ø160
Ø175
Ø215
Ø225

TEXT
Font: ROMANT
Height: 10

Ø75
Ø110
Ø130

40
30
50
20
60
10
70
0
80
90°

FLOW GAUGE

REFERENCE SIZES

4
18
12
6
27°

Ø10
25
45
8

ACTIVITY 33
Draw the given components and add all
dimensions, creating appropriate dimension
styles for the tolerance dimensions.

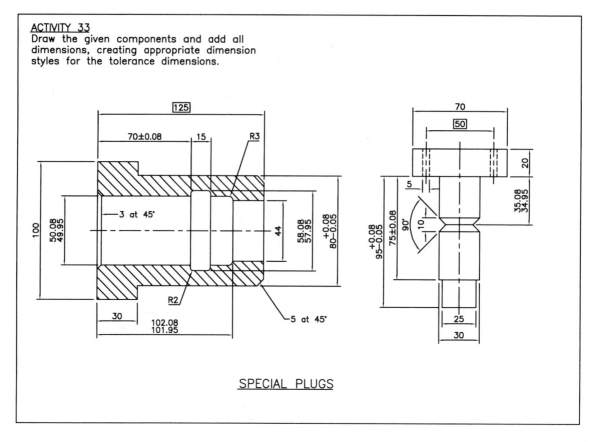

SPECIAL PLUGS

ACTIVITY 34
Draw the component shown and add the
dimensions and geometric tolerance.

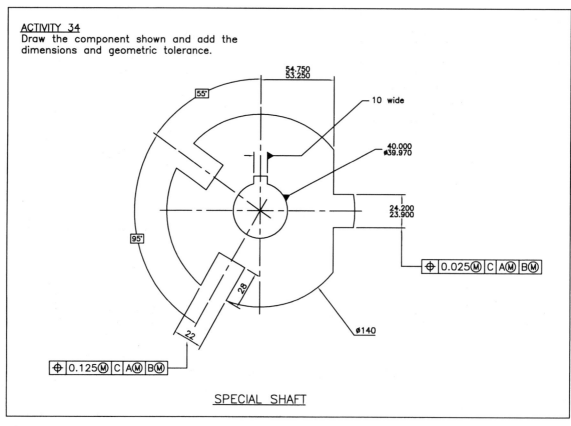

SPECIAL SHAFT

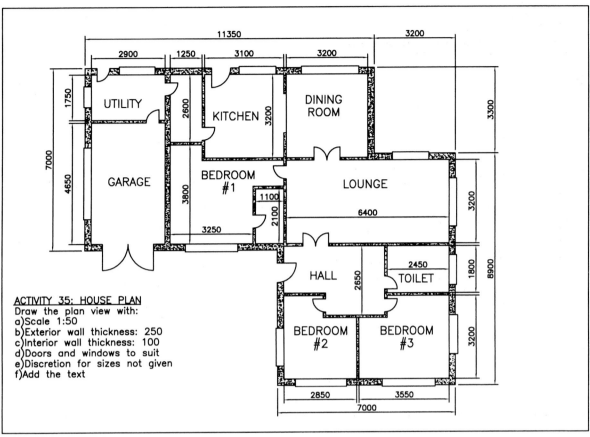

ACTIVITY 35: HOUSE PLAN
Draw the plan view with:
a) Scale 1:50
b) Exterior wall thickness: 250
c) Interior wall thickness: 100
d) Doors and windows to suit
e) Discretion for sizes not given
f) Add the text

ACTIVITY 36
1. Draw the two shapes with sizes given
2. Make blocks of the shapes with the suggested block names and insertion points
3. Insert the two blocks using the information given below using the information given in the assignment
4. Draw a 1 wide polyline through the roller centres then edit the polyline to a spline fit
5. With the DISTANCE command, find the vertical distance between the leftmost and rightmost roller centre points. MY distance was 14.98

REFERENCE

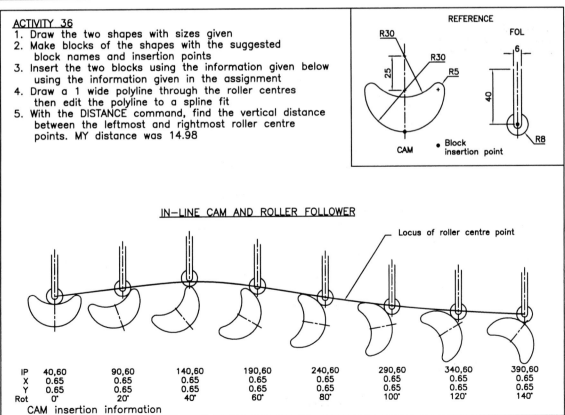

IN-LINE CAM AND ROLLER FOLLOWER

Locus of roller centre point

IP	40,60	90,60	140,60	190,60	240,60	290,60	340,60	390,60
X	0.65	0.65	0.65	0.65	0.65	0.65	0.65	0.65
Y	0.65	0.65	0.65	0.65	0.65	0.65	0.65	0.65
Rot	0°	20°	40°	60°	80°	100°	120°	140°

CAM insertion information

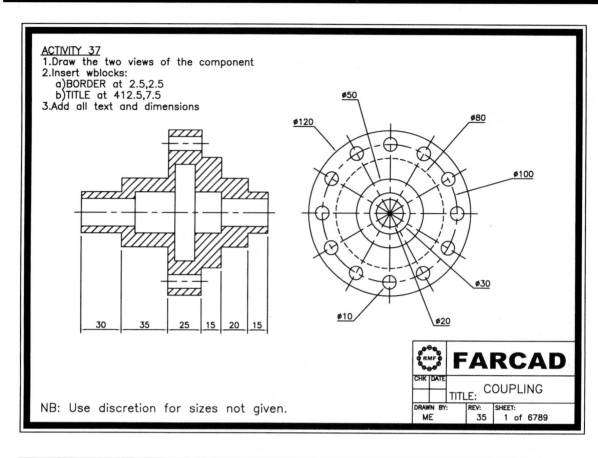

ACTIVITY 37
1. Draw the two views of the component
2. Insert wblocks:
   a) BORDER at 2.5,2.5
   b) TITLE at 412.5,7.5
3. Add all text and dimensions

Ø50
Ø120
Ø80
Ø100
Ø30
Ø10
Ø20

30 35 25 15 20 15

NB: Use discretion for sizes not given.

FARCAD
CHK | DATE
TITLE: COUPLING
DRAWN BY: ME | REV: 35 | SHEET: 1 of 6789

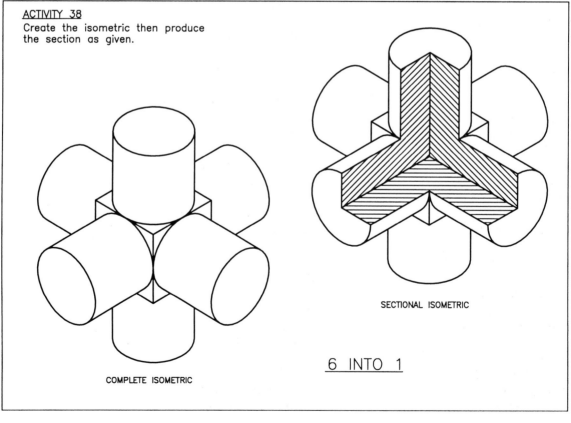

ACTIVITY 38
Create the isometric then produce
the section as given.

SECTIONAL ISOMETRIC

6 INTO 1

COMPLETE ISOMETRIC

Index